Smart Buildings

Related titles

Biopolymers and Biotech Admixtures for Eco-Efficient Construction Materials
(ISBN 978-0-08100-214-8)

Acoustic Emission and Related Non-destructive Evaluation Techniques in the Fracture Mechanics of Concrete
(ISBN 978-1-78242-327-0)

Eco-efficient Masonry Bricks and Blocks
(ISBN 978-1-78242-305-8)

Woodhead Publishing Series in Civil and
Structural Engineering: Number 69

Smart Buildings

Advanced Materials and
Nanotechnology to Improve
Energy-Efficiency and Environmental
Performance

Marco Casini

AMSTERDAM • BOSTON • CAMBRIDGE • HEIDELBERG
LONDON • NEW YORK • OXFORD • PARIS • SAN DIEGO
SAN FRANCISCO • SINGAPORE • SYDNEY • TOKYO
Woodhead Publishing is an imprint of Elsevier

Woodhead Publishing is an imprint of Elsevier
The Officers' Mess Business Centre, Royston Road, Duxford, CB22 4QH, UK
50 Hampshire Street, 5th Floor, Cambridge, MA 02139, USA
The Boulevard, Langford Lane, Kidlington, OX5 1GB, UK

Copyright © 2016 Elsevier Ltd. All rights reserved.

No part of this publication may be reproduced or transmitted in any form or by any means, electronic or mechanical, including photocopying, recording, or any information storage and retrieval system, without permission in writing from the publisher. Details on how to seek permission, further information about the Publisher's permissions policies and our arrangements with organizations such as the Copyright Clearance Center and the Copyright Licensing Agency, can be found at our website: www.elsevier.com/permissions.

This book and the individual contributions contained in it are protected under copyright by the Publisher (other than as may be noted herein).

Notices
Knowledge and best practice in this field are constantly changing. As new research and experience broaden our understanding, changes in research methods, professional practices, or medical treatment may become necessary.

Practitioners and researchers must always rely on their own experience and knowledge in evaluating and using any information, methods, compounds, or experiments described herein. In using such information or methods they should be mindful of their own safety and the safety of others, including parties for whom they have a professional responsibility.

To the fullest extent of the law, neither the Publisher nor the authors, contributors, or editors, assume any liability for any injury and/or damage to persons or property as a matter of products liability, negligence or otherwise, or from any use or operation of any methods, products, instructions, or ideas contained in the material herein.

British Library Cataloguing-in-Publication Data
A catalogue record for this book is available from the British Library

Library of Congress Cataloging-in-Publication Data
A catalog record for this book is available from the Library of Congress

ISBN: 978-0-08-100972-7 (print)
ISBN: 978-0-08-100640-5 (online)

For information on all Woodhead Publishing publications visit our website at https://www.elsevier.com/

Publisher: Matthew Deans
Acquisition Editor: Gwen Jones
Editorial Project Manager: Charlotte Cockle
Production Project Manager: Omer Mukthar
Designer: Maria Inês Cruz

Typeset by TNQ Books and Journals

Contents

Woodhead Publishing Series in Civil and Structural Engineering	ix
About the author	xiii
Acknowledgements	xv
Introduction	xvii

Part One   Smart buildings   1

1   Designing the third millennium's buildings   3
  1.1   Buildings as a key part of the energy and environmental system   3
  1.2   Smart, sustainable, and inclusive buildings   6
  1.3   Zero-energy buildings   10
  1.4   Green buildings   27
  1.5   Smart buildings   38
  1.6   Conclusions and future trends   49
      References   50

2   Advanced materials for architecture   55
  2.1   Building materials classification   55
  2.2   Nanotechnology   59
  2.3   Smart materials   84
  2.4   3D printing for architecture   94
  2.5   Conclusions and future trends   101
      References   102

Part Two   Smart insulation   105

3   Building insulating materials   107
  3.1   Heat transfer physics   107
  3.2   Classification and thermal properties   110
  3.3   Functional model and building facade applications   114
  3.4   Conclusions and future trends   124
      References   125

4   Advanced insulating materials   127
  4.1   Nanoporous insulating materials: aerogels   127
  4.2   Vacuum insulating panels   149

	4.3	Biobased insulating materials	160
	4.4	Transparent insulating materials	167
	4.5	Conclusions and future trends	174
		References	175
5	**Phase-change materials**	179
	5.1	Thermal mass and latent heat storage	179
	5.2	Classification and technical specifications	184
	5.3	Packaging and encapsulation methods	188
	5.4	Functional model and building design	193
	5.5	Building applications and products	197
	5.6	Conclusions and future trends	215
		References	216
6	**Advanced building skin**	219
	6.1	Cool roofs	219
	6.2	Green walls	229
	6.3	Environment-adaptive skin facades	239
	6.4	Conclusions and future trends	243
		References	244

Part Three   Smart windows   247

7	**Advanced insulation glazing**	249
	7.1	Advanced low-emission glazing	249
	7.2	Suspended film glazing	255
	7.3	Vacuum insulating glass	256
	7.4	Monolithic aerogel insulating glazing	257
	7.5	Advanced window frames	258
	7.6	Glazed double-skin facades	260
	7.7	Heating glazing	264
	7.8	Fire-resistant glazing	265
	7.9	ETFE transparent closures	266
	7.10	Conclusions and future development	274
		References	275
8	**Light and solar control glazing and systems**	279
	8.1	Antireflective glazing	279
	8.2	Self-cleaning glazing	280
	8.3	Light-redirection and optical systems	284
	8.4	Static solar protection glazing	291
	8.5	Advanced shading systems	296
	8.6	Conclusions and future development	302
		References	303

9	**Dynamic glazing**	**305**
	9.1 Passive dynamic glazing	305
	9.2 Active dynamic glazing	307
	9.3 Conclusions and future trends	323
	References	324
10	**Energy-generating glazing**	**327**
	10.1 Advanced photovoltaic glazing	327
	10.2 Bioadaptive glazing	347
	10.3 Conclusions and future trends	351
	References	352
Index		**355**

Woodhead Publishing Series in Civil and Structural Engineering

1 Finite element techniques in structural mechanics
   C. T. F. Ross
2 Finite element programs in structural engineering and continuum mechanics
   C. T. F. Ross
3 Macro-engineering
   F. P. Davidson, E. G. Frankl and C. L. Meador
4 Macro-engineering and the earth
   U. W. Kitzinger and E. G. Frankel
5 Strengthening of reinforced concrete structures
   Edited by L. C. Hollaway and M. Leeming
6 Analysis of engineering structures
   B. Bedenik and C. B. Besant
7 Mechanics of solids
   C. T. F. Ross
8 Plasticity for engineers
   C. R. Calladine
9 Elastic beams and frames
   J. D. Renton
10 Introduction to structures
    W. R. Spillers
11 Applied elasticity
    J. D. Renton
12 Durability of engineering structures
    J. Bijen
13 Advanced polymer composites for structural applications in construction
    Edited by L. C. Hollaway
14 Corrosion in reinforced concrete structures
    Edited by H. Böhni
15 The deformation and processing of structural materials
    Edited by Z. X. Guo
16 Inspection and monitoring techniques for bridges and civil structures
    Edited by G. Fu
17 Advanced civil infrastructure materials
    Edited by H. Wu
18 Analysis and design of plated structures Volume 1: Stability
    Edited by E. Shanmugam and C. M. Wang

19	**Analysis and design of plated structures Volume 2: Dynamics**
	Edited by E. Shanmugam and C. M. Wang
20	**Multiscale materials modelling**
	Edited by Z. X. Guo
21	**Durability of concrete and cement composites**
	Edited by C. L. Page and M. M. Page
22	**Durability of composites for civil structural applications**
	Edited by V. M. Karbhari
23	**Design and optimization of metal structures**
	J. Farkas and K. Jarmai
24	**Developments in the formulation and reinforcement of concrete**
	Edited by S. Mindess
25	**Strengthening and rehabilitation of civil infrastructures using fibre-reinforced polymer (FRP) composites**
	Edited by L. C. Hollaway and J. C. Teng
26	**Condition assessment of aged structures**
	Edited by J. K. Paik and R. M. Melchers
27	**Sustainability of construction materials**
	J. Khatib
28	**Structural dynamics of earthquake engineering**
	S. Rajasekaran
29	**Geopolymers: Structures, processing, properties and industrial applications**
	Edited by J. L. Provis and J. S. J. van Deventer
30	**Structural health monitoring of civil infrastructure systems**
	Edited by V. M. Karbhari and F. Ansari
31	**Architectural glass to resist seismic and extreme climatic events**
	Edited by R. A. Behr
32	**Failure, distress and repair of concrete structures**
	Edited by N. Delatte
33	**Blast protection of civil infrastructures and vehicles using composites**
	Edited by N. Uddin
34	**Non-destructive evaluation of reinforced concrete structures Volume 1: Deterioration processes**
	Edited by C. Maierhofer, H.-W. Reinhardt and G. Dobmann
35	**Non-destructive evaluation of reinforced concrete structures Volume 2: Non-destructive testing methods**
	Edited by C. Maierhofer, H.-W. Reinhardt and G. Dobmann
36	**Service life estimation and extension of civil engineering structures**
	Edited by V. M. Karbhari and L. S. Lee
37	**Building decorative materials**
	Edited by Y. Li and S. Ren
38	**Building materials in civil engineering**
	Edited by H. Zhang
39	**Polymer modified bitumen**
	Edited by T. McNally
40	**Understanding the rheology of concrete**
	Edited by N. Roussel
41	**Toxicity of building materials**
	Edited by F. Pacheco-Torgal, S. Jalali and A. Fucic

42 Eco-efficient concrete
    Edited by F. Pacheco-Torgal, S. Jalali, J. Labrincha and V. M. John
43 Nanotechnology in eco-efficient construction
    Edited by F. Pacheco-Torgal, M. V. Diamanti, A. Nazari and C. Goran-Granqvist
44 Handbook of seismic risk analysis and management of civil infrastructure systems
    Edited by F. Tesfamariam and K. Goda
45 Developments in fiber-reinforced polymer (FRP) composites for civil engineering
    Edited by N. Uddin
46 Advanced fibre-reinforced polymer (FRP) composites for structural applications
    Edited by J. Bai
47 Handbook of recycled concrete and demolition waste
    Edited by F. Pacheco-Torgal, V. W. Y. Tam, J. A. Labrincha, Y. Ding and J. de Brito
48 Understanding the tensile properties of concrete
    Edited by J. Weerheijm
49 Eco-efficient construction and building materials: Life cycle assessment (LCA), eco-labelling and case studies
    Edited by F. Pacheco-Torgal, L. F. Cabeza, J. Labrincha and A. de Magalhães
50 Advanced composites in bridge construction and repair
    Edited by Y. J. Kim
51 Rehabilitation of metallic civil infrastructure using fiber-reinforced polymer (FRP) composites
    Edited by V. Karbhari
52 Rehabilitation of pipelines using fiber-reinforced polymer (FRP) composites
    Edited by V. Karbhari
53 Transport properties of concrete: Measurement and applications
    P. A. Claisse
54 Handbook of alkali-activated cements, mortars and concretes
    F. Pacheco-Torgal, J. A. Labrincha, C. Leonelli, A. Palomo and P. Chindaprasirt
55 Eco-efficient masonry bricks and blocks: Design, properties and durability
    F. Pacheco-Torgal, P. B. Lourenço, J.A. Labrincha, S. Kumar and P. Chindaprasirt
56 Advances in asphalt materials: Road and pavement construction
    Edited by S.-C. Huang and H. Di Benedetto
57 Acoustic emission (AE) and related non-destructive evaluation (NDE) techniques in the fracture mechanics of concrete: Fundamentals and applications
    Edited by M. Ohtsu
58 Nonconventional and vernacular construction materials: Characterisation, properties and applications
    Edited by K. A. Harries and B. Sharma
59 Science and technology of concrete admixtures
    Edited by P.-C. Aïtcin and R. J. Flatt
60 Textile fibre composites in civil engineering
    Edited by T. Triantafillou
61 Corrosion of steel in concrete structures
    Edited by A. Poursaee
62 Innovative developments of advanced multifunctional nanocomposites in civil and structural engineering
    Edited by K. J. Loh and S. Nagarajaiah
63 Biopolymers and biotech admixtures for eco-efficient construction materials
    Edited by F. Pacheco-Torgal, V. Ivanov, N. Karak and H. Jonkers

64 **Marine concrete structures: Design, durability and performance**
Edited by M. Alexander

65 **Recent trends in cold-formed steel construction**
Edited by C. Yu

66 **Start-up creation: The smart eco-efficient built environment**
F. Pacheco-Torgal, E. Rasmussen, C. G. Granqvist, V. Ivanov, A. Kaklauskas and S. Makonin

67 **Characteristics and uses of steel slag in building construction**
I. Barisic, I. Netinger Grubesa, A. Fucic and S. S. Bansode

68 **The utilization of slag in civil infrastructure construction**
G. Wang

69 **Smart buildings: Advanced materials and nanotechnology to improve energy-efficiency and environmental performance**
M. Casini

About the author

Dr. Marco Casini is a leading academic in the green and smart building sector with over 20 years' experience in building sciences.

He is an environmental engineer with a PhD in environmental engineering, and is a research fellow in architecture technology at Sapienza University of Rome, Department of Urban Planning, Design, and Architecture Technology. Since 2002 he has been professor of Architecture technology and of Environmental certification of buildings at the Faculty of Architecture of Sapienza University, where he also teaches in several masters', PhD, and graduate schools on subjects pertaining to energy and environmental sustainability for buildings.

Dr. Casini's research activities cover a wide spectrum of topics within sustainable architectural design and energy efficiency of buildings, focusing on advanced materials and nanotechnologies for smart building envelopes and integrated renewable energy systems. He has worked as scientific coordinator on major projects, including the development of "Italian regional system for the certification of environmental sustainability of buildings—Protocollo ITACA Lazio" (2014) and the preparation of the Sustainable Energy Action Plan of Rome within the European Covenant of Mayors for Climate and Energy (2012).

Dr. Casini's professional activity includes scientific and technical consultancy on technological, environmental, and energy aspects related to the design and construction of complex building structures worldwide, as well as training on green building and smart cities strategies and policymaking for Italian public authorities (Prime Minister's Cabinet, Ministry of Internal Affairs, Regione Lazio).

He has been a member of several public technical working groups (Italian Environment Protection Agency, UNI, Bank of Italy, and the Italian technical body of the Conference of Regions and Autonomous Provinces, Regione Lazio) for the development of specific standards on environmentally sustainable construction.

Dr. Casini is the scientific director of the editorial board of the Italian scientific journal *Ponte* and a member of the editorial boards of several other international scientific journals in the fields of engineering and architecture. He has authored over 70 scientific publications on energy and environmental efficiency of buildings.

Acknowledgements

I would like to acknowledge Dr. Paolo Tisei for his collaboration in research activity and drawings preparation.

Introduction

The construction industry has been identified worldwide as one of the priority action areas in achieving the objective of a smart, sustainable, and inclusive development based on the efficient use of resources.

Because of their low energy efficiency, buildings are currently responsible for more than 30% of global world energy consumption and account for one-third of direct and indirect CO_2 and particulate matter emissions. Nevertheless, numerous studies show that the construction sector has the potential to improve energy efficiency with cost-effective interventions. An improvement of energy performance of buildings, along with reducing carbon emissions, would yield important benefits to building owners and occupants, such as improved durability, reduced maintenance, greater comfort, lower costs, higher property values, increased habitable space, increased productivity, and improved health and safety.

In this picture, the commitment of all countries to improving the energy efficiency of buildings has therefore greatly increased over the past decade, with the triple objective of significantly reducing the energy consumption of existing buildings, ensuring that all new buildings are characterized by high energy efficiency (very low-energy buildings), and using as much renewable energy as possible, instead of fossil fuels, to meet the energy needs of both new and existing buildings.

The challenge for the architecture of the new millennium focuses on several main areas of research in all phases of the building process, from planning through completion to final disposal of the building. Applications are in both new buildings and renovation or redevelopment of existing buildings.

The final objective is to provide a smart building, namely a building designed or renovated with intelligence (smart design) that uses the best typological (smart shape) and technological solutions, from both construction (smart envelope) and equipment (smart systems) points of view, able to interact intelligently with the environment and users (smart people) to provide, with a very low use of natural resources, an accessible, safe, comfortable, and healthy built environment, capable of improving the lives of all stakeholders involved (smart environment, smart living, smart economy, smart city).

In view of this strategic objective, the building envelope plays a key role and in recent years has undergone a thorough review of its features and requirements to find technological solutions that can guarantee continuous adjustment of environmental flows in relation to climatic conditions and other factors. The building envelope constitutes a complex system of barriers and filters that regulate the flow of heat, solar radiation, air, and steam, and can also convert radiation into energy (thermal and electrical), which is an essential element for the metabolism of the building.

The final goal is to provide a "smart envelope," capable of not only offering better performance compared to a traditional building shell, but also fulfilling new functions (power generation, light emission, image projection, air purification, self-cleaning surfaces, ability to self-repair, etc.) and adapting its characteristics in response to changes in external conditions (from transparent to opaque, from solid to liquid, from waterproof to permeable, etc.).

Thanks to huge progress in the field of materials science and nanosciences, technology solutions available today make it possible to accept this challenge of the new millennium architecture and take effective action, with more than satisfactory results, even in renovation of historical buildings subject to architectural restrictions.

Innovative insulation products (advanced insulating materials) that offer high thermal insulation performance with very low thickness along with dynamic glazing and building integrated renewable energy systems, are opening up important new possibilities in the fields of design and renovation of the building envelope to provide environment-adaptive skin facades. Examples are vacuum insulating panels (VIPs), nanoporous insulating materials (NIMs) like aerogel, transparent insulating materials (TIMs) that combine light transmission and thermal resistance in a few centimeters of thickness, special coatings to reflect infrared solar radiation (cool roofs), and active components and devices, so-called "smart materials," able to modify their characteristics in relation to different conditions imposed by climatic agents or users, such as phase-change materials (PCMs), chromogenic materials, and photocatalytic and organic photovoltaic materials.

The goal of this book is to provide readers with a state-of-the-art review of the latest advances in construction materials and building design. It takes into consideration both design and materials aspects, with particular focus on the next generation of construction materials and the most advanced products currently entering the market as high-priority, energy-efficient building envelope components.

The book is divided into three main parts.

In Part One, after an introduction about the issues concerning the design process in the third millennium and the differences between zero-energy buildings, green buildings, and smart buildings, the book addresses the issue of smart buildings, focusing on the envelope and how to make it "adaptive" thanks to the huge new possibilities offered by smart materials and nanotechnology.

Chapter 1 provides an overview of new issues related to building design, illustrating the design strategies to achieve maximum building efficiency, sustainability, and architectural quality. After an outline on the state of existing building stock and its role in energy consumption and global warming, an in-depth presentation of zero-energy, green, and smart building concepts and requirements is given, highlighting the main strategies and technical and technological design solutions. A new and inedited definition of smart buildings is introduced, responding fully to the needs of the architecture of the 21st century and in line with the new concept of a smart city. Particular attention is given to the envelope and how to make it "smart" by virtue of the new and extensive possibilities offered by nanotechnologies and smart materials. Lastly, an outlook on the importance of energy management systems and the "internet of things" to achieve a smart building is presented.

Chapter 2 gives an overview of the most advanced materials available today to reconcile the architectural features of buildings with the new challenges of energy and environmental efficiency. An in-depth analysis of nanotechnology and its application to the energy, environmental, and construction sectors is provided, focusing on innovative nanoproducts for architecture. The new class of highly innovative materials, so-called smart materials, is presented, addressing both property-changing and energy-exchanging materials, and illustrating their properties and application in the building sector. Lastly the chapter gives an overview of the enormous potential of three-dimensional (3D) printing technology for architecture, with particular focus on the realization of building components, structural elements, and entire buildings (3D house printing technology).

Part Two, devoted to the latest solutions for thermal insulation, the so-called smart insulation, presents the results of extensive and thorough research on the most innovative insulation materials, from nanotechnology to bioecological materials and PCMs, describing for each the technical characteristics, performance level, and methods of use. Also presented are the achievements in the field of green walls as a solution for upgrading the energy efficiency and environmental performance of existing buildings.

Chapter 3 focuses on the importance of the thermohygrometric characteristics of opaque closures in the energy efficiency of buildings. An overview of thermal insulation materials is presented, illustrating main thermal properties and heat transfer physics. Functional models and building application methods are described, highlighting the main differences in terms of energy performance, thermal comfort, and time and costs of realization.

Chapter 4 focuses on innovative solutions for thermal insulation of opaque closures in new and existing buildings, with an in-depth look at future prospect and developments. The most advanced thermal insulating materials, including NIMs such as aerogel and VIPs, are presented, providing a comprehensive analysis of different products available on the market, their advantages and disadvantages, their possible application in the building envelope, and their future developments. Different next-generation bio-based insulating materials (algae, mushrooms, wooden foam, etc.) are described, with particular attention to low environmental impact and reuse/recycle possibility. Main TIMs solutions and their applications are shown, comparing their performance to traditional insulating glass units.

Chapter 5 focuses on the importance of the heat capacity of the building envelope for energy efficiency and indoor thermal comfort, presenting the most innovative smart materials for latent heat storage in new and existing buildings. A thorough review of PCMs is given, explaining their classification, technical specifications, encapsulation methods, functional models, and application inside or outside the building, even combined with advanced insulating materials. Lastly, there is an in-depth look at future prospects and developments.

Chapter 6 provides an in-depth analysis of reflective coatings (cool roofs) available on the market, explaining their possible application for new and existing buildings and their potential to increase the energy efficiency of buildings and reduce urban heat island effect. Different technological solutions to integrate vegetable species as cladding systems for internal and external walls (green walls) are discussed, and the most innovative techniques are illustrated in detail. The last section looks at the state of research and future developments in environment-adaptive skin facades.

Finally, Part Three illustrates research on smart windows, with the assumption that transparent surfaces represent the most critical element in the energy balance of a building and simultaneously one of the most significant components of contemporary architectural quality. There is an extensive review of the technical features of transparent closures on the market or still under development, from so-called dynamic glazing up to bioadaptive and photovoltaic glazing, describing their esthetic potential and performance limits.

Chapter 7 focuses on the most innovative solutions to improve thermal transmittance of glazing, such as double and triple glazing, noble gas filling, low-emissivity coatings, suspended film glazing, vacuum insulated glass, monolithic aerogel, heating glazing, and glazed double-skin facades. It outlines different advanced insulating glazing solutions available on the market and their main characteristics and performance. Lastly, an outline of ETFE (ethylene tetrafluoroethylene) use for transparent closures is provided.

Chapter 8 illustrates the role of glazing in light and solar radiation control to achieve the maximum daylighting and energy efficiency levels by avoiding summer overheating and exploiting winter solar gain. Innovative solutions for enhancing and controlling visible light are shown, including self-cleaning glass, antireflective glazing, and light-redirection and optical systems. Main solutions for solar control are summarized, including advanced shading systems and static solar protection glazing, describing the main selective glazing available on the market and its performance and specifications.

Chapter 9 describes the most innovative high-performance dynamic glazing systems, aimed not only at reducing heat loss but also at controlling incoming solar radiation to maximize solar gain in winter and minimize it in summer, as well as ensuring the best natural lighting conditions with no glare. An analysis of the different types of dynamic glazing with both passive and active control is provided, illustrating their potential uses and the benefits achieved in terms of energy efficiency, environmental comfort, and architectural quality in both new constructions and existing building renovations. Particular attention is given to the different types of active dynamic glazing technologies (smart windows) available on the market or still in development, such as electrochromics, suspended particle devices, polymer-dispersed liquid crystals, and other emerging technologies. For each technology operation, performance, and feasibility are analyzed and compared.

Chapter 10 focuses on the role of glazed surfaces as an energy-generating component. The most advanced solutions to integrate photovoltaics into transparent surfaces are illustrated. Particular attention is paid to inorganic and organic semitransparent photovoltaic thin films, illustrating operation, performance, and feasibility for building applications. Emerging technologies such as spherical cell photovoltaic glazing, prismatic optical cell photovoltaic glazing, and transparent luminous solar collectors are also described. Lastly, innovative bioadaptive facades implemented with algae bioreactors are introduced, analyzing their operation, performance, and energy yield.

Part One

Smart buildings

Designing the third millennium's buildings

1.1 Buildings as a key part of the energy and environmental system

The construction industry is one of the global priority action areas for the achievement of "smart, sustainable, and inclusive" growth and a transition to a resource-efficient and low-carbon economy.

In fact, in 2012 buildings globally accounted for 32% (118.6 EJ) of final energy consumption, 53% of electricity consumption, and one-third of direct and indirect CO_2 and particulate matter emissions.[1–4] In the United States and Europe buildings account for more than 40% of total final energy use, reaching 80% in regions highly dependent on traditional biomass.[1,5]

Between 2000 and 2012 buildings' final energy consumption grew by 1.5% per year (from 102 EJ to 120 EJ),[4] a rate that has not slowed down despite the recent global economic crisis.[1]

As the global population is expected to increase by 2.5 billion people by 2050 and economic development and living standards are improving worldwide, energy use in the building sector is still set to rise greatly: with no improvements in the energy efficiency of the building sector, its energy demand is expected to grow by 50% by 2050.[2]

Nevertheless, numerous studies show that the construction sector has potential for improving energy efficiency, which can be exploited with interventions that are also effective in terms of costs.[1] After the energy industry itself, the buildings sector, including both residential and nonresidential, has the second-largest untapped and cost-effective energy-saving potential. Building emission reduction potential is also important, and standard new buildings can save as much of 80% of operational costs through integrated design methods, often at little, if any, extra cost over the building lifetime.[6]

Moreover, existing buildings have a huge unused potential of space available for the integration of renewable energy sources. In fact, 40% of total European Union (EU) electricity demand in 2020 would be met if all the roofs and facades of suitable European buildings were covered with photovoltaic (PV) panels.[7]

A wide global deployment of best available technologies and efficiency policies could yield annual savings in final energy use in buildings in the range of 53 EJ by 2050 (a 29% reduction in projected building energy consumption relative to a business-as-usual scenario), a value equivalent to the combined energy use of buildings in China, France, Germany, Russia, the United Kingdom, and the United States in 2012.[3,4]

The high energy consumption of buildings adds to the environmental impacts resulting from consumption of materials and drinking water, and waste from construction and demolition (the building sector is responsible for more than one-third of resource consumption globally, which equates to approximately 3 billion tonnes of raw materials annually, consumes more than 12% of the world's potable water, and accounts for about 40% of solid waste streams in developed countries).[8,9] In 2012 the cement sector alone accounted for 8.5% of total industrial energy use and 34% of direct industrial CO_2 emissions,[1] and concrete was the most widely used material on Earth (about 10 km^3/yr).[10] It is therefore apparent that any strategy aimed at reducing consumption of natural resources and emissions of fossil fuels and carbon must have as a priority objective the improvement of energy and environmental efficiency of both new and existing buildings. To meet the objective of limiting global temperature rise to 2°C, an estimated 77% reduction in total CO_2 emissions of the buildings sector by 2050 is required.[2]

Buildings consume energy for heating, cooling, interior ventilation, domestic hot water (DHW) production, lighting appliances, electrical equipment, people transport, and cooking, using gas (21%), electricity (30%), and biomass (29%) as the main energy carriers.[4] The incidence of each of the above items in total consumption varies depending on the type of building, its year of construction, and the climate zone in which it is located.

In 2012 space heating and cooling and DHW production accounted for nearly 60% of global energy consumption in buildings, representing the largest opportunity to reduce building energy consumption, improve energy security, and reduce CO_2 emissions, particularly due to the fact that space and water heating provision in some countries is still dominated by fossil fuel.[2] In the EU space heating is the largest end use in terms of final energy consumption, accounting for two-thirds of residential energy use and about 40% of services energy consumption, while in the United States these values are respectively 37% and 27%.[2,5]

Space cooling still accounts for less than 5% of final energy demand, but this value is dramatically rising worldwide, with an increase of 43% in the last 10 years compared to a 34% increase in building floor area and a 13% growth of world population.[4,11] Energy consumption for cooling is set to increase by almost 150% globally, and by 300–600% in developing countries, by 2050.[12] The European Commission predicts that the demand for cooling in buildings in the EU will increase by 72% by 2030.[13]

Energy consumption for appliances and other electrical equipment is also rising, due to the increasing number of appliances in homes and offices and their increased use in daily life, and represents the fastest-growing energy end use in buildings.[1]

Such high energy consumption in buildings is attributable to nonoptimal architectural choices (shape, orientation, windows-to-wall ratio, etc.), low thermohygrometrical performance of the building envelope (poor thermal insulation, low thermal inertia, presence of thermal bridges, ineffective screening systems, etc.), and low efficiency of HVAC (heating, ventilation, and air conditioning) and lighting systems, as well as a still low utilization of renewable energy sources.

Most of the existing building stock in Organisation for Economic Co-operation and Development (OECD) countries was in fact built before the introduction of specific energy performance requirements related to the building envelope or equipment. The first, not very stringent, energy codes for buildings were in fact introduced only in the 1970s in response to the oil crisis,[14] when 66% of the current EU building stock had already been built.[6] Among existing buildings, those dating from the 1945—80 period, built with the first industrial techniques and prior to the introduction of energy efficiency requirements, have the worst energy performance—even lower than those from before 1945, which were built with materials and techniques reflecting local conditions and taking into account bioclimatic considerations.[6]

Throughout history, several factors have constituted the key drivers of final energy use in buildings: population size, buildings sector size (eg, as measured by floor area or number of households in the residential subsector), economic activity (eg, as measured by gross domestic product—GDP), and building energy policies, as well as additional factors such as energy prices and climate. The amount each driver contributes to energy use differs from country to country, within countries themselves, and over time according to changes in social, economic, geographic, and demographic contexts, as well as in policy environments.[4]

An improvement of energy performance of buildings, along with reducing carbon emissions, would also yield important benefits to building owners and occupants, such as improved durability, reduced maintenance, greater comfort, lower costs, higher property values, increased habitable space, increased productivity, and improved health and safety. Multiple benefits of energy efficiency to governments often include reduced societal health costs, improved air quality, an improved tax base and lower budget variation, higher GDP, and enhanced energy security.[15] Utilities enjoy cost and operational benefits due to reduced customer turnover, reduced emissions, and reduced system capacity constraints. The pursuit of energy efficiency can also bring major benefits to the construction industry, whose performance, given the importance of the sector in terms of GDP, number of employees, and size of business and companies, can have a significant impact on the growth of the whole economy (the buildings industrial sector accounts for about 7% of the nonfinancial business economy in the 28 EU member states and provides 11.5 million direct jobs, about 8.8% of total employment in the nonfinancial business economy).[6]

In this picture, the commitment of all nations to improving the energy efficiency of buildings has therefore greatly increased over the past decade, with the triple objective of significantly reducing the energy consumption of existing buildings, ensuring that all new buildings are characterized by high energy efficiency (very low-energy buildings), and using as much as renewable energy as possible, instead of fossil fuels, to meet the energy needs of both new and existing buildings.

Concerning new construction, the recast EU Directive on the Energy Performance of Buildings (2010/31/EU Directive) stipulates that by 2020 all new buildings constructed within the EU should reach nearly zero-energy levels (ie, buildings for which on average the energy need for heating and cooling is less than 30 kWh/m^2 yr).[4] The Building Technologies Program of the US Department of Energy has instead the strategic goal of achieving "marketable zero energy homes in 2020 and commercial zero energy buildings in 2025," while countries of the Asia—Pacific Economic Cooperation, such as Japan and Korea, have issued policies and set up clear and aggressive objectives for nearly zero-energy buildings, and established financial and taxation policies to stimulate development.[4] This means that in less than a decade all new buildings will be required to demonstrate very high energy performance and their reduced or very low energy needs will be significantly covered by renewable energy sources.

This objective is certainly an important goal (in Europe more than a quarter of the 2050s' future building stock is yet to be built), but considering that more than 50% of the current global building stock will still be operative in 2050 (in OECD countries this figure reaches or exceeds 75%), it is clear that the reduction of energy consumption and carbon emissions associated with buildings cannot be achieved without deep and rapid energy upgrading of the existing building stock (known as deep energy renovation, with a minimum reduction of energy consumption of 50—60% compared to prerenovation levels).[4,12,16,17]

Spurred in part by expanding efficiency-targeted policies, global energy efficiency investment in buildings is rising more rapidly than the overall growth of the construction sector: it was estimated to have reached US$90 billion (±10%) in 2014 and is expected to increase to over US$125 billion (excluding appliances) by 2020, with significant potential for additional profitable investments.[11] With improved and more widely implemented energy efficiency codes, standards, and programs, efficiency investment in each building is set to increase across most OECD national building markets. However, this level is still much lower than the estimated investment needed—US$215 billion by 2020—for the buildings sector to meet the International Energy Agency 2 degrees Celsius scenario (2DS).[11]

An effective energy upgrade of the existing building stock goes through increasing the rate, quality, and effectiveness of building renovation, as the current annual renovation rate is only 1% (1.2—1.4% in Europe, with a demolition rate of 0.1% per year).[12,18] This requires a major commitment on all regulatory, economic/financial, technological, and cultural fronts to increase minimum standards, reduce renovation costs, minimize disturbance for occupants, and address specific barriers, including socioeconomic ones and those linked to the building ownership structure. Effective solutions need to be demonstrated and widely replicated to help increase renovation to minimum rate of 2—3% per year.[12]

To separate aggregate building energy use from its historical drivers, several energy policy measures must act together. These include improved building energy codes for both new constructions and existing buildings undergoing retrofit, improved standards and labels for appliances, whole-building rating, labeling and disclosure programs, educational programs, and capacity building, as well as improved data availability and quality to allow for informed policy design and implementation.

1.2 Smart, sustainable, and inclusive buildings

The demand expressed by society in the third millennium is for a building system that can finally leave its consumer role and assume that of an energy producer. This new concept, which is a real cultural revolution compared to the past, cannot be realized without profound and important changes to the design canons that guided the architectural forms of the last century.

It is necessary in the construction sector to take a holistic approach to design that follows the principle of "life-cycle thinking," taking into account the environmental, social, and economic impacts of design choices along the entire life cycle of the building: from site identification, to construction, operation, and the final decommissioning and recovery of materials ("from cradle to cradle").

The aim of this approach is to promote a building industry based on knowledge and innovation, more resource-efficient, greener, and more competitive, with a high employment rate and favorable to social and territorial cohesion. A building sector is understood as:

> *a process in which all stakeholders (clients, financers, engineers, builders, material manufacturers, authorities, etc.) apply functional, economic, environmental and quality considerations to produce and renovate buildings and create a built environment that can be: attractive, durable, functional, accessible, comfortable and healthy; efficient in terms of resources (especially in terms of energy, materials and water); respectful of the context and local culture and heritage; competitive in terms of costs, especially in a long-term perspective.*[19]

With the objective of a "smart, sustainable, and inclusive" building industry, the aim is thus to overcome the age-old conflict between so-called "traditional architecture" and "bioarchitecture" or "green architecture," identifying a single architecture designed to meet the needs, both expressed and implied, of all present and future stakeholders.

Pursuing sustainability in the construction industry means interpreting sustainable development in the light of three closely linked aspects—economic, environmental, and social (triple bottom line)—while always matching it with specific and essential technical, functional, and quality requirements. In fact, the final quality of building work resides not only in meeting the needs of direct promoters and users, but also in providing a larger value for the community through the effects that any work of territory transformation has in cultural, economic, environmental, and social terms.

As regards environmental efficiency, the building system must encourage the use of renewable energy sources, exploitation of rainwater, proper treatment and recovery of waste water, and use of materials compatible with the environment that can be easily reused and recycled, do not contain dangerous substances, and can be safely disposed.

From the construction point of view, the need for building envelope configurations capable of reducing the energy demand is increasingly pushing architectural design toward the integration of building and equipment technologies, leading to the adoption of a single integrated building-equipment system which can change its performance over time in relation to the variations of environmental context and user needs.

At the same time, technological progress is constantly bringing to the market materials, components, technical-constructive and equipment solutions that can reduce the environmental impact, including health and safety aspects, in every phase of the building process.

We are therefore facing a dynamic inventory of technical and technological solutions that, if well orchestrated, may result overall in new and exciting architectural languages in continuous evolution. For this reason the energy efficiency challenge cannot and should not be seen as a restriction to the freedom of the designer in proposing solutions in response to the formal request of the client, but rather as an incentive to explore new fields of scientific and technological research in the fields listed above.

This challenge to architecture from third-millennium society to balance the world system of sustainable development requires a "global awareness" by all involved stakeholders, and will certainly lead to structural changes in the production sphere and an expansion of the regulatory framework which governs the construction industry. However, as stated by Italo Calvino, creativity and imagination are stimulated by respect for explicit rules, regulations, and constraints (*contraintes*). The constraint is a creative tool which boosts the chances of achieving original and bizarre solutions; paradoxically, it is "a celebration of the freedom of invention."[20,21]

A modification of the design process is thus required as it becomes more complex due to the introduction of the new "efficiency" parameter, necessitating the use of new skills along with new assessment and calculation tools. It is in fact necessary to control the three environmental, typological, and constructive levels simultaneously to achieve the best harmonization of the compositional, structural, and technological choices.

In this new scenario, the design process must necessarily go through a careful analysis of the context with which the intervention will have to interact during its entire life cycle. The reading of the programmatic framework in which the intervention is inserted, knowledge of the approval procedures, and consideration of all actors involved in the construction process are essential to make project choices, and thus technology solution choices, according to precise objectives of environmental compatibility, architectural and functional quality, energy efficiency, and economy. From the results of this analysis will flow all design inputs necessary to ensure the technical and economic feasibility of the work to be carried out and the meeting of the various needs of construction quality.

According to architect Jean Nouvel:

> *the modernity of today architecture lies in connection with the context. The context is the network of constraints that a building must comply with: economic, environmental, cultural, symbolic and technical. Constructing generic buildings, to be placed anywhere, not specific to urban areas, is doing things with no value.*[22]

The elements found in the course of the analysis will constitute neither the solution to the design choices nor a limitation to the design process. They rather represent a stimulus to creative design, "the grammar and syntax" of the architectural language that can reconcile formal and space quality with other forms of building quality.

In this context, the use of building information modeling (BIM) systems for computable representation of construction projects has become imperative for achieving the objectives of building quality required by contemporary architecture and reducing design, construction, and building management costs. BIM virtual modeling has opened up new and interesting perspectives to the design process in terms of creativity, verification, and optimization of alternative solutions, process control, and costs. Through BIM virtual modeling it is possible to integrate in real time the formal architectural solution (three dimensions) with the effects in terms of energy and environmental performance (four dimensions) and of construction, operation, and maintenance costs (five dimensions), controlling the entire project up to the construction phase from the formal, technological, and economic points of view.

For these reasons the EU, in its Directive 2014/24/EU of February 26, 2014 on Public Procurement, expressly invited member states to provide for public works contracts "to use specialized electronic tools such as, in particular, Building Information Modeling software" (Art. 22, para. 4), starting from 2016. Today this is already required in several European countries, such as England, the Netherlands, Denmark, Finland, and Norway.

The challenge for architecture in the new millennium therefore focuses on some key research areas that apply to every stage of the building process, from planning and construction through to the disposal of the building, and find their application in both new constructions and the restructuring and/or upgrading of existing buildings.

- A first research area regards the study of building shape and orientation, the positioning and sizing of openings, and the distribution of inner spaces to optimize the relationship with the environmental context and facilitate aspects such as solar gain, natural daylight and ventilation, acoustic comfort, and reduced consumption of resources and impacts on the environment (smart building shape).
- A second area concerns the building envelope, and focuses on the development of materials, products, and building systems that will improve the energy and environmental performance of the building and the comfort level for occupants through the use of advanced thermal insulation (smart insulation) and transparent closures (smart windows), as well as specific surface treatments or special coatings for building materials (smart surfaces). This area also includes the study of prefabricated components and the relationships between production processes and organization of the construction yard, up to the three-dimensional printing of products, individual components, or entire buildings, and the robotization of construction or demolition activities (smart envelope).
- A third area deals with the achievement of maximum efficiency of mechanical systems and equipment, the integration of renewable energy sources and their relationship to the grid, and the introduction of automation and control systems for buildings (a BMS—building management system) designed to manage data from all the building's equipment centrally to optimize energy management, microclimate, and security (smart energy systems).
- The last research area regards the study of user behavior to reduce energy waste and bring theoretical standards closer to the real end consumption (smart people).

It is apparent that the challenge of energy and environmental efficiency in buildings also requires new and effective tools to support the design activity. These tools shall be able to simulate, in a dynamic and integrated way, the building performance in terms of comfort, energy consumption, impacts on the environment, and time and costs of realization, to allow the optimization, from the very first project phases, of the different choices in relation to objectives, context, and available budget (building performance optimization and simulation tools). A review of available building performance simulation tools shows that in fact up to now most developed tools used at different stages are not proper design tools, but rather tools to evaluate the performance of a design (building performance simulation): out of 406 such tools listed on the US Department of Energy website in 2012, only 19 allowed building performance optimization.[23]

1.3 Zero-energy buildings

1.3.1 Definition and concepts

The term "zero-energy buildings" (ZEBs) is intended generally to indicate a category of buildings with very high energy performance, characterized by a very low or almost zero annual energy requirement that is entirely or very significantly covered by renewable energy, including energy from renewable sources produced on site or nearby.[4,24]

These buildings are thus characterized by an extremely low global energy consumption ($EP_{gl,tot}$) and, through the use of energy from renewable sources taken from the grid or produced on site or in the neighborhood ($EP_{gl,ren}$), almost zero consumption of nonrenewable energy ($EP_{gl,nren}$):

$$EP_{gl,tot} = EP_{gl,nren} + EP_{gl,ren} \approx \left(EP_{gl,ren}\right) \approx 0 \qquad [1.1]$$

As buildings are normally connected to an external network (grid-connected) for energy supply (electricity and gas), and in view of the discontinuous nature of renewable energy sources (available only at certain times of the day), what is actually equal to zero is not the consumption of nonrenewable primary energy, but the balance between primary energy supplied and the self-produced energy consumed directly or fed into the grid. They are thus more commonly and correctly referred to as "net ZEBs": their excess energy produced is fed into the electricity distribution grid, which in turn supplies the ZEBs with the energy required at times of insufficient self-production (at night or in conditions of poor wind or sunlight).

As consumption of nonrenewable energy is associated with both economic costs for the purchase of fuel and electricity and environmental costs due to emissions of carbon dioxide into the atmosphere, ZEBs are also referred to as "net zero-energy-cost buildings" or "net zero-energy-emissions buildings."[25] In fact, generally the primary energy use of a building accurately mirrors the depletion of fossil fuels and is proportional to CO_2 emissions. Proportions are distorted only if nuclear electricity is involved.

Buildings in which renewable energy production covers energy needs to a very significant extent, although without achieving zero net consumption of nonrenewable energy, are referred to as "nearly zero-energy buildings"; and when the production of renewable energy actually exceeds the needs of the buildings, they are called net plus-energy buildings or net positive-energy buildings.

Renewable energy can be produced by equipment located on the roof or integrated into the building envelope (generation on building footprint), or via ground systems (on-site generation from on-site renewables). It can be produced on site from outside sources such as biomass (on-site generation from off-site renewables), or in the neighborhood, as in the case of common shared facilities built in conjunction with a larger group of buildings (nearby generation), or produced off site and available through the network (off site in the grid).[26–28]

Renewable energy produced anywhere in the grid has the advantage of giving any building the opportunity to reach zero CO_2 emissions, considering that buildings in cities typically are not able to produce enough renewable energy, on site or nearby, to achieve carbon neutrality.

The definition of a ZEB does not cover investments in plants powered by renewable energy sources located in other places (off-site generation) or the purchase of green energy (off-site supply), although these are positive in terms of reducing global greenhouse gas emissions.

The energy performance of a ZEB is usually determined based on the amount of annual primary energy actually consumed, or expected to be required to meet the different energy needs for standard use of the building, such as lighting, winter and summer air conditioning, DHW production, ventilation, and, for the tertiary sector, lifts and escalator systems (energy and transport services).

The energy performance therefore refers to a standard use of the building (under certain conditions of interior temperature, light comfort, etc.) and not to the real use, as this reflects user behavior and could lead to measured energy consumption values significantly different from the theoretical values.

Global energy performance as well as that of individual building energy (space heating, summer air conditioning, DHW production, lighting) and transport (elevators and escalators) services are generally measured by performance indexes (EP_{tot}) expressed in total annual primary energy per area unit ($kWh/m^2 yr$), or sometimes, for tertiary-sector buildings, per heated volume unit ($kWh/m^3 yr$).

The use of the total primary energy consumption value for measuring building energy performance is necessary to compare energy consumption values of different natures and thus allow comparison with legal requirements or between different buildings. With an equal energy demand, buildings can indeed consume very different primary energy quantities according to the way that the energy is produced (Table 1.1).

The calculation of renewable, nonrenewable, and total primary energy consumption is carried out using specific primary energy conversion factors (f_p), provided and constantly updated by government agencies depending on the energy source used, taking into account the losses of extraction, processing, storage, transport, conversion or transformation, transmission or distribution, and everything else needed to deliver the supplied energy to the border of the system (see Table 1.2).

The energy performance of the building in terms of primary energy can either be measured only during the use phase (the most common case) or refer to a building's broader life cycle, thus including the phases of manufacture of materials (embodied energy), transportation, and construction.

The current definition of a ZEB excludes energy consumption related to the production stages of materials and building systems (embodied energy), transport to the yard, and construction and demolition of the building itself.

From an energy point of view, in fact these phases currently have a modest relevance compared to consumption in the use phase of the building: in a 50-year scope heating, cooling, lighting, and hot water production on average total more than 90% of the overall energy consumption throughout the entire life cycle of the building.[30,31] Furthermore, due to insufficient consistency in results from different life-cycle analysis (LCA) tools, it may be too early to require LCA information as part of regulatory requirements.

Nevertheless, as energy consumption during the use phase decreases, operation and maintenance of building systems and equipment, energy consumption during construction and disposal phases, and occupant usage patterns become more important and should be considered in the near future.

Table 1.1 Energy production and use level definitions

Energy production and use	Definition
Primary energy	Energy deriving from a source present in nature and not from transformation or conversion of any other form of energy. This classification includes both renewable energy sources (such as solar, wind, hydro, geothermal, biomass) and depletable sources, such as directly usable fuel (crude oil, natural gas, coal) or nuclear energy.
Secondary energy	Energy not directly available in nature and produced from a primary energy source. Examples are gasoline, a result of chemical refining, electric energy, and hydrogen. The production process, as a conversion from one source to another, inherently implies a loss of efficiency in the process, with part of the initial energy irretrievably lost.
Primary energy conversion factor	Dimensionless ratio that indicates the amount of primary energy consumed to produce one unit of supplied energy for a given energy carrier. It takes account of the energy required for extraction, processing, storage, transport, and, in the case of electricity, the average yield of the system of generation and the average transmission losses of the national electricity system or the average losses of distribution network in the case of district heating.
Energy consumption	Energy supplied to the final consumer's energy boundary.
Useful energy	Energy available to consumers after the last conversion made in the consumer energy conversion equipment, thus final energy consumption minus conversion losses.

In fact, LCAs of energy consumption of existing low-energy buildings show that the influence of embodied energy on the overall lifetime energy consumption of these buildings is much higher than for an inefficient building.[25,27] Similarly, usage patterns become increasingly important as buildings become more efficient, as the manner in which buildings are operated becomes a larger factor of overall energy consumption. Consequently, as designers move toward more energy-efficient buildings through energy sufficiency, energy efficiency, and clean energy, policymakers look to additional aspects such as energy management, embodied energy, and usage patterns to extract further energy savings.

1.3.2 Strategies

Regardless of the reference model, design and construction of a ZEB require a comprehensive approach to promote rational use of energy by reducing requirements (energy sufficiency), installing high-efficiency systems (energy efficiency), and using

Table 1.2 **Primary energy conversion factors of different energy carriers used in Italy**[29]

Energy carrier	$f_{P,nren}$	$f_{P,ren}$	$f_{P,tot}$
Natural gas	1.05	0	1.05
LPG (liquefied petroleum gas)	1.05	0	1.05
Diesel and bunker fuel	1.07	0	1.07
Coal	1.10	0	1.10
Solid biomass	0.20	0.80	1.00
Liquid and gaseous biomass	0.40	0.60	1.00
Grid supplied electricity	1.95	0.47	2.42
District heating	1.50	0	1.50
Solid urban waste	0.20	0.20	0.40
District cooling	0.50	0	0.50
Thermal energy from solar collectors	0	1.00	1.00
Electricity produced by PV, miniwind, and minihydro systems	0	1.00	1.00
Thermal energy from outside environment—free cooling	0	1.00	1.00
Thermal energy from outside environment—heat pump	0	1.00	1.00

renewable energy sources (clean energy) through an optimal balance of passive (affecting mainly the architectural, morphological, and building technology variables) and active (concerning the operation of technological systems) measures to achieve the best comfort conditions, proper use of resources, and reduced impacts on the receptor ecosystem.

It is apparent that the pathways to reach the net zero-energy objective are numerous and diverse, but should be carefully customized for the particular building type, use and operation, and climate conditions. The key to achieving the net zero-energy goal is to place energy at the forefront of the design process, along with more conventional considerations such as architectural shape, function, and costs, but with comfort at least equally important.

In particular, the goal of reducing energy consumption must be pursued at four different levels.

- The typological level, concerning the shape and orientation of the buildings, the positioning and sizing of the windows, and interior layout (smart shape).
- The technical and construction level, regarding the thermohygrometric characteristics of the building envelope (smart envelope).
- The technological level, inherent to the choices of equipment and use of renewable energy sources (smart energy systems).
- The sociocultural level, through education and involvement of users (smart people).

For new buildings it is possible to intervene at all four of these levels, whereas for existing buildings design solutions are normally concerned only with the technical construction (building envelope), technology (equipment), and social levels.

Typological and technical-construction levels allow reducing the energy requirements of the building (energy sufficiency measures). The so-called "passive house" approach reflects these concepts for cold and moderate climates. The best performance on building envelopes being achieved today is compliant with the Passivhaus specifications established by the Passivhaus Institut in Darmstadt in Germany in 1990, which provide an energy demand for heating and cooling lower than 15 kWh/m^2 yr and allow only marginal use of air conditioning.[32]

The technological level allows, for the same energy requirement, reducing both total primary energy consumption through greater equipment efficiency (energy efficiency measures) and nonrenewable primary energy consumption, thanks to the use of renewable sources (clean energy measures). Finally, the sociocultural level allows reduction of energy waste, bringing real end consumption closer to the theoretical standards (consumer behavior measures).

The objective is to achieve the energy performance level which gives the lowest cost along the estimated economic life cycle of the building, determined considering investment costs related to energy and maintenance, operating, and eventual disposal costs (cost-effective optimal level).

1.3.2.1 Typological-level measures

At the typological level, the reduction of energy requirements should be primarily pursued through a design mindful of the environment context (see Table 1.3) and aimed at architectural solutions in which building shape and orientation, distribution of rooms, and placement of windows have a profound effect on energy behavior, favoring natural lighting and ventilation, winter daylight, and exploitation of renewable energy to reduce energy losses in winter and overheating in summer (passive solar design). Correct orientation, an appropriate building shape, and rational spatial and functional organization of the interiors can yield significant energy savings (30–40%) without additional costs and significantly improve the conditions of thermal, sound, and visual comfort and air quality inside buildings and most used spaces.[31,33]

For tall buildings, a careful study of both shape and surface materials can reduce wind-induced load, with consequent savings in the amount of material used in the support structure. The boxy shape used until the end of the last century to minimize the costs of construction and maximize leasable areas of a building has thus been replaced by tapered or twisted shapes, or characterized by large openings able to reduce resistance to wind and oscillation, with consequent savings in the use of structural materials.

In this picture, the objective of the design process is to mediate between shape, function, and efficiency to achieve the highest "building quality" in its modern meaning of formal-spatial, technological, technical, environmental, maintenance, operational, and useful quality: "form follows sustainability."

Table 1.3 Environmental context influence on design and technological choices

Climate parameters	Monitoring parameters	Design choices	Design objectives
Air temperature	Daily and monthly average temperature Monthly average of minimum and maximum daily temperatures Degree days	Building form factor and orientation Room distribution Building envelope thermal resistance Shading systems Mechanical equipment sizing Outdoor space characterization	Thermohygrometric comfort Reduction of energy consumption
Wind	Prevailing direction sector for each month Average speed of prevailing wind (m/s) Presence in prevailing sector (%) Wind calm (%) Average monthly wind speed (m/s)	Building form factor and orientation Room distribution Size and position of openings Building envelope thermal resistance Renewable energy sources Noise and pollution barriers	Thermohygrometric, physiologic, respiratory/smell comfort Exploitation of renewable energy sources and reduction of energy consumption Structure cost reduction
Humidity	Monthly average of minimum and maximum relative humidity	Thermohygrometric characteristics of building envelope (humidity condensation) Ventilation needs Mechanical equipment sizing Outdoor spaces	Thermohygrometric, physiologic, respiratory/smell comfort Structure maintenance
Rain	Annual and monthly rainfall average Average distribution of rainy days, days when soil is covered by snow, thick fog days, stormy days, and dry days	Roof type and inclination, rainwater harvesting and disposal, choice of vegetation	Rainwater recovery and reduction of water consumption Correct disposal of waste water Vegetation management
Sun path	Available sun hours over course of a year for each elevation orientation and roof Radiation energy related to sunshine hours affecting building Ratios with external air temperature Sun ray penetration into interior spaces	Building ground plan and orientation Sizing, orientation, and position of glazed surfaces and shading systems Inclination of sun-collecting surfaces Room distribution Thermal inertia of building envelope	Improvement of hygienic and thermohygrometric well-being Reduction of building energy consumption Exploitation of available solar energy via passive and active systems
Sky conditions	Monthly average of fair weather, overcast, and cloudy days	Sizing, orientation, and position of glazed surfaces and shading systems Average reflecting factor of indoor surfaces	Improvement of visual comfort conditions Reduction of building energy consumption

In particular, the design choices at the typological level aimed at reduction of energy requirements concern:

- optimization of the ratio between the dispersing surface and the heated volume of the building (ie, the form factor), favoring shapes able to optimize thermal exchanges with the external environment in both winter and summer;
- orientation of the building to maximize winter solar gain, reduce it during summer, and favor natural lighting of interiors;
- positioning and sizing of the openings relative to the orientation of the facades and intended use of the interior space to increase winter solar gain, reduce it during summer, and ensure a high level of natural lighting and ventilation;
- distribution of rooms and organization of internal spaces in relation to the different orientations, to maximize the conditions for thermal and visual comfort;
- choice of solutions for improvement of the microclimate, such as a green system, vegetable barriers, reduction of impermeable surfaces, etc.

A thorough analysis of meteorological parameters of the intervention is thus particularly important in this design phase, including analysis of climate and microclimate, study of sun path and natural lighting, and survey of the availability of renewable energy sources.

Climate aspects have the most influence on the design choices at the building scale: the climatic characteristics of the place will in fact dictate building shape and orientation, type of materials, characteristics of the housing, internal distribution of rooms, and equipment systems solutions.

The building's exposure to climatic loads not thoroughly considered during the design phase can also lead to premature decay of the building envelope, in particular perimeter walls and roofs, directly affected by the action of wind and rain, temperature variations, and water vapor diffusion, consequently increasing maintenance costs.

1.3.2.2 Technical and construction-level measures

The importance of the building envelope in determining energy performance should not be underestimated, as space heating and cooling account for over 30% of all energy globally consumed in buildings, rising to as much as 50% in cold-climate countries. In the residential sector, the share of energy used for heating and cooling in cold countries is over 60%.[12]

As regards the technical-construction level, design choices should relate to the optimization of the thermohygrometric characteristics of the building envelope, concerning both the opaque components (to maximize inertia and thermal insulation characteristics), and the transparent components (to maximize thermal insulation, light transmission, and protection from summer solar radiation).

In particular, the design choices at the technical-construction level aimed at the reduction of energy requirements concern:

- high levels of insulation in walls, roofs, and floors to reduce heat losses in cold climates, optimized using life-cycle cost assessment;
- high levels of thermal inertia and highly reflective surfaces in roofs and walls to reduce summer overheating in hot climates;

- high-performance windows with low thermal transmittance and climate-appropriate solar heat-gain coefficient (SHGC), and solar shading systems to favor solar gain in winter and reduce it during summer;
- properly sealed structures to ensure low air infiltration rates with controlled ventilation for fresh air (air sealing alone can reduce the need for heating by 20−30%);[1,12]
- minimization of thermal bridges (components that easily conduct heat/cold), such as high thermal conductive fasteners and structural members.

The most advanced materials and technologies available on the market for improving building envelope energy performance include the following.

- Advanced insulating materials (fiber-reinforced aerogel blankets, vacuum insulating panels) with extremely reduced thickness, able to allow architectural integration, minimal encumbrance, thermal bridges reduction, and piping application.
- Transparent insulating materials to combine daylight with energy efficiency.
- Phase-change materials to raise thermal inertia of the envelope or internal partitions and provide thermal heat storage for solar passive heating without increasing mass.
- Reflective coatings to increase thermal insulation or reduce overload solar radiation (cool roofs).
- Self-cleaning and antireflective glazing to maximize luminous transparency.
- Advanced thermal insulation glazing (monolithic aerogel insulating glass units) and heating glazing to reduce thermal loss from windows in cold climates and improve thermal comfort.
- Dynamic glazing (smart windows) to allow dynamic control of solar radiation and HVAC integration.
- PV glazing to reduce incoming solar radiation and simultaneously produce electricity.
- Living green walls to reduce incoming heat flow during summer and improve urban air quality.
- Building envelope solutions integrated with operation of mechanical equipment (transpired solar collectors, ventilated walls, double-glazed skin facades, passive solar systems).

1.3.2.3 Technological-level measures

At the technology level design choices shall aim to identify systems solutions capable of allowing the best conditions of interior comfort and at the same time maximizing the reduction of primary energy needed to satisfy all requirements (winter and summer air conditioning, DHW production, ventilation, and lighting) by increasing the efficiency of all equipment systems and using renewable energy sources. The goal is to realize positive-energy buildings (net plus-energy or net positive-energy buildings) able to self-produce on site, through renewables, more energy than is needed to meet the different needs.

For this reason, it is essential that the design of equipment systems goes hand in hand with architectural design, and therefore starts at the very moment in which the work is conceived and the needs of the client are compared with constraints and resources. Equipment choices may provide different solutions which, to achieve a high-quality construction, must integrate perfectly with the architectural choices and offer a concrete response to the requirements that the shape alone may not be able to satisfy fully.

Evaluation of the best technological and equipment solutions for improving the energy performance of a building must take the following main aspects into account.

- Architectural integration.
- Global and nonrenewable primary energy savings achieved by the proposed solution.
- Reduction of the environmental impact resulting from direct (fuel consumption) and indirect (electricity consumption) air emissions, with particular reference to CO_2 (the main climate-altering gas produced in energy transformations), as well as possible noise and electromagnetic emissions.
- Economic viability of the proposed solution (life-cycle costing).

In particular, design choices at the technological level aimed at reducing global primary energy consumption concern:

- the method of heat and electricity production and the exploitation of renewable energy sources;
- equipment of a high energy efficiency class, with particular reference to heat pumps for space heating, cooling, and hot water production, focusing on the latest heat pump technology using CO_2 as working fluid, and on combined heat and power (CHP) and combined heat power and cooling, along with the introduction of variable-flow systems;
- energy recovery systems (condensing boilers, heating recovery ventilation systems, dual compressor heat pumps);
- thermal storage including sensible (hot water, underground storage), latent (ice storage, phase-change materials), and thermochemical storage;
- electricity storage systems (lithium ion batteries, fuel cells);
- flow regulator power systems to reduce energy consumption of electrical installations through adjustment and stabilization of the energy flow;
- hybrid natural cooling systems to precool ventilated air through ground pits, air exchange, water, and night sky radiative cooling;
- low-temperature conditioning equipment, such as floor, wall, or ceiling radiating panels integrated with thermal solar, heat pumps, or condensation boilers;
- highly efficient systems and appliances for interior space lighting (Light emitting diodes (LED), organic LEDs, quantum dots LEDs);
- building/home energy management systems (BEMS/HEMS);
- "internet of things" devices for user awareness and behavior improvement.

Among the different possible sources of nonfossil energy (wind, solar, aerothermal, geothermal, hydrothermal, tidal, hydropower, biomass, landfill gases, residual gases from purification processes, and biogas), those that can more easily be applied in today's architecture are undoubtedly solar technologies, vertical axis wind technologies,[34] biomass energy exploitation, and energy harvesting systems such as in-pipe minihydro systems[35] and piezoelectric systems for pavements (Fig. 1.1).

Renewable energy can be used in various forms in thermal systems, such as through the capture of solar heat gains for space or water heating, the use of biomass for heating, or heat pumps (thermal, geothermal, or hydrothermal) for space and water heating as well as space cooling.

Electricity can be produced by PV, wind, minihydro, and piezoelectric systems. The use of cogeneration (CHP) or trigeneration (cogeneration of heat, cooling, and power) systems using gas, biomass, or hydrogen with fuel cells allows combining the

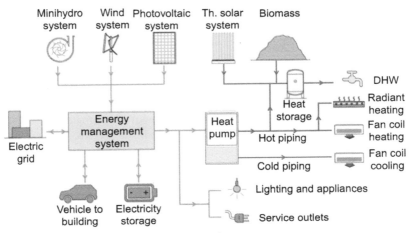

Figure 1.1 Building energy equipment system scheme.

production of electrical and thermal energy, ensuring greater conversion efficiencies compared to separate production.[36] District heating and cooling systems driven by energy from renewable sources are also viable.

These technologies shall be accurately mixed, integrated, and operated according to the specific climate and local conditions, including cost and incentives.

Plug-in hybrid and electric vehicles can be integrated into the shared energy concept, serving as both electrical storage and load management devices, but also providing backup power to houses during emergency situations such as natural disasters (vehicle to building and vehicle to grid).

To ensure energy-efficient operation, building energy parameters must be adequately controlled and monitored by building service systems. It is fundamental to provide interoperable energy control, monitoring, and automation tools for efficient heating, cooling, ventilation, DHW production, lighting, shading, energy storage and generation, and other building services while ensuring high levels of indoor environment quality.

Suboptimal control of energy-consuming systems such as heating, cooling, ventilation, and lighting is in fact endemic in existing buildings, with spaces heated when it is not necessary, lighting left on, ventilation operating continuously at maximum capacity, etc. As a result large quantities of energy are wasted, and thus there is considerable potential for savings.

Several analyses agree that building automation technologies and building automation control systems (BAT/BACS) solutions are among the cost-optimal measures able to yield economically viable energy savings almost regardless of the quality of the building envelope and the occupant user profile.[37] Assessments based on EU standards and official cost-optimal building performance requirements show how intelligent controls and BATs produce consistent energy savings in residential dwellings. In service buildings, energy consumption reductions of more than 50% are also reported,[2] allowing a great improvement in energy performance classification.[38]

In addition to being a cost-optimal measure, BAT/BACS require relatively low investments and can be implemented without extensive work causing the building to be taken out of service. Building owners who find themselves forced by reasons of occupation or budget constraints to postpone major energy-related renovations can quickly benefit from investing in improving the operating efficiency of their energy equipment. Moreover, BAT/BACS will still provide valuable energy savings even if at a later stage the building envelope is renovated. BAT/BACS provide a yet more cost-optimal alternative when the building envelope cannot be upgraded to state-of-the-art performance levels, for example when cultural or historic factors prevent intrusive actions.

Furthermore, BAT/BACS can also complement the extra functionality provided by smart meters, improve the use of renewable energy sources, and increase overall grid stability by providing buildings with massive load-shifting and storage management capabilities. By integrating demand-response-enabling elements into energy management systems, it is possible to achieve the best building energy system interaction and optimize energy consumption, production and storage at the building level considering the availability and price of energy supplied via the grid. In fact, grid-tied net ZEBs still have considerable upstream impacts on the environment and energy supply infrastructure because of their diurnal and/or seasonal dependence on a centralized energy supply when renewable self-production is insufficient. Future research and development must recognize the complex interactions between individual buildings, the community, and the larger scale (eg, urban and grid-wide).[39]

1.3.2.4 Sociocultural-level measures

The energy transition toward ZEBs will require not only the implementation of new technological solutions but also a change in user behavior with respect to uptake of energy efficiency solutions and increasing use of renewable energy sources. Households, public authorities, large and small enterprises, and other energy users should therefore be considered at the heart of the energy system and become active market players.[39]

For instance, energy efficiency or demand response solutions could produce changes in user energy consumption behavior, providing flexibility as well as new opportunities to produce renewable energy at local level. In this context, consumers should be directly engaged in the energy system through better understanding, information, and market transformation. Innovative technologies, products, and services should also help consumers to manage their energy demand and supply actively.[40]

In recent years important market changes have occurred, with the emergence of smart meters, wireless sensors, small-scale renewable energy systems, connected smart appliances, home networks, smart energy infrastructures, and local energy storage systems, setting the path toward smart systems. These new technologies, products, and services are already helping consumers in better understanding and managing their energy demand (eg, by detecting malfunctions or optimizing intelligent control and automation) and energy supply.

Moreover, the net zero-energy objective places greater emphasis on ensuring that a building performs as expected after occupancy. Therefore it is often essential to have the designers involved in the commissioning and first months or years of operation, because many features may require some fine tuning and occupants may need some education on the building's proper functions, as energy performance, comfort, and building operation—both passive and active—are more than ever closely linked. In cases of building energy renovations it is fundamental to involve the building users across the whole energy retrofit process and instruct them about correct energy behaviors.

1.3.3 Building-integrated photovoltaics

PV technologies are expected to be the main resource for on-site generation of electricity in net ZEBs.[41-47] In fact they have huge potential in terms of their ability to integrate with the building envelope to reach the definition of a new architectural language in which solar-energy-capturing surfaces can become an expressive and compositional element. For this reason, PV systems are to be considered as just one more element in the building that is being designed, and therefore choosing their proper location is important, not only from a technical point of view but also from the esthetic point of view to ensure that the module and its supports have the maximum visual and efficiency quality. PV technologies should be interpreted as an element in architecture that contributes to the formal quality of the whole.

Integration may be limited to the application of panels on the outer surface of the building envelope coplanar with the supporting surface without the replacement of the materials that constitute the supporting surfaces (building-adapted PV systems—BAPV). BAPV modules are used solely for generating energy, even as part of the architectonical composition of the building, but do not replace or function as a construction component. Such is the case for example in retrofit applications, with PV modules over sloped roofs, arranged with a given angle on flat roofs, or assembled on facades or shading devices.

But integration can go further, by replacing the actual architectural elements of the building with PV ones (building-integrated PV systems—BIPV). In BIPV, PV construction elements are part of the building envelope, acting not only as an energy generator but also performing an envelope function on roofs, facades, atria, or shading elements (Figs. 1.2 and 1.3).[7,48-51]

The design approach can aim at "invisible application" in which the solar modules are located out of sight or intentionally designed not to be distinguished from the rest of the building envelope, or push at enhancing or even determining to a significant extent the image of the building (PV highlighted and leading application).[48]

In fact, every element of the building surface, from the roof to the perimeter walls, the transparent closures, and the shading screens, is suitable for the installation of PV systems. Their integration into the building organism allows, in addition to energy production, the provision of other major functions such as sound insulation, thermal insulation, and solar radiation protection, resulting in reduced air-conditioning load during summer.

Figure 1.2 DSSC building-integrated photovoltaic facade, SwissTech Convention Center, Geneva.
Courtesy of Solaronix SA ©.

Figure 1.3 Semitransparent a-Si building-integrated photovoltaic facade, GenYo Center, Granada.
Courtesy of Onyx Solar LLC ©.

Integration can give sensible economic benefits: by replacing the building material with a solar module, the value of the first is deducted from the total cost of the system; furthermore, if the system is really an integral part of the building, the cost of its support structure and the lot on which it is placed are already accounted for.

Currently, several products for the integration of PV technologies in the building are available on the market, characterized by advanced design, flexibility, attention to detail, and variety in color, shape, and transparency degree, to meet any requirement. Thanks to both inorganic (second generation) and organic (third generation) thin-film technology, PV systems can now be integrated on any surface and in any building element, from roof tiles to flat or curved windows, ethylene tetrafluoroethylene (ETFE) cushions, shading elements, pavements, and street furniture elements.

Current PV modules available on the market yield electrical power of 80—150 Wp per square meter of surface area exposed to the sun, depending on the type of solar cells used (c-Si monocrystalline and polycrystalline, a-Si, CdTe, and CIGS thin film).

A typical crystalline silicon (c-Si) PV module consists of a glass sheet for mechanical support and protection, and laminated encapsulation layers of ethylene vinyl acetate for ultraviolet (UV) and moisture protection. The module typically comprises 60—96 individual solar cells, each 15 cm^2 and capable of producing 4—5 W under peak illumination (Wp). The module is completed by a fluoropolymer backsheet for further environmental protection and an aluminum frame for mounting. Common module dimensions are 1 m × 1.5 × 4 cm, and peak power ratings range from 260 to 320 W.[7,42]

For a given installed power, produced electricity depends on available sunlight, and thus on latitude, the state of the sky, the temperature, the orientation and inclination of the module, and its ability to capture direct as well as diffused solar radiation. Under optimal orientation and inclination, electricity production can range according to different technologies from 87 to 187 kWh/m^{2}yr for a latitude of 55°41'0", corresponding to the city of Copenhagen, Denmark with solar irradiation of 1025 kWh/m^2 per year, and from 167 to 359 kWh/m^{2}yr at a latitude of 38°1'0", corresponding to the city of Trapani, Italy with solar irradiation of 1963 kWh/m^2 per year (see Table 1.4).

Over time the efficiency of solar cells and modules has greatly increased, with an equally important cost reduction of installed Wp.[52] In the last 10 years the efficiency of average commercial wafer-based silicon modules has increased from about 12% to 16%, while CdTe module efficiency increased from 9% to 13%. In the same time span, in Germany prices for a typical 10—100 kW$_p$ PV rooftop system dropped from €5000/kWp to €1300/kWp, a net price regression of about 74% over 10 years. The experience curve (also called the learning curve) started in 1980 shows that module price decreased by about 20% with each doubling of the cumulative module production due to economies of scale and technological improvements.[52] Solar model aging and performance degradation has also improved over time, and today available products come with guaranteed performance of 90% of original power output after 10 years of weather exposure and 80% after 20—25 years.

Furthermore, testing on PV products manufactured in the 1980s with lower standards than today showed >80% performance in a vast majority of samples.[52]

Today, the main BIPV products (see Fig. 1.4) available on the market are:

- unconventional PV modules specifically designed for architectural applications, such as flexible modules, thin-film strips on a rigid support, PV roofing tiles, and transparent PV modules for facades, windows, and roofs conveniently built and installed to allow the passage of light inside the building envelope;
- special components for roofing, opaque vertical surfaces, or ventilated facades, formed by the assembly and integration of unframed laminated PV modules in place of conventional building materials, and fitted with a European-patented fixing system.

Table 1.4 Comparison between main PV technologies available on the market: efficiency and energy yield

Technology	Crystalline silicon		Thin film		
	Mono	Poly	CdTe	CIGS	a-Si/μc-Si
Product	Sun power X21	Sharp Nd-RC260	First solar	Solar frontier SF 170-s	3sun
Module efficiency (%)	21.5	15.8	14.7	13.8	10.0
Area (m^2/kW_p)	4.7	6.3	6.8	7.2	10.0
Maximum energy yield at latitude 55°N (1025 kWh/m^2yr)					
Production for installed power (kWh$_{el}$/kW$_p$yr)	871				
Production for installed surface (kWh$_{el}$/m^2yr)	187	138	128	120	87
Maximum energy yield at latitude 41°N (1737 kWh/m^2yr)					
Production for installed power (kWh$_{el}$/kW$_p$yr)	1476				
Production for installed surface (kWh$_{el}$/m^2yr)	317	233	217	203	148
Maximum energy yield at latitude 38°N (1963 kWh/m^2yr)					
Production for installed power (kWh$_{el}$/kW$_p$yr)	1669				
Production for installed surface (kWh$_{el}$/m^2yr)	359	264	245	230	167

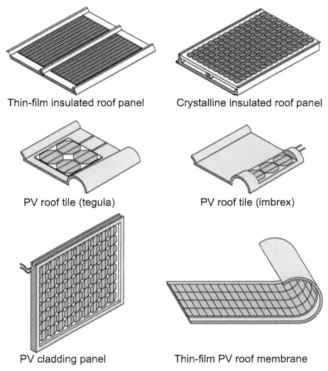

Figure 1.4 Building-integrated photovoltaics.

These elements must be designed and manufactured industrially to provide, in addition to electricity generation, basic architectural features such as:

- thermal insulation of the building, ensuring the maintenance of energy requirements thanks to thermal transmittance values comparable to those of the architectural component replaced;
- waterproofing and the subsequent impermeability of the underlying building structure;
- mechanical resistance, which should be comparable with the building element replaced.

Semitransparent PV (STPV) products use different PV cell technologies, from spaced opaque crystalline silicon cells to thin-film transparent PV cells, and can replace conventional windows and skylights in both commercial and residential buildings.

In most commercial buildings, where reduction of cooling energy costs is important, instead of resorting to tinted or ceramic fritted outer glass panes in insulating windows to reduce solar transmittance, STPV glazing may be used to reduce solar heat gains and generate solar electricity, while providing adequate daylight and vision. In residential buildings, STPV modules can be used in skylights or adjacent conservatories and greenhouses.

PV systems can also be integrated with heat production systems (BIPV with heat recovery or BIPV/thermal—BIPV/T), with the advantage of requiring less surface area to produce the same amount of thermal and electrical energy compared to side-by-side BIPV and solar thermal technologies (Fig 1.5). Depending on the

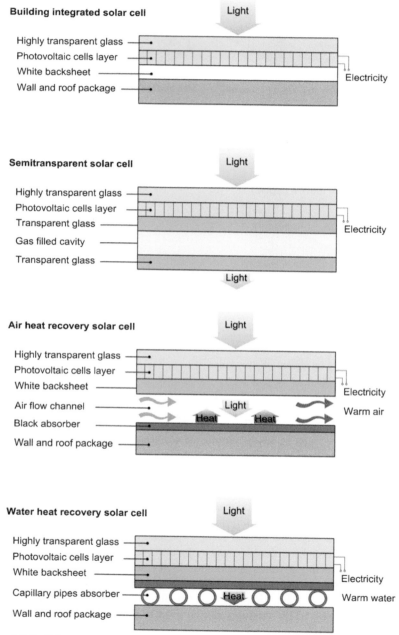

Figure 1.5 Building-integrated photovoltaic (BIPV) and BIPV/thermal systems.

climate, operating conditions, and design, BIPV/T systems can produce 2—4 times more heat than electricity.

In these systems, the absorbed solar energy that is normally wasted as heat is instead recovered either actively, using a fan or pump, or passively by a heat removal fluid (HRF) flowing on the rear side of the PV layer. As the fluid circulates behind the PV module, it extracts heat from the cells through convection, reducing the PV cell temperature and increasing the electrical efficiency. The recovered heat can be used either directly or indirectly for water and space heating (eg, indirectly using a heat pump).

BIPV/T systems can be classified as air-based or water-based depending on the HRF used. Water systems have typically higher thermal efficiencies than air-based systems due to the better thermophysical properties of water compared to air, and do not need additional heat transfer enhancement measures such as fins or roughness elements added at the back of the PV module.[23] Air-based systems, however, do not require any significant maintenance because leakage is not as critical as in liquid systems (where the liquid also needs to be periodically replaced or drained).

Another classification depends on the HRF temperature, identifying low-temperature ($<60°C$) or high-temperature ($>60°C$) BIPV/T systems. Low-temperature open-loop BIPV/T—air systems are normally used to reduce space heating energy consumption by preheating the fresh air entering the HVAC system in commercial, multiresidential, or industrial buildings. Closed-loop BIPV/T—air or BIPV/T—water high-temperature systems are employed for DHW heating, where higher outlet temperatures are needed. However, as PV cell temperatures are generally higher than in an open-loop system, lower electrical performance and accelerated PV module deterioration are to be taken into account.

Commercial hybrid PV/thermal systems for building-adapted installation are already available or about to enter the market. UK-based Naked Energy, for example, is introducing in 2016 its Virtu hybrid panel, which combines high-efficiency vacuum-tube technology to harvest solar heat with electricity production by a crystalline silicon solar cells array covering the coaxial water tube. Combined heat and electricity output efficiency reaches 95%, due to the self-rotation of tubes and arrays which increases energy yield by 20—45%, and the system is able to favor either heat or electricity generation according to the need.

1.4 Green buildings

The term "green building" indicates a building which has a lower environmental impact compared to an identical building constructed in accordance with local regulations and current practice, as evaluated along its entire life cycle: from raw material extraction to product manufacturing and transport, construction, and management and maintenance of the building until its decommission.[8,53–55]

Environmental aspects taken into account in this assessment concern relations with the context, resource consumption, diffusion of pollutants into air, water, and soil, waste, and indoor well-being.

Compared to a ZEB, the differences are on the one hand in the performance requirements, and on the other hand in which phase of the life cycle these requirements are assessed.

As regards the first aspect, in addition to the energy efficiency requirements that characterize ZEBs, a green building must be able to offer its users the optimum conditions of thermal, visual, acoustic, electromagnetic, and psychological comfort and indoor air quality through efficient use of all resources (including water and material resources), a low level of emissions, and the use of materials compatible with the environment, possibly locally sourced (so-called zero km material), which can guarantee both durability and reuse and recyclability of the construction components at the end of operating life. In particular, acoustic comfort is often neglected in the design of standard and net ZEBs as it may conflict with design for optimal daylight and natural ventilation. In fact, recent postoccupancy evaluations of low-energy buildings show they often score highly for all categories of occupant satisfaction except acoustic quality and privacy.[23]

Concerning timing of assessments, unlike ZEBs, whose energy performance is measured only in the operative phase of the building, the performance of a green building is measured along its entire life cycle (*from cradle to cradle*), including production (*from cradle to gate*), transport (*from gate to site*), construction, and demolition phases (*from construction to cradle*), to reduce the net environmental imbalance between the benefits and burdens to the environment during the different phases of the building process.

Over the last 20 years the green building has arisen as the most important and progressive trend in the building industry. It is being recognized as a long-term business opportunity, as shown by the significant increase in the number of green building technology patents filed since the beginning of the new century, covering both constructional and architectural elements and equipment systems (renewable energy integration, energy-efficient HVAC and lighting technology, etc.).[56–59]

But while design, construction, and operation of buildings have steadily progressed, these forward steps are still minute when compared with the rate of change that is required to avoid the worst effects of climate change and other global environmental challenges.

Nevertheless, the growth of green buildings is unmistakable, and today 51% of building designers, contractors, and owners expect more than 60% of their work to be green by 2015, up from 28% of firms in 2012. For instance, in 2012 green building construction in the United States was already worth US$85 billion, and 20% of residential and 44% of nonresidential building projects worldwide qualified as green building activity, with green building project activity forming 38% of total construction investment.[11]

1.4.1 Green building products

In the concept of sustainable construction, building materials play a key role throughout the entire life cycle of the building in several respects.

The overall life of a building begins with the extraction of the raw materials needed for production of construction elements, which are then transported to the site and used for its realization, and ends with the management of demolition waste coming from its decommission. The extraction and processing of raw materials in building products imply the consumption of materials and energy and the release into the environment of pollutants and waste related to manufacture, transportation from production sites to the construction yard, and installation methods.

Looking at *design for resource conservation* and *design for low-impact materials, cleaner production, and reuse and recycling*, green building designers are increasingly focused on choosing environmentally friendly products, certified (Eco-label) or with an environmental product declaration, and environmentally friendly materials that are of vegetable or animal origin (wood, bamboo, cork, cardboard, cellulose fiber, hemp, flax, straw, corn starch, wheat stalks, sheep wool, seaweed, mushrooms, poultry feathers), recycled (recycled plastic products, cardboard, glass, metal, stone and bricks), or even recovered and reused after selective demolition (including the reuse of containers),[60,61] free of hazardous substances, whose processing cycles have low environmental impact, and possibly supplied by companies holding an ISO 14001 or EU EMAS Regulation certified environmental management system, or that participate in an extended producer responsibility program or are directly responsible for extended producer responsibility.[62,63]

In this picture, using vegetation for constructing living walls or green roofs (Figs. 1.6–1.8) is increasingly widespread, with the vegetable species forming a real technological component of the building envelope. In summer the vegetation reduces heat flows entering through the envelope thanks to its shading, the absorption of radiant energy needed for photosynthetic processes, and the thermal role of the evaporation−transpiration processes, in addition to the greater thermal inertia offered by the mass of the culture medium. The vegetation also improves air quality, thanks to the dust-filtering action of the plants, and reduces the visual impact of walls or roofs where vegetation can become a natural habitat for the development of biodiversity.[64,65]

Figure 1.6 Green roof in Chicago.
Photo by the author.

Figure 1.7 Green roof in Chicago.
Photo by the author.

Figure 1.8 Living wall, Oasis d'Aboukir, Patrick Blanc, Paris.
Courtesy of Patrick Blanc.

The search for more sustainable construction methods and the growing awareness of the importance of embodied energy in the global environmental impact of buildings have also brought designers' attention back to the structural use of wood, bamboo, and engineered wood-derived products, so-called "massive timber"—such as cross-laminated timber made from solid wood layers set at 90° orientations, laminated strand lumber made from a matrix of thin chips, and laminated veneer lumber made from thin laminations of wood, similar to plywood but much larger in scale—which can be used in buildings up to 40 stories ("playscrapers") even in areas of high seismic activity.[66–68]

The main construction method for playscrapers uses a central core of solid wood hosting staircases and lifts, and wood floor slabs and steel beams, providing a system ductile enough to withstand wind and seismic loads. Construction costs are on par with comparable concrete or steel buildings, with great environmental benefits added. In fact, a 20-floor wooden building removes up to 3100 tonnes of carbon per year from the atmosphere, compared with an output of up to 1200 tonnes of CO_2 by a concrete building of similar size; the difference of 4300 tonnes is equivalent to the carbon emissions of 900 cars over the course of a whole year.

Moreover, according to the Canadian Wood Council:

> *compared to wood design, the steel and concrete designs embody 26% and 57% more energy, emit 34% and 81% more greenhouse gases, release 24% and 47% more pollutants into the air, discharge 400% and 350% more water pollution, produce 8% and 23% more solid waste, and use 11% and 81% more resources (from a weighted resource use perspective).*[69]

Green building design also pays special attention to resource supply methods, according to the principles of sustainable management (for instance, certified wood from forests managed according to the sustainability principles of programs such as the Forest Stewardship Council (FSC) Programme for the Endorsement of Forest Certification Schemes, the Sustainable Forest Initiative, or bio-based products compliant with the Sustainable Agriculture Network's Sustainable Agriculture Standard, tested using ASTM (American Society of the International Association for Testing and Materials) Test Method D6866 and legally harvested).[70]

The objective of construction sustainability also requires the designer to focus on the one hand on the use of local materials and products (zero km material; Table 1.5), and on the other hand on adoption of lighter building materials and promotion of space optimization strategies through the choice of small-size prefabricated elements to be assembled on site, or even directly realized on site via three-dimensional printing processes (*design for efficient distribution*).

Table 1.5 Zero km materials requirements (Living Building Challenge 3.0)[71]

Living Building Challenge
• 20% or more of materials construction budget must come from within 500 km of construction site.
• An additional 30% of materials construction budget must come from within 1000 km of the construction site or closer.
• An additional 25% of materials construction budget must come from within 5000 km of the construction site.
• 25% of materials may be sourced from any location.
• Consultants must come from within 2500 km of the project location.

The use of nonlocal components is only justified when it brings advantages in building performance, and energy efficiency in particular, contributing to a significant reduction in operating costs. In this case, transport methods with less environmental impact, such as rail transport, are preferred.

The goal of reducing energy consumption and environmental impacts in the use and maintenance phase of the buildings (*design for use and green maintenance*) is facilitated by the progressive introduction of advanced nanotechnology materials, whose physical and mechanical properties greatly exceed those of traditional materials, and of active components and devices, so-called "smart materials," able to modify their own characteristics in relation to different operating conditions determined by climatic agents or users, allowing the realization of dynamic facades.

Thanks to nanotechnology it is now possible to improve the performance of existing materials and synthesize new classes of materials with properties not previously conceivable. Innovations include high-performance nanoporous insulating materials; ceramic materials with greater hardness and toughness; metals with greater hardness, higher tensile strength, and special electrical properties; and polymers with higher mechanical strength, resistance to heat, chemical attacks, atmospheric agents, and aging, better gas-exchange barrier properties, and improved transparency and electrical conductivity. Surfaces of existing materials can be treated to make them, for instance, nonslip, scratch resistant, corrosion resistant, nonwettable, self-cleaning, antibacterial, oil repellent, sterile, reactive, UV resistant, insulating, or heat or light radiation reflective, and able to modify the level of transparency (electrochromic, photochromic, or thermochromic) or produce electric or luminous energy (piezoelectric, PV, photoluminous).

To mitigate and reduce air pollution, particularly interesting are photocatalytic products such as paints, mortars, plasters, and coatings containing titanium dioxide (TiO_2), a photocatalyst element able to change the speed of a chemical reaction through the action of light. By using UV light energy, photocatalysts induce the formation of strongly oxidizing reagents, which are able to decompose by oxidation organic (benzene, dioxins) and inorganic (NO_x, SO_x, CO, NH_3) substances present in the air to turn them into harmless products. These photocatalytic products may be used in outdoor applications, for construction of sidewalks, squares, roundabouts, coatings of facades, and porches, and indoors in kitchens, bathrooms, schools, and hospitals. Among recent applications of TiO_2-based photocatalytic products on building envelopes, particularly interesting are the Vanke and Palazzo Italia pavilions at Milan 2015 Expo (Figs. 1.9 and 1.10). The latter featured a 9000 m^2 air-purifying building shell composed of 900 panels made of "i.active Biodinamic" cement based on TX Active.

Equally interesting is the "low-tech biotech" innovative building approach, which combines engineering progress with the most advanced biological technologies. An notable example is the Hy-Fi project by The Living (Fig. 1.11), winner of the Young Architects Program for constructing the installation at the New York Museum of Modern Art that provided exhibition spaces for "Warm Up Summer Music" in June–September 2014. Three towers were built with special biological bricks manufactured from a revolutionary combination of corn husk and fungi mycelium, applying a

Figure 1.9 Palazzo Italia, Studio Nemesi, Expo 2015 Milan. Photo by the author.

Figure 1.10 Vanke pavilion, Daniel Libeskind, Expo 2015 Milan. Photo by the author.

construction process invented by New York start-up Ecovative. Each organic brick was "grown" starting from living raw material supplied from the fungi mycelium, then dried with corn husks. This combination produces a material with excellent thermal performance, perfect for green architecture, volatile organic compounds-free, flame resistant, and 100% ecological. Biological bricks placed higher in the towers are coated with a reflective surface developed by 3M which transmits daylight within the structure without compromising the thermal performance; the porosity of these materials also favors natural ventilation, ensuring an environment rich in colors and light.

Since the choice of materials will influence the environmental impact of the building at the end of its life cycle, linked to the possibility or lack thereof to recover used materials and safely dispose of them, thereby reducing the production of solid waste, it

Figure 1.11 Hy-Fi, The Living, New York.
Courtesy of Ecovative Ltd.

is crucial to use products that can be easily disassembled into their components and reused or recycled at the end of life. In particular, materials must be capable of recovering their original performance characteristics and not imply particularly polluting or highly energy-consuming treatment processes, and technical components that contain materials which may be incompatible with one another in terms of recycling should be avoided.

It should be noted, however, that the real potential for recovering building materials is linked not only to the recyclability of individual products but also to the construction techniques used and the method of demolition at the end of the building life. To close the waste cycle, the green building must be designed with a focus on the recovery of its components: an unbundled building made with recoverable components which are constituted by construction products recovered or made from recycled materials (*design for disassembly, reuse, and recycling*).

While selecting technology solutions in view of their efficiency, it is important to remember that there are no ecological materials in absolute terms, as the life of a building component is inextricably linked to the life of the building, influencing and in turn being influenced by it. The overall sustainability of the life cycle of products is therefore dependent on the distance between the place of manufacture and that of installation, the characteristics of the context in which the intervention will be carried out, the application methods of the product into the building, the use conditions, and the

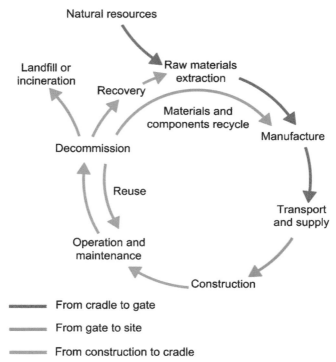

Figure 1.12 Life cycle of building materials and products.

method of disposal. All these variables are not related to the scale of the product, and are definable only from time to time on the basis of the context, building characteristics, and project requirements.

Underlying this complex assessment is technical information provided by manufacturers, which should be as responsive as possible to the demands of the designer and allow her/him to verify the impacts of the product associated with the three main phases of the process leading to the realization of the building (Fig. 1.12): the life cycle of the product off site (*from cradle to gate*); the transport of the product (*from gate to site*); and the life cycle of the building (*from construction to cradle*).

1.4.2 Green building rating systems

The environmental quality of a green building is normally assessed by a process of performance evaluation and certification carried out by independent third parties and accredited according to correspondence with specific requirements (green building rating systems).[72] The evaluation criteria concern the building's integration into the environment, efficient use of resources, atmospheric emissions, environmental compatibility of materials, quality of indoor environment, and use of innovative technology solutions.[71,73]

Popular assessment and certification systems include, at the international level, the Leadership in Energy & Environmental Design system, developed by the US Green Building Council in 1996 and used since 2000, the Building Research Establishment Assessment Method, introduced in 1990 by the British Building Research Establishment and now widely used in the UK and Northern European countries, and the Living Building Challenge program, created in 2006 by the nonprofit International Living Future Institute and defining the most advanced sustainability measures in the built environment possible today. Unlike the other systems, Living Building Challenge certification is based on actual, rather than modeled or anticipated, performance, with buildings operational for at least 12 consecutive months prior to evaluation.

Other national building environmental certification schemes are the Green Mark in Singapore, the UAE Estidama Pearl Rating system, the DGNB in Germany, the Green Star in Australia and South Africa, and Protocollo ITACA in Italy.

More and more buildings around the world boast certified high environmental performance. Among the most sustainable recently built buildings with environmental certification, in the services sector particularly noteworthy is the Bullitt Center in Seattle by the Miller Hull Partnership (Figs. 1.13–1.15), voted "Sustainable Building of the Year 2013" in the famous international competition organized by WAN (*World Architecture News*) and certified according to both Net Zero Energy standards and Living Building Challenge systems, the latter achieved in 2015 through a 12-month assessment of its operative phase starting from January 2014. All materials used for its construction are FSC certified and come exclusively from areas adjacent to the construction site, encouraging the local economy and avoiding unnecessary CO_2

Figure 1.13 Bullitt Center, Miller Hull Partnership, Seattle.
Courtesy of Nic Lehoux/Bullitt Foundation ©.

Figure 1.14 Bullitt Center, photovoltaic roof.
Courtesy of John Stamets/Bullitt Foundation ©.

Figure 1.15 Bullitt Center, photovoltaic roof.
Courtesy of John Stamets/Bullitt Foundation ©.

emissions. In addition, to comply with the Living Building Challenge certification standards, all potentially hazardous or noxious substances and materials, as clearly identified on the "Red List" of the certification system, have been banned from the building. Rainwater is recovered and used to supply toilets, while a series of raingardens around the building will avoid flooding phenomena and prefilter water before it is dispersed in the soil. Featuring an intelligent, highly efficient building envelope, the Bullitt Center is able to produce on site 100% of its net energy demand from renewable sources, through the PV roof and solar facades which together deliver an output of about 230,000 kWh/yr.

Regardless of the assessment and certification system chosen, the construction of a green building requires a design approach even more attentive to the environmental conditions and the characteristics of the intervention site. In addition to the climate, particular attention should be paid to the geological, hydrological, and vegetation context and the historic, cultural, and landscape scope to assess their possibilities and potentials of transformation and ensure the harmonization of the intervention with the context characteristics, preserving habitats and natural resources and minimizing impacts on the environment.

1.5 Smart buildings

The literature already includes several definitions of smart, or intelligent, buildings, all of which broadly refer to the presence inside the building of more or less complex building management systems, providing active systems and controls that allow the motorized action of specific subordinate functions and appliances.[74–80]

In this book it is instead assumed that the smart building is based on a very different paradigm from what is conventionally understood, considering it as the evolution of both the ZEB and the green building (Table 1.6).

This evolutionary leap was made possible on the one hand by the availability of new materials, new technologies in the energy and information sector, and increasingly advanced information technology tools to support the design process, and on the other hand by the need to address the new challenges of contemporary society that require a different way to design, build, and upgrade buildings.

The term "smart" therefore refers both to the quality of the construction process (planning, design, and construction) and to the quality of its final product, highlighting for both the ability to keep up with the evolving needs of clients, users, and community.

The wide use of the term smart in the new century (smartphones, smart grids, smart city, smart objects, etc.) is designed to underline a ground-breaking difference from all that precedes it in terms of the ability to respond through performance to the expressed and implied needs of users and all stakeholders involved.

In the smart building, the established characteristics of energy and environmental efficiency evaluated along the entire life cycle add those of architectural quality, intelligence, and resilience through continuous interaction in the formal and functional aspects with the context and with users.

The first form of interaction that distinguishes a smart building concerns the esthetic and emotional aspect, and the deep relationship between art and architecture.

A smart building is in fact primarily a building of high formal and space quality, able to stimulate actively those who live or work there, promoting culture, well-being, participation, and creativity.[81] A smart building aims at being a real icon of the 21st century, able to combine innovative design and technological solutions with the highest performance in terms of comfort, energy efficiency, and social and environmental sustainability.

The second form of interaction is with the environmental context (indoor microclimate parameters, outdoor climate, weather forecast) and the built environment (power

Table 1.6 **Building quality objectives of zero-energy, green, and smart buildings**

Building quality objective	Zero-energy building	Green building	Smart building
Building process phase	Building operation	Entire life cycle	Entire life cycle
Functionality			
Site sustainability		■	■
Esthetics			■
Usability			■
Accessibility			■
Ease to equip			■
Safety			■
Security			■
Integrability			■
Building management system			■
Sustainable mobility		■	■
Resilience			■
Indoor well-being			
Sunshine	■	■	■
Daylight	■	■	■
Thermohygrometric values	■	■	■
Ventilation and air changes	■	■	■
Indoor air quality		■	■
Acoustic climate		■	■
Electromagnetic fields		■	■
Psychological well-being		■	■
Dynamic control of environmental parameters			■
Dynamic anisotropy of building envelope			■
Efficiency and environmental impacts			
Reduction of energy consumption	■	■	■
Use of renewable energy sources	■	■	■
Saving of water resources and rainwater recovery		■	■
Waste water disposal		■	■

Continued

Table 1.6 Continued

Building quality objective	Zero-energy building	Green building	Smart building
Reduced consumption and environmental impact of building products and materials		■	■
Building waste management		■	■
Reduction of soil, water, and air emissions		■	■
Competitiveness			
Simplification of realization procedures			■
Choice of durable and inexpensive materials		■	■
Choice of technical and technological solutions with easy maintenance	■	■	■
Reduction of management and maintenance costs	■	■	■

grid, water supply, information from the web and social networks) to acquire data and information useful for optimizing its internal operation, always aimed at the utmost efficiency and effectiveness in use of environmental and economic resources, and the ability to withstand even extreme weather events with remarkable resilience.[82]

This last property, resilience, is acquiring more and more importance in architecture as a qualifying element of a smart building, as shown by the Sure House project by Stevens Institute of Technology—a solar-powered home with an uncanny hurricane-resistant design that is also able to act as a community base in the event of a natural disaster—which won the 2015 US Solar Decathlon competition, and by the recent introduction of a specific certification system, "RELi", which provides a comprehensive listing of resilient design criteria for the latest integrative processes in developing communities, neighborhoods, buildings, homes, and infrastructure of the next generation (Resiliency Action List by C3 Living Design Project).[83]

A resilient design aims at achieving buildings, and therefore communities, that are shock resistant, healthy, adaptable, and regenerative through a combination of diversity, foresight, and the capacity for self-organization and learning.

The third interaction is with the users and their needs, in a process of continuous exchange of information characterized by mutual respective learning.

Ultimately, a smart building is a thinking building, perfectly integrated into its environment, connected to the global network, and able to receive, process, and take advantage of data and information to communicate with its users and share its infrastructure with that of the city and the transport systems, acting as founding element in the achievement of the future smart city.

A smart building (Table 1.7) is thus a building designed, or renovated, with intelligence (smart design) that uses the best typological (smart shape) and technological solutions, from both the construction (smart envelope) and equipment (smart systems)

Table 1.7 **Smart building constituents and advantages**

Smart building components		Smart building elements	Smart building advantages
Smart design	Smart shape	Smart context integration	**Smart environment**
	Smart envelope	Smart insulation	Lower materials, energy, and water consumption
		Smart windows	Lower emissions and wastes
		Smart surfaces	**Smart living**
	Smart systems	Smart HVAC and lighting	Maximum accessibility
		Smart renewable energies	Higher comfort
		Smart BEMS/HEMS and "internet of things"	Better air quality
			Security and safety
	Smart people	Smart management	Architectural quality
		Smart user behaviors	**Smart economy**
			Lower operating costs
			Lower maintenance costs
			Higher property value
			Lower vacancy rates and additional revenue expectations
			Resilience
			Increased productivity
			Smart city
			Grid integration
			Media integration
			Mobility integration

points of view, able to interact intelligently with the environment and with users (smart people) to provide an accessible, safe, comfortable, and healthy built environment, capable of improving the lives of all stakeholders involved (smart environment, smart living, smart economy, smart city).

In full compliance with the aim of promoting a "smart, sustainable, and inclusive building industry" based on knowledge and innovation, more efficient in terms of resources, greener, and more competitive, a smart building aims to satisfy in its performance the expressed or implied needs of all parties involved: quality for the end user of the building, related to use constraints and specific needs; quality for the clients and managers of the building, in both technical (maintenance, safety) and economic (marketability of the property, return on invested capital) aspects; quality for the community, resulting from environmental integration and use of resources; and quality for the entire chain of operators involved in the construction process, starting from the design stage.

From a construction and technological point of view, a smart building is characterized by the presence of an adaptive building envelope (smart envelope) and an advanced building automation system (smart BEMS/HEMS) for monitoring internal and external environmental parameters and dynamic adjustment of all equipment (thermal, electrical, water, security, renewable energy, etc.) in coordination with the different configurations assumed by the building envelope to optimize both user well-being and consumption of environmental and economic resources. Smart BEMS/HEMS are also able to learn from the needs and behavior of users to anticipate solutions or provide recommendations, to reduce management costs and improve comfort conditions.

1.5.1 Smart envelope

Nowadays, the building envelope has acquired a leading role in the process of formal and technological recognition of architecture because it acts as an urban catalyst and a diffuser of the quality of architecture more than any other architectural element. The widespread diffusion of smartphones and social networks has further accentuated this role, freeing buildings from their geographical context and enabling a new level of involvement of citizens, who are becoming today more and more active partners in the design process.

Images shared on the web and via social media allow access to architecture with unprecedented immediacy, creating the conditions for a global conversation on buildings and their impact, and pushing designers to seek solutions that are more responsive to contemporary social and environmental issues and to make buildings ever more comfortable, environmentally friendly, and intelligent.

Specifically, the aim of meeting the growing demand for thermal and environmental comfort, associated with the pressing need to improve the energy efficiency of buildings up to the standard of "carbon-neutral" buildings or ZEBs, is demanding a major overhaul of the approach to building design by shifting the focus from the form to the function, and from the structure to the envelope.

In fact, the building envelope plays a strategic role in the energy and environmental performance of the building, significantly affecting the levels of indoor comfort. It consists of a complex system of environmental barriers and filters, potentially able not only to regulate the flows of heat, solar radiation, air, and steam, but also to convert solar, wind, and mechanical energy into heat and electricity to meet different demands of the building.

In this picture, the building envelope has become the main object of regulations and also the main research, development, and testing field for all agents involved in the building process, in both new constructions and existing retrofit interventions. Advanced building envelope design can reduce the capital costs of heating and cooling systems, since required heating and cooling loads can be reduced by 40—60%.[5]

The objective is to obtain an intelligent building envelope which, in analogy with the function of the skin in the human body, is able to adapt fully to the environmental conditions in a dynamic way, ensuring, in full synergy with the systems and equipment, efficient, continuous, and automatic management of all matter and energy flows in accordance to climate, user behavior, and market conditions of energy.[31,76,80]

A "smart envelope" is thus a building envelope with advanced performance, designed with intelligence (smart design) that employs, in relation to the climate, the best technological solutions, from both construction (smart insulation, smart windows, smart surfaces, and smart facade systems) and equipment (smart management systems) points of view, to provide users with the best conditions of safety, indoor well-being, and efficient use of energy and environmental resources.

The term "smart" therefore relates both to the choice of technological solutions able to optimize the life cycle of the building in relation to the site of intervention, and to the behavior of the envelope, capable of not only offering better performance compared to a traditional building shell, but also fulfilling new functions (power generation, light emission, image projection, air purification, self-cleaning surfaces, ability to self-repair, etc.) and adapting its characteristics in response to changes in external conditions (from transparent to opaque, solid to liquid, waterproof to permeable, etc.).

The adaptive behavior of the casing (climate-adaptive building shell) may comprise both the macro scale, with the presence of moving mechanical parts, and the micro scale, through a change of physical or chemical properties of the materials.[84,85]

The first case regards the kinetic envelope, and dynamic response primarily involves sun screening, natural ventilation, and solar energy capture systems capable of orientating, opening, or closing automatically according to the need. The movement may also involve the entire building through sophisticated rotation mechanisms.[86-88,91]

The second case regards more properly the intelligent skin, and the response to environmental stimuli involves the internal structure of materials able to modify their optical, electrical, or thermal properties or their geometry, or to convert energy from one form to another (smart materials).[89,90,92-94]

The adaptive behavior of the building envelope can be passive, without direct control by the users, or active, with manual or automatic control, even remote, or a combination of the two. An active operating behavior requires the integration of sensors, control systems, and actuators in the components of the building envelope.

These variations of the characteristics of the building envelope may or may not be in communication with HVAC and lighting systems; in the former case, equipment is able to adjust its operation automatically in relation to the materials and energy flows actually entering the building within a single integrated building—equipment system, managed with advanced building automation programs (smart buildings).

The evolutionary path of adaptive capacity of the smart envelope on the one hand leads to the building's ability to self-learn from user behavior, anticipating desired configurations in accordance with changing environmental conditions, and on the other hand brings the chance to educate users themselves, positively modifying their actions through a continual and continuously evolving conversation. This communication between different building elements and between the building itself and the users and the outside world, in a sort of global connection, reflects the current trend of the social model in which individuals are increasingly interconnected through the internet and social media in a continuous exchange of information. In turn, user behavior data made available by buildings can provide information useful on a larger scale (big data) to guide territorial development as well as energy and environmental sustainability policies.

Today available solutions for the construction or renovation of high-performance building envelopes, both opaque and transparent, are varied and stem from the profound evolution that materials, products, and construction systems have experienced over the last decade, spurred by energy efficiency and environmental protection objectives.

In particular, recent years have seen the gradual introduction in the construction industry of advanced materials and nanotechnology products with physical and mechanical properties greatly exceeding those of conventional materials, as well as the availability of active components and devices, so-called "smart materials" and "smart devices," able to change their characteristics in relation to different operating conditions determined by climatic elements or users.

These new products and systems may be used for thermal insulation of the building, giving superior performance with extremely reduced thickness (smart insulation), for glazed components allowing dynamic performance (smart windows), for surface treatment of building materials to improve their characteristics or introduce new ones (smart surfaces), and for facade systems to allow adaptive behavior and maximum integration with building equipment (smart facade systems).

In any case, it is important to stress that a predetermined optimal building envelope does not exist, but rather a repertoire of technological solutions are available, to be employed in an appropriate manner in relation to the type of users, the complexity of the building, the environmental (heating or cooling dominated), economic (or developed undeveloped), and local cultural context, and whether it is a new construction or a retrofit intervention (Table 1.8).

1.5.2 Smart systems

A peculiar element that distinguishes a smart building from other constructions is the presence of an advanced BEMS.[37,38] BAT/BACS are identified by several studies as a missing link for the successful implementation of ZEBs and the deployment of an effective smart grid.

Table 1.8 Recommended most effective advanced technologies for smart buildings related to climate and type of intervention

Climate	Smart envelope materials and technologies	
	New construction	Retrofit
Hot	• Very low SHGC glazing or dynamic glazing • Dynamic shading systems • BIPV shell and PV glazing • Cool roofs • Phase-change materials (thermal inertia) • Green roofs and walls • Self-cleaning glazing and surfaces • Ventilated roof and walls • Adaptive skin facades • Nano electro-mechanical systems (NEMS)	• Very low SHGC glazing or dynamic glazing • Solar control films for windows • Cool roofs and reflective coatings • PV shading systems • Ventilated walls • Green walls • Phase-change materials (thermal inertia) • Self-cleaning glazing and surfaces
Cold	• Advanced insulation glazing • Glazed double-skin facades • Dynamic shading systems • Antireflective glazing • Light redirection and optical systems • Heating glazing • Bioadaptive glazing • ETFE envelope • BIPV shell • Self-cleaning glazing and surfaces • Transparent insulating materials • Phase-change materials (passive solar gain) • Advanced insulating materials for thermal bridge correction and equipment insulation • Adaptive skin facades • NEMS	• Advanced insulation glazing • Dynamic glazing • Antireflective glazing • Suspended film glazing • Light redirection and optical systems • Heating glazing • BIPV/BAPV shell • Self-cleaning glazing and surfaces • Advanced insulating materials for interior or exterior insulation and equipment insulation • Phase-change materials (passive solar gain or internal insulation) • Green walls • Reflective thermal coatings for interior walls

In fact, most optimistic scenarios estimate that in the EU alone more widespread adoption and improved operation of BAT/BACS could yield savings of up to 150 Mtoe (1745 TWh) per year by 2028, equal to 22% of all building energy consumption and around 9% of total final energy consumption of the entire EU, corresponding to an abatement potential of up to 419 $MtCO_2$ per year.[37] Even in more realistic depictions of potential savings, these ramp up progressively to reach 13% of building energy consumption by 2035—still over 5% of the EU's entire energy consumption.[37]

For these reasons, adoption of modern BAT and management systems is expected to increase from 26% of all service-sector floor area today to 40% by 2028 even without further policy intervention.[37] In the residential sector, penetration of HEMS is also projected to reach 40% of homes by 2034, up from 2% today, without additional intervention.[37]

Most advanced BMSs allow central management of data from all equipment systems in the building to optimize energy efficiency, microclimate conditions, and security systems, achieving savings of up to 50%.

For European service buildings, BMS installation yields an assumed net average energy saving of 37% for space heating, water heating, and cooling/ventilation, and 25% for lighting.[38] In residential buildings, proper installation and use of the best BAT/HEMS available are estimated to save, on average, 25% of heating and hot water energy compared to a typical control arrangement that satisfies current requirements in building codes.[38] If instead the comparison is to the average default installed control systems in the building stock, savings will be over 30%.[38]

All individual building subsystems (electrical distribution, air conditioning, water, lighting, data network, access control, elevator control, fire and security, renewable energy and energy storage, electric mobility) can become part of a single central system, also able to learn users' needs and behavior, anticipate solutions, and provide use recommendations (Table 1.9). These systems make use of forecasting, optimization, and evaluation algorithms that acquire real-time data from smart sensors and meters placed in strategic points of the building, capable of detecting internal microclimate parameters, space, information/communication technology infrastructure use, attendance, weather data, and energy quality. Based on the analysis of input data, the system suggests actions to support building management optimizing energy consumption; users, on the other hand, as an active part of the system can monitor consumption instant by instant, by tablet or smartphone, and improve their behavior accordingly, becoming themselves actors in the energy saving.

Table 1.9 Building management system operation

Monitored parameters	Equipment and systems controlled
Meteorological parameters	Air-conditioning system
Indoor air temperature	Lighting system
Humidity	Shading systems
CO_2 concentration	Water grid
Lighting levels	Electricity grid
Electricity consumption	Data network
Building access and presences	Fire prevention system
Elevators use	Security systems
Energy produced from renewable sources	Heat and electricity storage

Depending on the type of system used, smart building appliances are able to perform partially autonomous (following fixed and predetermined responses to environmental parameters) or fully autonomous programmed functions, based on responses to environmental parameters directed by dynamic programs that can be generated or improved by self-learning.

BEMS functions should include effective integration and interoperability with energy grids, as well as demand response investigation and the energy management of individual customers.[39] In a smart building, the building automation system (BAS) can in fact take advantage of energy price data from the smart grid, weather forecast, and occupancy patterns to determine the best energy management strategy through interventions in the charge or discharge of energy storage devices, temperature setpoints, the charging or discharging schedules of electric vehicles, or appliance usage.[95-98]

The BAS will also be able to decide whether it is more convenient to sell the power generated by building-integrated renewables or use it internally. From this point of view, the so-called "vehicle-to-building" or "vehicle-to-grid" technology is particularly interesting, as it allows energy exchange between buildings and vehicles aimed at reducing energy consumption and managing emergency situations.[99] For instance, the building can take advantage of energy contained in vehicle batteries at times when electricity is most expensive and recharge them when energy costs less, or use parked vehicles as an emergency power supply in case of shortages. Promising experiments have already been launched in Japan and Europe by Nissan with the Leaf cars and in the United States by BMW with a program "paying people not to plug in," and this particular market is set to expand rapidly in the coming years, expecting to reach nearly $21 million by 2024.[100]

Already available on the market are several models of smart meter, which permit two-way, near real-time communication between the building and the grid and are considered as one of the enabling technologies for migration toward advanced load management strategies in commercial and residential buildings. Unlike conventional meters, which record only cumulative values, smart meters provide real-time pricing signals for the BAS, thus giving clear information about incentives and most favorable tariffs. Electric or thermal energy storage technologies are another major driver for load management, as they allow easy planning of the use of energy resources over time: increasing the energy storage capacity significantly adds to the potential of advanced control strategies to improve system performance.

Among the most innovative intelligent integrated platforms, noteworthy is the Digital Building Operating Solution (Di-BOSS), jointly developed by Finmeccanica, Columbia University, and the real estate firm Rudin Management Company. It is already operating in four skyscrapers in New York (3 Times Square, 345 Park Avenue, 355 Lexington, and 415 Madison Avenue) and it is planned to be installed on 12 other buildings; in just six months after its activation it achieved about $500,000 in savings, or approximately 10% of total energy expense.

Among the best-performing buildings featuring BMSs, of particular interest is the Hong Kong International Commerce Center by Kohn Pedersen Fox (2013), elected tall building with the best energy performance in the world in 2014 by the Council on Tall

Buildings and Urban Habitat. Thanks to a sophisticated system of monitoring and control of environmental and energy parameters managed by an advanced BMS, the building had in 2013 an efficiency 90% better than all commercial tall buildings with high energy efficiency.

1.5.2.1 Internet of things

In addition to traditional building-integrated home automation systems, recent market introductions include a number of devices applicable to individual components of the domestic technological system: by exploiting wireless transmission of data, these give users, literally, the house at hand without the need for any wiring or upgrade of existing plant.[101]

Dubbed the "internet of things" (IoT), this revolution of the large network of web-connected, smartphone-controlled everyday objects is an ever-expanding sector, intended to grow to 26 billion units installed in 2020, representing an almost 30-fold increase from 0.9 billion in 2009, most of which are in residential dwellings.[102] IoT devices can be integrated in all energy-consuming equipment (air-conditioning systems, electrical switches and outlets, lamps, appliances, plumbing, etc.) or in building envelope elements such as doors and windows, offering users the ability to optimize energy efficiency, micro-climatic conditions, safety, and security.

These systems are applicable to any type of building and are able to monitor intelligently indoor environmental parameters or energy consumption of individual appliances or electrical outlets, providing users with a stream of real-time information on their smartphones, anywhere in the world they may be. Services also include sending email or SMS notifications, and alarm or equipment failure warnings if necessary.

In addition to data reading, some of these systems allow real-time operation, always via smartphone, of household equipment and appliances either by programming their on–off schedules or adjusting activity levels as needed. It is also possible to control objects using voice commands, such as "hey computer, lights on" or "hey computer, lights red."

Objects become self-recognizable and acquire intelligence due to their ability to communicate information about themselves and gain access to aggregate information from other devices. Alarm clocks anticipate themselves in the event of traffic, plants tell the watering system when it is time to be watered, smart thermostats track the comings and goings of occupants to adjust the temperature when necessary, smart lighting self-switches on after a prolonged absence for security purposes, an indicator warns you when your shower is taking too long, the heat produced by appliances is stored to produce energy to recharge the phone, windows close autonomously when outside air is too polluted, chairs encourage regular little walks between more sedentary activities, and the bed adjusts the temperature according to our sleep cycle, turns the lights off, and in the morning commands the coffee pot to prepare coffee. Furthermore, access to the house can be provided by facial recognition or smartphone via smart connected locks.

All objects can take a proactive role by connecting to the net. The IoT could prove particularly useful during energy demand peak times, with the ability to manage power consumption in a more rational way. Inside the building, all devices can talk to each other, interact with the network, and generate information on the amount of energy they are consuming. Once information is gathered, the data could be used to take decisions to manage the overall system better. Smart appliances, for example, could independently make decisions on when to start their cycle based on when the network is freer and more energy is available, or when the cost is lower.

The goal of the IoT is to make the electronic world draw a map of the real world, giving an electronic identity to things and places from the physical environment. Objects and places equipped with radio frequency identification labels or quick response codes already communicate information to the network or to mobile devices such as smartphones.

This growing phenomenon has prompted the technology giants Apple (Home Kit), Google (Nest), and Samsung (SmartThings) to develop IoT solutions for buildings, and in 2015 pilot smart homes were demonstrated by Ikea (Smart 10) and Intel (Tiny House) to show the potential of the IoT for improving users' quality of life while at the same time reducing waste and consumption.

However, to turn this scenario into a proven reality, substantial problems still need to be addressed, starting with the "dialogue" between various applications and devices manufactured by different companies to reach the definition of a common standard at an international level. For this purpose, recently different nonprofit consortia have been established to define wireless connectivity requirements of IoT devices, and thus ensure interoperability between systems regardless of form factor, operating system, or manufacturer.

1.6 Conclusions and future trends

The responsibility to achieve smart sustainable and inclusive buildings requires considerable efforts from all sectors of the building community. To make smart buildings a reality, designers will need the tools to design and apply better-integrated equipment, manufacturers will need to produce high-efficiency equipment and materials and develop the know-how to integrate them into buildings, and both will have to monitor occupant needs carefully to provide comfortable living conditions.

Market-oriented initiatives to foster adoption of ZEB technologies and support ZEB marketing activities should be pursued, including energy and environmental certification enforcement.

Achieving a smart building and converting the existing building stock from an energy waster to an energy producer will also require new technologies for both building envelopes and equipment, including renewable energy sources tailored to each specific construction period, building type, and climatic zone.

It is apparent that traditional materials, products, and technologies are not adequate to meet this challenge of upgrading the energy efficiency of buildings. What is necessary instead are advanced solutions able to combine energy performance with building architectural quality and the highest livability, ensuring the achievement of the optimal cost-effective level.

Some of the technologies needed to transform the buildings sector are already commercially available and have proven cost-effective, with payback periods of less than five years. Others are more costly, and will require government intervention if they are to achieve a wider market uptake.

To date, the construction industry has found it difficult to integrate these key technologies effectively into its operations to achieve sustainable long-term competitiveness, and such integration should also be promoted.

The knowledge of how to optimize the use of these technologies is underdeveloped, and little authoritative design guidance is available. Technical information is still poor and nonhomogeneous, and no official price lists have been released. Furthermore, applications in real building retrofit interventions are still too few and too recent to allow the comparison of laboratory data with real-life performance, which is characterized by interlinked and interdependent technologies and influenced by physical attributes of the building, local climate, and user behavior. At the same time current available energy simulation software is not up to date with these new and useful technologies, and is thus unable to calculate real improvements of retrofit interventions accurately. Moreover, no early-stage decision-making tools are available to assist designers and constructors in identifying during the first design phases the best solutions to adopt to improve the energy efficiency of existing buildings.

Training designers and workforces on the installation and use of integrated solutions is a prerequisite to ensure a wider diffusion of these technologies.[103]

Future research and development must recognize the complex interactions between individual buildings, the community, and the larger scale (eg, urban and grid-wide). New advanced BEMS functions should therefore include effective integration and interoperability with energy grids, as well as demand response investigation and the energy management of individual customers, using next-generation innovative technologies enabling smart grids, storage, and energy system integration with an increasing share of renewables. These technologies and services should provide for advanced solutions able to ensure the needed stability and security in the context of an increasing share of variable renewable energy sources in the electricity grid.[39]

Further development finally concerns IoT and BEMS integration, and the definition at an international level of a common standard dialogue between various applications and devices manufactured by different companies.

References

1. IEA (International Energy Agency). *Tracking clean energy progress 2015*. Paris: OECD/IEA; 2015a.
2. IEA. *Transition to sustainable buildings: strategies and opportunities to 2050*. Paris: OECD/IEA; 2013a.
3. IEA. *Energy technology perspective 2015*. Paris: OECD/IEA; 2015b.
4. IEA. *Building Energy performance metrics*. Paris: OECD/IEA; 2015c.
5. US DOE (United States Department of Energy). *R&D roadmap for emerging window and building envelope technologies*. Washington: DOE; 2014.

6. JRC. (Joint Research Centre Institute for Energy and Transport). *Energy renovation: the trump card for the new start for Europe*. Brussels: JRC; 2015.
7. EPIA. European Photovoltaic Industry Association. *Solar generation 6: solar photovoltaic electricity empowering the world*. Brussels: EPIA; 2011.
8. UNEP (United Nations Environment Programme). *Building design and construction: forging resource efficiency and sustainable development*. Nairobi: UNEP; 2012.
9. ECORYS. *Resource efficiency in the building sector*. Rotterdam: ECORYS; 2014.
10. Pacheco-Torgal F, Diamanti MV, Nazari A, Goran-Granqvist C, editors. *Nanotechnology in eco-efficient construction: materials, processes and applications*. Cambridge: Woodhead Publishing; 2013.
11. IEA. *Energy efficiency market report 2015*. Paris: OECD/IEA; 2015d.
12. IEA. *Technology roadmap: energy efficient building envelopes*. Paris: OECD/IEA; 2013b.
13. European Commission. *Cleaner cooling – meeting the rapid growth in transport and urban cooling demand*. Brussels: Committee of the Regions Building; June 18, 2015.
14. IEA. *Modernising building energy codes*. Paris: OECD/IEA; 2013c.
15. IEA. *Capturing the multiple benefits of energy efficiency*. Paris: OECD/IEA; 2014.
16. BPIE (Belgium Buildings Performance Institute Europe). *A guide to developing strategies for building energy renovation*. Brussels: BPIE; 2013.
17. GBPN (Global Buildings Performance Network). *What is a deep renovation definition?* Paris: GBPN; 2013.
18. EEFIG (Energy Efficiency Financial Institutions Group). *Energy efficiency – the first fuel for the EU economy. Part 1: buildings (Interim report)*. Brussels: European Union; 2014.
19. European Commission. *Towards a thematic strategy on the urban environment, COM(2004) 60 final*. 2004. Brussels.
20. Barenghi M, editor. *Italo calvino, Saggi 1945–1985*. Milano: Mondadori; 1995.
21. Calvino I. *Romanzi e racconti*. Milano: Mondadori; 1994.
22. Casamonti M. *Jean nouvel*. Milano: Motta Architettura; 2008.
23. Athientis A, Obrien W, editors. *Modeling, design and optimisation of net-zero energy buildings*. Berlin: Ernst & Sohn; 2015.
24. Directive 2010/31/EU of the European Parliament and of the Council of 19 May 2010 on the energy performance of buildings (recast). *Off J Eur Union* 18.06.2010;**L 153**:13.
25. NREL (US National Renewable Energy Laboratory). *Zero energy buildings: a critical look at the definition*. Oak Ridge: NREL; 2006 (CP-550–39833).
26. NREL. *Net-zero energy buildings: a classification system based on renewable energy supply options*. Oak Ridge: NREL; 2010 (TP-550–44586).
27. BPIE. *Principles for nearly zero-energy buildings*. Brussels: BPIE; 2011.
28. ECOFYS. *Towards nearly zero-energy buildings: definition of common principles under the EPBD*. Cologne: Ecofys; 2013 (BESDE10788).
29. Italy Decreto del Ministero dello Sviluppo Economico del 26 Giugno. *Applicazione delle metodologie di calcolo delle prestazioni energetiche e definizione delle prescrizioni e dei requiisti minimi degli edifici*. 2015 (G.U. n. 162, 15.07.2015, S.O. n. 39).
30. UNEP. *Buildings and climate change. Status, challenges and opportunities*. Kenya: UNEP; 2007.
31. Casini M. *Costruire l'ambiente. Gli Strumenti e i metodi della progettazione ambientale*. Milano: Edizioni Ambiente; 2009.
32. PHI (Passivhaus Institut). *The independent institute for outstanding energy efficiency buildings*. Darmstadt: PHI; 2013.
33. Casini M. *Progettare l'efficienza degli edifici. La certificazione di sostenibilità energetica e ambientale*. 2nd ed. Roma: DEI; 2013.

34. Casini M. Small vertical axis wind turbines for energy efficiency of buildings. *J Clean Energy Technol* 2016;**4**(1):56−65.
35. Casini M. Harvesting energy from in-pipe hydro systems at urban and building scale. *Int J Smart Grid Clean Energy* 2015;**4**(4):316−27.
36. Lo Basso G, de Santoli L, Albo A, Nastasi B. H_2NG (hydrogen-natural gas mixtures) effects on energy performances of a condensing micro-CHP (combined heat and power) for residential applications: an expeditious assessment of water condensation and experimental analysis. *Energy* 2015;**84**:397−418.
37. Debusscher D, Waide P. *A timely opportunity to grasp the vast potential of energy savings of building automation and control technologies*. Brussels: European Copper Institute; 2015.
38. Waide Strategic Efficiency. *The scope for energy and CO_2 savings in the EU through the use of building automation technology*. UK: Waide Strategic Efficiency; 2014.
39. European Commission Decision C. 6776 of 13 October 2015. *Secure Clean Effic Energy Horizon 2020 Work Programme* 2015: 2016−2017.
40. JRC. *Strategic energy technology SET plan. Towards an integrated roadmap research and innovation challenges and needs of the EU energy system*. Brussels: JRC; 2014 (93056).
41. Perez R, Perez M. A fundamental look at energy reserves for the planet. *IEA SHC Sol Update* 2009;**50**:2−3.
42. MIT (Massachusetts Institute of Technology). *The future of solar energy*. Boston: MIT; 2015.
43. Solar Power Europe (European Photovoltaic Industry Association). *Global market outlook for solar power 2015−2019*. Brussels: Solar Power Europe; 2015.
44. European Commission. *Renewable energy progress report. COM(2015) 293 final*. 2015. Brussels.
45. IPPC. (International Panel of Climate Change). *Renewable energy sources and climate change mitigation, intergovernmental panel on climate change*. Geneva: IPPC; 2011.
46. JRC. *Renewable energy in Europe for climate change mitigation*. Brussels: JRC; 2015.
47. IEA. *Renewables information 2015*. Paris: OECD/IEA; 2015e.
48. Cronemberger AJ, Almagro Corpas M, Cerón I, Caamano-Martín E, Sánchez SV. BIPV technology application: highlighting advances, tendencies and solutions through Solar Decathlon Europe houses. *Energy Build* 2014;**83**:44−56.
49. Luque A, Hegedus S. *Handbook of photovoltaic science and engineering*. West Sussex: Wiley; 2011.
50. Mcevoy A, Markvart T, Castaner L. *Practical handbook of photovoltaics, fundamentals and applications*. Waltham (MA): Architectural Press; 2012.
51. Roberts S, Guariento N. *Building integrated photovoltaics a handbook*. Basel: Birk Hauser; 2009.
52. Fraunhofer Institute for Solar Energy Systems. *Photovoltaics report*. Freiburg: Fraunhofer ISE/PSE AG; 2015.
53. Howe JC, Michael Gerrard M. *The law of green buildings: regulatory and legal issues in design*. Washington: American Bar Association; 2010.
54. EPA. *Green building*. 2008. Available at http://www.epa.gov/greenbuilding/pubs/about.htm.
55. ISO 15392:2014-*Sustainability in building construction − general principles*.
56. Dodge Data and Analytics. *World green building trends 2016: developing markets accelerate global green growth 2015*. Bedford: Dodge Data and Analytics; 2015.
57. Dodge Data & Analytics. *Green and healthier homes: engaging consumers of all ages in sustainable living smart market report 2015*. Bedford: Dodge Data & Analytics; 2015.
58. EPO-UNEP. *Climate change mitigation technologies in Europe − evidence from patent and economic data*. Nairobi-Munich: Unep/EPO; 2015.

59. McGraw Hill Construction. *World green building trends: business benefits driving new and retrofit market opportunities in 60 countries*. New York: McGraw Hill; 2013.
60. Slawik H, Bergmann J, Buchmeier M, Tinney S. *Container atlas: a practical guide to container architecture*. Berlin: Gestalten; 2010.
61. Kotnik J. *New container architecture: design and sustainability*. Barcelona: LinksBooks; 2013.
62. Wines J. *Green architecture*. Colonia: Taschen; 2000.
63. Jodidio P. *100 contemporary green buildings*. Taschen; 2013.
64. Cantor SL. *Green roofs*. New York: Norton & Company; 2008.
65. Blanc P. *The vertical garden: from nature to the city*. New York: Norton & Company; 2012.
66. Green M. *The case for tall wood buildings*. Vancouver: Wood Enterprise Coalition; 2012.
67. Skidmore O, Merril. *Timber Tower research project*. Chicago: SOM; 2013.
68. Karacabeyli E, Lum C. *Technical guide for the design and construction of tall wood building in Canada*. Vancouver: FP Innovations; 2014.
69. Canadian Wood Council. Energy and environment in residential construction. In: *Sustainable building series*, vol. 1. Ottawa: CWC; 2013a.
70. Canadian Wood Council. Certified wood products. In: *Sustainable building series*, vol. 10. Ottawa: CWC; 2013b.
71. International Living Future Institute. *Living building challenge 3.0-documentation requirements*. Seattle: International Living Future Institute; 2014.
72. ITU (United Nations specialized agency for information and communication technologies − ICTs). *Sustainable buildings*. Geneva: ITU; 2012.
73. USGBC (US Green Building Council). *LEED 2009 for new construction and major renovations rating system*. Washington: USGBC; 2009.
74. Buckman AH, Mayfield M, Beck SBM. What is a smart building? *Smart Sustain Built Environ* 2014;3(2):92−109.
75. Wonga JKW, Lia H, Wang SW. Intelligent building research: a review. *Automation Constr* 2005;14:143−59.
76. Clements-Croome DJ. Sustainable intelligent buildings for people: a review. *Intell Build Int* 2011;3(2):67−86.
77. Sinopoli J. *Smart building systems for architects, owners and builders*. Oxford: Butterworth-Heinemann; 2010.
78. Holden J. *Introduction to intelligent buildings: benefits and technology*. Walford: BRE-Global; 2008.
79. Gadakari T, Mushatat S, Newman R. Intelligent buildings: key to achieving total sustainability in the built environment. *J Eng Proj Prod Manag* 2014;4(1):2−16.
80. Wigginton M, Harris J. *Intelligent skins*. Oxford: Elsevier Architectural Press; 2006.
81. Clements-Croome DJ. Creative and productive workplaces: a review. *Intell Build Int* 2015; 7(4):1−20.
82. JRC. *The challenge of resilience in a globalised world*. Brussels: JRC; 2015.
83. C3 Living Design Project. *RELi resiliency action list + credit catalog*. 2015 [rev 05.02.15].
84. Loonen RCGM, Trčka M, Cóstola D, Hensen JLM. Climate adaptive building shells: state of the art and future challenges. *Renew Sustain Energy Rev* 2013;25:483−93.
85. Hargave J. *It's alive. Can you imagine the urban building of the future?*. London: Arup; 2013.
86. Pesenti M, Masera G, Fiorito F, Sauchelli M. Kinetic solar skin: a responsive folding technique. *Energy Procedia* 2015;70:661−72.

87. Miao Z, Li J, Wang J. Kinetic building envelopes for energy efficiency: modeling and products. *Appl Mech Mater* 2011;**71−78**:621−5.
88. Razaz ZE. Sustainable vision of kinetic architecture. *J Build Apprais* 2010;**5**(4):341−56.
89. IEA ECBCS Annex 44. In: Heiselberg P, editor. *Integrating environmentally responsive elements in buildings − expert guide Part 1: responsive building concepts.* IEA; 2009.
90. Fox M, Kemp M. *Interactive architecture.* New York: Princeton Architectural Press; 2009.
91. Schumacher M, Schaeffer O, Vogt M. *Move: architecture in motion-dynamic components and elements.* Basel: Birkhäuser; 2010. 2010.
92. Gruber P. *Biomimetics in architecture − architecture of life and buildings.* Wien: Springer-Verlag; 2011.
93. Velikov K, Thün G. Responsive building envelopes: characteristics and evolving paradigms. In: Trubiano F, editor. *Design and construction of high-performance homes: building envelopes, renewable energies and integrated practice.* London: Routledge; 2012.
94. Sala M, Romano R. Innovative dynamic building component for the Mediterranean area. In: *International Scientific Conference CleanTech for sustainable buildings from nano to urban scale CISBAT 2013, Lausanne, Switzerland, 4−6 September 2013*; 2013.
95. Kiliccote S, Piette MA, Ghatikar G, Hafemeister D, Kammen D, Levi BG, et al. Smart buildings and demand response. In: *AIP conference proceedings*, vol. 1401; March 5−6, 2011. Berkeley.
96. Shaikh PH, Nor NBM, Nallagownden P, Elamvazuthi I, Ibrahim T. A review on optimized control systems for building energy and comfort management of smart sustainable buildings. *Renew Sustain Energy Rev* 2014;**34**:409−29.
97. Wong J, Li H, Lai J. Evaluating the system intelligence of the intelligent building systems. *Autom Constr* 2008;**17**(3):284−302.
98. Wong JKW, Li H. Development of intelligence analytic models for integrated building management systems (IBMS) in intelligent buildings. *Intell Build Int* 2009;**1**(1):5−22.
99. Wang Z, Wang L, Dounis AI, Yang R. Integration of plug-in hybrid electric vehicles into energy and comfort management for smart building. *Energy Build* 2012;**47**:260−6.
100. Navigant Research. *Vehicle grid Integration.VGI applications for demand response, frequency regulation, microgrids, virtual power plants, and renewable energy integration.* Navigant; 2015.
101. Casini M. Internet of things for energy efficiency of buildings. *Int Sci J Archit Eng* 2014; **2**(1):24−8.
102. Gartner. *Forecast: the internet of things, worldwide, 2013.* Gartner; 2013.
103. Losasso M. *Percorsi dell'innovazione. Industria edilizia, tecnologie, progetto.* Napoli: Clean; 2010.

Advanced materials for architecture

2.1 Building materials classification

Building materials can be classified in different ways according to the objective of their classification.[1–5]

Specifically, main classification systems used in construction concern composition, technical properties, environmental sustainability, the role played in the building system, or the size scale (see Table 2.1).

As regards their composition, in particular, building materials are usually divided into the following main classes.

- *Metallic materials*: inorganic substances composed of one or more metallic elements (iron, copper, aluminum, etc.), which may also contain some nonmetallic elements (carbon, nitrogen, oxygen, etc.), and are characterized by a crystalline structure in which atoms are regularly arranged in space. Metallic materials (metals and alloys) are commonly divided into two classes: ferrous metals/alloys which contain a large percentage of iron (cast iron and steels), and metals/alloys that contain little or no iron (aluminum, copper, nickel, zinc, titanium, lead, tin, gold, silver, etc.).

Table 2.1 **Main classification systems for construction materials**

Category	Subcategory	Scale of engineering
Composition	• Metallic materials, ceramic materials, polymeric materials, composite materials	Macromaterials Micromaterials Nanomaterials
Technical properties	• Mechanical, thermal, electrical, optical, acoustic, magnetic, chemical, radioactive	
Sustainability	• Renewability, recycled content, embodied energy, absence of toxic substances, compostability, degradability, recyclability, reusability, durability, easy maintenance, global warming potential, ozone depletion potential, acidification potential	
Function	• Binders and aggregates for mortars and concrete, reinforcing materials, materials for structural components, load-bearing elements, windows (frame, glazing, and shades), thermal and acoustic insulation materials, water- and vapor-proofing materials, materials for floors, walls covers, and ceilings, materials for systems and equipment, paints, primers, solvents, and adhesives	

- *Ceramic materials*: inorganic materials (mainly oxides and/or silicates) consisting of metallic and nonmetallic elements chemically bound to each other and characterized by an amorphous or mixed crystalline structure. This category covers stone materials, including masonry, sands, gravels, clay, conglomerates such as concrete, and mineral binders such as aerial and hydraulic lime, cements, and mortars; and materials that have undergone heat processing, including bricks, ceramics, porcelain, and glass.
- *Polymeric materials*: materials consisting of long chains or networks of organic molecules (containing carbon). They either have a natural origin, such as wood and natural fibers (wool, cotton, silk, etc.), or are synthetic, such as thermoplastic materials (polyvinyl chloride, polypropylene, polyethylene), thermosetting materials (resins), elastomers (gums), technical textiles, and bituminous materials.
- *Composite materials*: materials consisting of a fibrous phase (reinforcing fibers with high mechanical properties) incorporated in a continuous phase (matrix). Depending on the type of matrix, composites are classified as polymeric, metallic, or ceramic. Most used in buildings are polymer composites (fiber-reinforced polymers—FRPs); after fiberglass and aramid, most interesting applications today relate to carbon FRP materials capable of offering excellent performance in the consolidation and static reinforcement of existing building structures.

Regardless of the classification system used, advanced materials are defined as those (both new and deriving from the modification of existing materials) specifically designed to possess new or improved technical properties (structural or functional) or environmental features compared to materials traditionally used to perform the same functions.[6,7]

The introduction of advanced materials (Fig. 2.1) in the construction industry is contributing substantially to the achievement of energy and environmental efficiency

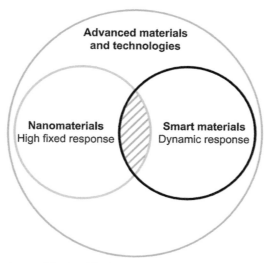

Figure 2.1 Advanced materials classification.

improvements of buildings throughout their life cycle and leading to profound changes in the whole construction process, from design to implementation.

Advanced materials are in fact a key enabling technology for a number of other technologies, and have a main role in enabling advanced/high-value manufacturing and addressing society's key socioeconomic grand challenges like climate change and improved efficiency in using resources, supporting and enabling the development of new products through radically different and unique functionality, including low-carbon technologies.

As a matter of fact, advanced materials, in addition to being included among key enabling technologies by the European Union (EU), are one of the great eight technologies identified by the UK government.

Advanced materials can introduce new functionalities and improved properties (Table 2.2), while adding value to existing products and processes, in a sustainable approach oriented to resource conservation, low-impact materials, cleaner production, efficient distribution, green use and maintenance, reuse, remanufacture, disassembly, recycling, and safe disposal. This is particularly relevant in the building sector,

Table 2.2 Advanced materials for architecture: enhanced and new properties

Enhanced properties	
• Mechanical resistance	• Insulation properties
• Hardness (abrasion resistance)	• Thermal conductivity
• Shock resistance	• Light transmission or reflection
• Weldability	• Thermal radiation reflection
• Quick compacting and curing time	• Surface energy
• Corrosion and oxidation resistance	• Electrical conductivity
• Ultraviolet (UV) resistance and control	• Photoelectric effect
• Water protection	• Energy storage
• Fire behavior	
New properties	
• Self-healing	• Biocidal activity
• Air purifying	• Electromagnetic radiation block
• Easy to clean	• Light emission
• Antigraffiti and antistain	• Chromogenism (dynamic glass)
• Scratch resistant	• Thermotropism (PCMs)
• Antireflection	• Real-time structural health monitoring
• Antiicing and fogging	

characterized by high environmental impacts in terms of resource consumption and waste production.

Advanced materials in architecture can produce a fixed high-performance response or dynamic behavior in response to external stimuli; the latter characterizes the so-called *smart materials*.

These new or improved properties are the result of an engineering process at the nanoscale (nanomaterials) or at larger-scale levels (micromaterials or macromaterials). In particular, the nanometric scale offers higher potential in developing innovative materials and products and improving performance of existing ones.[7–10]

The fundamental properties of matter at the nanoscale level can be significantly different from the same material in its bulk state. These nanoscale property variations can potentially be exploited by embedding the materials in others at either the micro or macro level, improving the performance of the bulk material to the desired extent in real-life applications. Applications of nanomaterials are various; in particular they have great potential in the field of thermal energy. Their high surface-to-volume ratio allows them to impart sensible thermophysical property changes in bulk materials and thereby enhance their energy performance.

Advanced materials may be applied by themselves as end products or be integrated in other products or devices, both during the manufacturing process and in the final unit to enter the market.

Concerning the construction sector, in particular, most interesting and innovative developments in the field of advanced materials regard nanocomposites and coatings (thin films, nanocoatings, nanopaints)—thin coverings deposited on a base material to provide new or improved surface qualities or appearance.

The application of advanced coatings improves mechanical, physical, and chemical performance of building materials and components, and glazing in particular, by imparting new properties, both fixed (antireflection, scratch resistance, etc.) and dynamic (chromogenics, self-cleaning effect, shape-memory alloys, etc.).

To achieve zero-energy buildings and proceed with a deep and steady energy retrofit of the existing building stock, research in the sector of advanced materials is now focusing on the development of:

- modular "plug-and-play" components and systems fully integrated with advanced three-dimensional (3D) surveying techniques;
- innovative insulation solutions to address cold bridges and improve the airtightness of the building envelope (nanoporous insulating materials, transparent insulating materials—TIMs, vacuum insulating panels—VIPs);
- dynamic glazing and coatings for solar control of windows;
- highly efficient thermal energy storage systems (phase-change materials—PCMs);
- highly efficient lighting systems (quantum dot light-emitting diodes—QLEDs and organic light-emitting diodes—OLEDs);
- highly efficient integrated energy generation systems based on renewable sources (organic and dye-sensitized photovoltaic (PV) solar cells);
- energy systems and controls for better monitoring of the energy performance of a building (nano electrical mechanical systems—NEMS);
- prefabricated and 3D printed components.

Many of these technologies are already available, but diffusion across different countries is nonhomogeneous, due to their high cost and a lack of awareness by main market actors about the convenience potential of the best available technologies. This is especially true for combined solutions that consider a whole-system approach. Workforce training on the installation and operation of integrated solutions is mandatory to ensure better diffusion of these technologies.

Regarding 3D printing, this rapid prototyping technique is determining a profound revolution in the sector of manufacturing and production of building components, greatly reducing processing, transportation, and installation costs. 3D printing makes the realization of single customized objects as cheap as large-scale manufacturing, subverting scale economics. This new technology could have a revolutionary impact on the industrial sector, on par with the printing machine in 1450, the steam motor in 1750, and the transistor in 1950, changing all fields of industry including the building sector.[11]

2.2 Nanotechnology

The history of nanotechnology is generally understood to have begun in December 1959 when physicist Richard Feynman gave a speech, "There's Plenty of Room at the Bottom," at an American Physical Society meeting at the California Institute of Technology in which he identified the potential of nanotechnology.[12] Feynman said it should be possible to build machines small enough to manufacture objects with atomic precision, and that if information could be written on an atomic scale, "all of the information that man has carefully accumulated in all the books in the world can be written … in a cube of material one two-hundredths of an inch wide—about the size of the smallest piece of dust visible to the human eye." He claimed that there were no physical laws preventing such achievements, while noting that physical properties would change in importance (eg, gravity becoming less important), and surface phenomena would begin to dominate materials' behavior.

In 1974 Norio Taniguchi first used the word "nanotechnology," in regard to an ion sputter machine, to refer to "production technology to get the extra-high accuracy and ultrafine dimensions, i.e., the preciseness and fineness on the order of one nanometer"[13] In the 1980s Eric Drexler authored the landmark book on nanotechnology, *Engines of Creation*, in which the concept of molecular manufacturing was introduced to the public at large. It is due to Drexler that public imagination has been captured by the potential of nanotechnology and nanomanufacturing.[14] With the invention of the scanning tunneling microscope in 1981, followed by the atomic force microscope in 1986, nanoscience and nanotechnology really started to take off.

Today nanoscience and nanotechnology hold great promise for future innovation and transformation, as recognized by most of the industrial world and confirmed by government funding allocated to this area.[15] Hundreds of products based on nanoscience and nanotechnology are already in use across many sectors, and analysts expect related markets to grow to hundreds of billions of euros in the next few years. In 2014 the world market for nanotechnology products was worth €23 billion, and it is expected to grow to €59 billion in 2019—an annual growth rate of 19.8%.[16,17]

The EU considers nanotechnology as a key enabling technology to foster radical breakthroughs in vital fields such as healthcare, energy, environment, and manufacturing, and to allow European industries to retain competitiveness and capitalize on new markets. Many new small and medium-sized enterprises are being established, and the number of spin-off and start-up companies in this high-tech sector is growing at a steady and quick pace. For this reason, the Horizon 2020 EU funding program for 2016–17 aims at bridging the gap between nanotechnology research and the wider market by scaling up laboratory experience to the industrial scale and demonstrating the viability of a variety of manufacturing technologies to leverage large-scale market introduction of innovative, safe, and sustainable nano-enabled products for different applications.[18]

A number of barriers still need to be addressed, such as costs and health and environmental issues, to understand both the benefits and the hazards of the nanoscale across all phases of product life cycle, from manufacture to use and final disposal.[19]

In fact, data on exposure and effects of manufactured nanoparticles in the environment is still scarce and incomplete. Their presence is difficult to monitor due to their diminutive size and low concentration levels; and it is equally difficult to distinguish between particles from manufactured nanomaterials, naturally occurring ones, and those produced incidentally, even more so in complex matrixes such as cosmetics, food, waste, soil water, or sludge.

In Europe, risks related to nanomaterials have been investigated since 2004 by the Scientific Committee on Emerging and Newly Identified Health Risks (SCENIHR), the Scientific Committee on Consumer Products, the European Food Safety Authority, and the European Medicine Authority. In 2009 SCENIHR concluded that "the hypothesis that smaller means more reactive, and thus more toxic, cannot be substantiated by the published data. In this respect, nanomaterials are similar to normal chemicals/substances in that some may be toxic and some may not. As there is not yet a generally applicable paradigm for nanomaterial hazard identification, a case-by-case approach for the risk assessment of nanomaterials is still warranted." Furthermore, SCENIHR's opinion in 2010 was that "it should be stressed that 'nanomaterial' is a categorization of a material by the size of its constituent parts. It neither implies a specific risk nor does it necessarily mean that this material actually has new hazard properties compared to its constituent parts."

Thus not all nanomaterials may have negative or toxic effects on human beings or the environment, as demonstrated by synthetic nanomaterials such as carbon black or titanium dioxide which have long been used and show little toxicity. It is evident that each nanomaterial's behavior in the environment, including uptake, distribution within living organisms, and interaction with other pollutants, may be understood only on the basis of specific physiochemical information, as its distinctive footprint stems from its chemical composition, shape, and structure and its behavior may differ according to the environment media.

2.2.1 Definition and concepts

Nanosciences constitute a whole new approach to research and development, aimed at controlling the structure and fundamental behavior of matter at the atomic and molecular levels.

In particular, nanotechnology is defined as the application of scientific knowledge to manipulate, control, and synthesize matter in the nanoscale (from approximately 1—100 nm), where properties and phenomena dependent on size and structure can emerge, distinct from those associated with individual atoms and molecules or with bulk materials.

The prefix "nano" in SI units indicates 10^{-9} as a billionth of the unit (=0.000000001). A nanometer (nm) thus equals a billionth of a meter, a dimension tens of thousands times smaller than the thickness of a human hair and corresponding to the length of a small molecule (see Table 2.3). To better understand nanometric scale proportions, consider that the ratio between a 1 nm diameter nanoparticle and a soccer ball (22 cm) is the same as that between the soccer ball itself and the planet Earth (12.742 km).

The lower limit of 1 nm is introduced to avoid single and small groups of atoms being designated as nanoobjects or elements of nanostructures (the smallest atoms are those of hydrogen, which are approximately a quarter of a nm diameter). Proper nanotechnology must in fact build its devices from atoms and molecules. The upper limit of 100 nm is arbitrary, but is around the size at which phenomena not observed in larger structures start to become apparent and can be exploited in nanodevices.

Table 2.3 **Sizes of nanoscale objects**[20]

Object	Diameter (nm)
Hydrogen atom	0.1
Buckminsterfullerene (C60)	0.7
Carbon nanotube (CNT; single wall)	0.4—1.8
Six carbon atoms aligned	1
DNA	2
Proteins	5—50
CdSe quantum dot (QD)	2—10
Ribosome	25
Virus	75—100
Semiconductor chip features	90 or above
Mitochondria	500—1000
Bacteria	1000—10,000
Capillary	8000
White blood cell	10,000
Hair	40,000—100,000
Pinhead	1,000,000

Although nanotechnology refers to the intentional human manipulation of matter at the nanoscale, there are numerous examples of naturally occurring nanostructures, including ribosomes responsible for protein synthesis, chloroplasts which carry out vegetable photosynthesis, the structure of a gecko's feet soles, the surface of a lotus leaf which provides self-cleaning capacity (the lotus effect), the shell of the abalone, the adhesive used by mussels to anchor to solid surfaces underwater, the spider web, and the rhinoceros horn.[9]

The appeal of nanotechnology stems on the one hand from the growing need for miniaturization (electronics, medicine, etc.), and on the other hand from the possibility of exploiting the new properties provided by nanomaterials.

At the nanometric scale, materials gain new and surprising properties which can apply to all technology sectors. Bulk properties of materials, such as strength and electrical, chemical, and optical characteristics, may indeed change dramatically when miniaturized to nanoscale dimensions: at this scale (<100 nm) they no longer depend on chemical composition alone, but also on size and form. This occurs for two main reasons.

First, a nanomaterial has a greater specific surface area compared to the same mass of material produced in a macro-scale form. This enhances their chemical reactivity (in some cases materials that are inert in their larger form are reactive when produced in a nanoscale form), and affects their strength or electrical properties.

Second, quantum effects begin to dominate the behavior of matter at the nanoscale, particularly at the lower end (<50 nm), affecting the optical, electrical, and magnetic behavior of materials.

The so-called quantum size effects describe the physics of electron properties in solids with the smallest particle size. They do not come into play by moving from macro to micro dimensions, but become dominant when the nanometer size range is reached. That is because all bulk properties of any material are merely the average of all the quantum forces affecting all the atoms that make up the material. As the scale reduces, eventually a point is reached in which the averaging no longer works and the specific behavior of individual atoms or molecules has to be dealt with—this behavior can indeed be very different from when the same atoms are aggregated into a bulk material.

By operating at the molecular level, nanotechnology opens up new possibilities in material design. In a world where quantum physics rules, objects may change color, shape, and phase much more easily than when in the macro scale. Fundamental properties like strength, surface-to-mass ratio, conductivity, and elasticity can be designed to create dramatically different materials.

For instance, normally opaque substances become transparent (copper); inert materials become catalysts (platinum); stable materials become combustible (aluminum); solids turn into liquids at room temperature (gold); and insulators become conductors (silicon).

Composites made from ceramic or metal nanoparticles can suddenly become much stronger than predicted by existing macroscale materials science models: metals with a grain size of around 10 nm can be as much as seven times harder and tougher than their ordinary counterparts with grain sizes in the hundreds of nanometers.

Properties gained by materials at the nanoscale may be static (*high-performance fixed response*) or dynamic (*dynamic performance* or *smart behavior*). In the latter case, nanomaterials become part of the category of so-called smart materials, able to vary their own characteristics according to external stimuli (chromogenics, thermotropics, etc.)

2.2.2 Classification

The term "nanomaterials" includes all materials having at least one of the three dimensions confined to the nanoscale range <100 nm (nanoobjects), as well as those with an internal or surface structure at the nanoscale (nanostructured materials or bulk nanomaterials), even if they have none of the three dimensions at the nanoscale (zero-dimensional or 0D nanomaterials).[21] Nanomaterials are not restricted to newly discovered materials but include nanoscale forms of well-known materials such as gold, silver, platinum, iron, and others.

According to the number of physical dimensions in the nanoscale, nanoobjects are classified into nanoplates (one dimension in the nanoscale, 1D); nanofibers (two dimensions in the nanoscale, 2D) further divided into nanotubes (hollow nanofibers), nanorods (solid nanofibers), and nanowires (electrically conducting or semiconducting nanofibers); and nanoparticles (all three dimensions in the nanoscale, 3D) (Fig. 2.2 and Table 2.4).[22] In addition to size and shape, several other parameters relate to the function and phenomena exhibited by nanoobjects. Those include composition, morphology, crystalline structure, and surface features, and can all have a major influence on the key nanoscale phenomena exhibited by nanoobjects such as magnetic, optical, catalytic, electronic, and other properties.

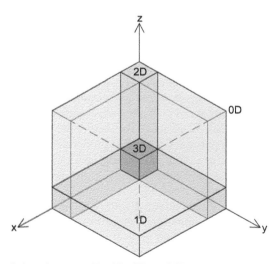

Figure 2.2 Space relations between 0D, 1D, 2D, and 3D nanomaterials.

Table 2.4 **Nanoproduct classification**

Nanomaterials ISO/TS 80004-1:2015	
Nanoobject Material with one, two, or three external dimensions in the nanoscale (1–100 nm)	**Nanostructured material** Material having internal nanostructure or surface nanostructure
3D: Nanoparticle Nanoobject with all external dimensions in the nanoscale where the lengths of the longest and the shortest axes of the nanoobject do not differ significantly (eg, QDs, nanospheres, buckyballs) *2D: Nanofiber* Nanoobject with two external dimensions in the nanoscale and the third dimension significantly larger (eg, nanorods, nanowires, single or multiwalled CNTs) *1D: Nanoplate* Nanoobject with one external dimension in the nanoscale and the other two external dimensions significantly larger (eg, graphene, monolayers, thin films)	*Nanostructured powder* Powder comprising nanostructured agglomerates, nanostructured aggregates, or other particles of nanostructured material *Nanocomposite* Solid comprising a mixture of two or more phase-separated materials, one or more being nanophase *Solid nanofoam* Solid matrix filled with a second gaseous phase, typically resulting in a material of much lower density with a nanostructured matrix, for example, having nanoscale struts and walls, or a gaseous nanophase consisting of nanoscale, or both *Nanoporous material* Solid material with nanopores *Fluid nanodispersion* Heterogeneous material in which nanoobjects or a nanophase are dispersed in a continuous fluid phase of a different composition

2.2.2.1 Nanoplates

1D nanoobjects (nanoplates), characterized by an overall exterior thickness with nanoscale dimension (<100 nm), may or may not have a nanoscale internal grain structure, but are anyway classified as 1D nanomaterials.

1D nanomaterials are generally deposited on a substrate or support with dimensions above the nanoscale, and the overall thickness is the sum of the film's and substrate's thickness. In this case, 1D nanomaterials can be considered a nanocoating. In contrast, when the substrate thickness also has nanoscale dimensions or when multiple layers with nanoscale thickness are deposited sequentially, the nanomaterial can be classified as a multilayer 1D nanomaterial, independently from internal structure dimensions (Fig. 2.3).

1D nanomaterials may also be patterned with physical features such as channels, grooves, and raised lines at various scales, mostly required for electronic applications. These patterns can have different geometries and dimensions at the nanoscale or above, and, in the case of multilayered nanomaterials, can be etched into any layer.

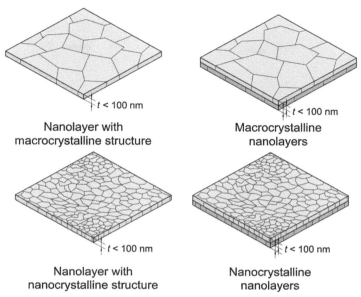

Figure 2.3 Different nanolayer structures.

2.2.2.2 Nanostructured materials

Nanostructured materials are not confined to the nanoscale in any dimension (0D nanomaterials; Fig. 2.4), but are instead characterized by an internal or surface structure at the nanoscale. They include nanostructured powders, nanocomposites, solid nanofoams, nanoporous materials, and fluid nanodispersions.

In both nanopowders and fluid nanodispersions, the nanoobjects (or their aggregates or agglomerates) are arranged in a nonrandom distribution (identifying a short-range order, thus a true structure).

Nanocomposite materials can be composed of a multiple arrangement of nanosize crystals (nanocrystalline materials), most typically in different orientation, or can contain dispersions of nanoparticles, bundles of nanowires and nanotubes, or multiple nanolayers. Nanocomposites are formed by two or more materials with distinctive properties acting synergistically to obtain properties that could not be achieved by each material alone. The nanocomposite matrix, which can be polymeric, metallic, or ceramic, has dimensions larger than the nanoscale, whereas the reinforcing phase is commonly at the nanoscale. Distinctions are based on the type of reinforcing nanomaterial added, such as nanoparticles, nanowires, nanotubes, or nanoplates.

2.2.3 Manufacturing processes

The operative and methodological path that leads to the realization of nanoscale materials is today based on two alternative methods: top-down or bottom-up approach (Fig. 2.5).[9,23]

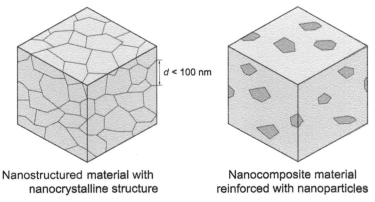

Figure 2.4 0D nanomaterials.

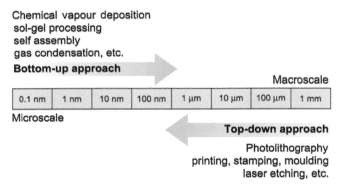

Figure 2.5 Nanotechnology manufacturing approaches.

The top-down approach obtains nanosized structures by starting from the processing of a solid with discrete dimensions. The most common process is UV photolithography, which is expensive and extremely complex, as well as ill-adapted to sub-100 nm processing due to difficulty in focusing light beams. Further developments are electron beam lithography and X-ray lithography, with improved nanoscale precision but still slow and cumbersome in operation. More recent lithography methods instead employ mechanical processes such as printing, stamping, and molding, which are more versatile and require cheaper equipment. These use a silicon matrix etched by traditional lithography to reproduce it on a polymer base, which can be subsequently used for printing molecular inks on gold or silicon, with 50 nm resolution, or as mold for stamping and casting polymer solutions with a resolution up to 10 nm.

The bottom-up approach, or "atomic technology," is based on synthesis processes and aims at creating new structures starting from the atomic and molecular level, completely revolutionizing current manufacturing processes. In this approach the nanostructure is generated by the subsequent addition of atoms, with techniques based on the activation of chemical processes, such as chemical vapor deposition (CVD) or sol—gel processing, or physical processes (dip pen lithography or self-assembly). Bottom-up methods allow control of the structure dimensions from the atomic and molecular level, and are proven to be preferable to the initial top-down methods for the production of large quantities of high-quality nanostructured materials with a good possibility of translating to large-scale production.

Among chemical bottom-up methods, the most used is certainly CVD (Fig. 2.6). This consists of the controlled realization of nanoobjects though a process of decomposition and deposition on a solid substrate of specific molecular precursors in gaseous form. Transport of the precursor is carried out by a gas (oxygen, argon, nitrogen, hydrogen, etc.), which also takes decomposition by-products away. The most common chemical reactions involved in CVD processes are pyrolysis, reduction, and oxidation.

The CVD technique is normally used to produce nanoplates (1D) such as graphene, nanowires or nanorods (2D) such as CNTs—by using specifically etched substrates—or even nanoparticles (3D) up to 5 nm in size.[9]

A further development of the CVD technique is molecular beam epitaxy, which allows accurate growing of thin layers of highly pure crystals by depositing each component in quasivacuum conditions (10^{-10} torr) to avoid any collision between atoms and molecules and the supporting substrate. The atomic layer epitaxy technique instead allows alternate deposition of single atomic layers on the same surface, each time saturating the substrate with the right chemical precursor and allowing the film to grow with atom-sized precision.

Both techniques allow the manufacturing of 1D nanoplates.

Chemical methods that work in the liquid phase are sonochemical processing, which can synthesize particles as small as 2 nm in size, and sol—gel processing,

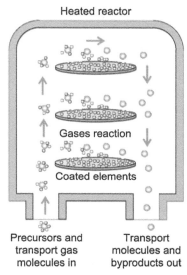

Figure 2.6 Chemical vapor deposition process.

a versatile technique that allows manufacture of 3D nanoparticles, 1D nanoplates and nanomembranes, and 0D nanoporous materials such as aerogel.

Among physical bottom-up methods, most interesting is certainly the self-assembly process, in which atoms and molecules self-arrange automatically between themselves following the least energy-dispersive scheme, and create the final product without any direct intervention from outside. This process takes advantage of molecular chemistry and thus requires each specific molecule to be designed to self-arrange as planned according to its energy bonds. Extremely miniaturized and defect-free structures can be obtained, with the possibility of easily integrating organic and inorganic elements within the same system. Self-assembly processing can produce 3D nanoparticles, 2D nanoobjects such as CNTs, and 1D nanoplates.

Other bottom-up physical processes employ probe scanning microscopes, able to investigate surfaces not with optical methods, unviable due to diffraction caused by the excessive wavelength of light compared to details examined (<100 nm), but with nanoscale scanning tips able to map the surface by measuring its atomic force (atomic force microscopy) or by the tunnel effect (scanning tunneling microscopy). In addition to enabling accurate investigation of nanometric components, these devices allow users to modify them by moving and assembling single atoms. Among these methods is dip pen nanolithography, which uses the probe of an atomic force microscope, able to move with nanometric precision thanks to piezoelectric actuators, to deposit a variety of molecular inks according to desired schemes by a tip printhead with resolution up to 50 nm.

Physical bottom-up processes include inert gas condensation and inert gas expansion, in which atoms are vaporized and dispersed in a controlled environment, favoring their aggregation in increasingly large clusters to obtain 3D nanoobjects.

Research is currently focused both on devising new synthesis methods for processing nanoobjects and improving established ones to reduce costs and improve the quality of output materials. Small defects at the atomic and molecular level are still common, especially in high-end electronics-targeted nanoobjects such as monolayer semiconductors.

A possible method for obtaining defect-free monolayers for use in LED displays, solar cells, or transistors, has been devised by researchers at Berkeley University and National Laboratory.[24] The desired nanofilm, in this case a molybdenum disulfide (MoS_2) semiconductor monolayer, is dipped in an organic superacid solution, chemically treating it by removing contaminants and also filling in missing atoms through a protonation reaction. This way it is possible to achieve a defect-free end product with much improved performance: the MoS_2 monolayer photoluminous effect increased from 1% to 100% after treatment.

2.2.4 Applications in energy, environmental, and construction sectors

From medical applications to information technology, from aerospace to the automotive industry, from building construction to energy production and storage or environmental protection, nanotechnology is proving able to help solve many issues of contemporary society, allowing the creation of products and processes for more specific uses, with higher performance and lower environmental impact throughout their entire life cycle (material follows function) (Table 2.5).[25] For these reasons, nanoscience is often referred to as "horizontal," "key," or "enabling," since it can reconnect different scientific disciplines and benefit from interdisciplinary and converging approaches.[26]

The main applications of nanotechnology in the various sectors are as follows:

- Creation of nanostructured materials by the addition of free or embedded nanoobjects within the composition of traditional materials and products, both solid (nanocomposites) and liquid (nanosuspensions), such as cement, metals, ceramics, plastics, textiles, paints, adhesives, and lubricants, as well as the modification of their internal structures to make them nanocrystalline.
- Use of nanocoatings for surface treatment of traditional products such as glass, wood, ceramics, textile materials, and PV cells.
- Modification of surface structures of traditional products to make them nanostructured.
- Creation of new nanoporous materials (nanoporous insulating materials, nanofilters, nanomembranes, etc.).
- Development of nanodevices such as NEMS.

In the energy sector nanotechnology can provide important benefits across the entire supply chain, from production to distribution (power transmission and heat transfer), storage (batteries, supercapacitors, PCMs, and adsorptive storage with nanoporous materials), and end use (thermal insulation, lighting, and appliances), including the development of renewable energy, hydrogen generation, fuel cells, and smart grids with nanosensors for intelligent and flexible grid management capable of managing highly decentralized power feeds.[27,28]

Table 2.5 EU nanotechnology patents classification[17]

B82Y	Nanotechnology
B82Y5/00	Nanobiotechnology or nanomedicine, eg, protein engineering, drug delivery
B82Y10/00	Nanotechnology for information processing, storage, or transmission, eg, quantum computing, single-electron logic
B82Y15/00	Nanotechnology for interacting, sensing or actuating, eg, QDs as markers in protein assays or molecular motors
B82Y20/00	Nanooptics, eg, quantum optics, photonic crystals
B82Y25/00	Nanomagnetism, eg, magnetoimpedance, anisotropic magnetoresistance, giant magnetoresistance or tunneling magnetoresistance
B82Y30/00	Nanotechnology for materials or surface science, eg, nanocomposites
B82Y35/00	Methods or apparatus for measurement or analysis of nanostructures
B82Y40/00	Manufacture or treatment of nanostructures
B82Y99/00	Subject matter not provided for in other groups of this subclass

In particular, concerning renewable energies, nanotechnologies can improve:

- PVs, with nanooptimized cells (polymeric, dye sensitized, QD, thin film, multiple junction) and antireflective coatings;
- wind energy, with nanocomposites for higher and stronger rotor blades, protection from wear and corrosion, nanocoatings for bearings and power trains, etc.;
- geothermal energy, with nanocoatings for wear-resistant drilling equipment;
- hydrotidal power, with nanocoatings for corrosion and oxidation protection;
- biomass energy with nano-based precision farming (nanosensors, controlled release and storage of pesticides and nutrients).

In the environment protection sector, aimed at preventing water, air, and soil pollution, nano-based technologies offer sensible help in achieving environmental objectives and targets by improving many conventional processes or introducing altogether new approaches. In particular, benefits provided by nanotechnology regard the environmental monitoring of pollutants, processes for water and air cleaning and purification, water desalinization, and soil remediation methods.

Chemical sensors based on nanotubes and nanowires have already proven effective, as have NEMS serving as sensors, controllers, analyzers, and communication devices allowing uses in many situations where larger-scale devices were not viable.

New nano-based catalytic, filtration, and adsorbent methods are greatly improving the efficiency of purification processes of water, air, and soil. Recently, advanced MoS_2 nanomembranes with extremely tiny "nanopore" holes proved able to filter up to 70% more water than graphene, and to be more cost-effective and energy-efficient in removing salt from seawater than standard desalination methods relying on reverse-osmosis membranes.[29]

Soil depuration may also take advantage of nanopolymers designed to attract hydrophobic pollutant agents and trap them to ease their removal. This concept was studied to isolate and extract pesticides and endocrine interferents such as bisphenol A from water and soil via a nanoparticle dispersion of polyethylene glycol, a polymer commonly used in pharmaceuticals, and biodegradable polylactic acid plastic. Thanks to their hydrophobic core and hydrophilic shell, these particles attract and absorb pollutants on their surface and then stabilize and aggregate under UV radiation, allowing removal via filtration or sedimentation.[30]

Oil-absorbent nanoporous polymers and aerogels are also proving extremely effective in purifying water, for instance, after oil spills: in particular, absorbent boron nitride powder (so-called "white graphene") performance is being further enhanced by processing it into nanosheet aerogels able to soak up oil and organic solvents up to 33 times their own weight.[31]

Finally, exterior surfaces and pavements enhanced with nanocoatings or nanomaterials can potentially help in reducing air pollution (photocatalytic effect).[32]

In the construction sector, the use of nanotechnology can bring many benefits by improving energy and environmental performance of buildings, with application potential in all technical elements, from structures to opaque and transparent closures, internal partitions, systems, and appliances.[8,33]

Several applications are now available for this specific sector to increase the mechanical, physical, and chemical properties of materials, products, and equipment, as well as their durability over time, with the aim of improving environmental comfort, safety, and energy efficiency of buildings and at the same time reducing operating and maintenance costs and environmental impacts.[9,33–35]

In particular, the use of nanomaterials in architecture may allow important improvements in:

- performance of load-bearing structures, with use of lightweight nanocomposite construction materials (CNTs, metal matrix composites, nanocoated light materials, ultra-high-performance concrete (UHPC), polymer composites, etc.);
- thermal and lighting control, with use of high-performance nanoporous insulating materials (aerogels, VIP, TIM), highly reflective coatings (cool roofs), and low-e, antireflective, passive and dynamic solar control and self-cleaning nanocoatings for glazing;
- surface characteristics of building elements, with the application of specific nanocoatings or nanopaints (self-healing, air purifying, self-cleaning, antibacterial, antigraffiti and antistain, photoluminous, scratch resistant, antireflection, antiicing, antifogging, fouling resistant, oxidation and corrosion resistant, UV resistant, fire resistant, etc.);
- energy efficiency in lighting (LEDs, QLEDs, OLEDs);
- more efficient and integrated renewable energy systems (polymeric, dye sensitized, QDs, thin-film and multijunction PV solar cells, fuel cells, etc.);
- building and environmental monitoring and control, with use of nano-embedded sensor, actuation, and control systems (NEMS).

2.2.5 Nanoproducts for architecture

To obtain products with new or improved properties, nano-based technology can now find wide application in all main construction materials, elements, and systems of buildings (Tables 2.6 and 2.7): structural materials as steel, concrete, or wood; plasters and screeds; construction ceramics; plastics and polymers; glazing; insulating materials; paints; adhesives; lubricants; energy storage systems; renewable energies; lighting systems; air and water depuration; building and environmental monitoring and control systems.[36,37]

Within these construction materials, elements, and systems, nanomaterials can be integrated in the form of solid nanocomposite materials, solid materials enhanced with nanocoatings or characterized by nanostructured surfaces, nanoporous materials, and fluid nanodispersion in nanopaints, nanoadhesives, and nanolubricants.

More specifically, in the nanocomposite class of materials one or more phases at nanoscale dimensions (3D, 2D, and 1D) are embedded in a polymeric, metallic, or ceramic matrix. Nanoobjects embedded in the matrix (usually a few percent by weight) can be made of various primary materials, and their addition can provide the material with unique and enhanced properties compared to conventional composite materials.

Nanocoatings or nanofilms are thin films which incorporate nanomaterials, or have a thickness at the nanoscale (<100 nm), deposited on a substrate material to enhance

Table 2.6 **Main building material properties improved through nanotechnology**

Enhanced properties	Nanotechnology	Building products
Durability	Al_2O_3, SiO_2 (silicon dioxide), and ZnO particles	Concrete
Mechanical resistance and crack prevention	Polymers, SiO_2, FeO, and $CaCO_3$ particles, CNTs	Concrete
	Cu particles	Steel
	$CaSO_4$, SiO_2 particles	Gypsum drywalls and ceilings
Hardness (abrasion resistance)	CNTs, FeO particles	Concrete
Shock resistance	V, Mb particles	Steel
	Al_2O_3 particles	Concrete
Weldability	CuO, Ca, Mg	Steel
Formability	Nanopolymers	Concrete
Quick compacting	TiO_2 (titanium dioxide) particles, CSH nanocrystals	Concrete
Curing time	TiO_2, $CaCO_3$ particles, CSH nanocrystals	Concrete
Corrosion resistance	SiO_2 particles	Paints
	Al_2O_3, SiO_2 coating	Steel
Oxidation resistance	CuO particles	Steel
UV resistance	ZnO, CeO_2 coating	Wood
Moisture resistance	$C_6H_{10}O_5$	Concrete
Water protection	Fatty acids coating	Wood
	SiO_2 particles	Concrete
Fire behavior	SiO_2 particles	Glass
	TiO_2, particles	Concrete
Insulation properties	SiO_2 aerogel, SiO_2 particles	Insulating materials
	Hydro-NM-oxide particles	Paints
Thermal conductivity	CNTs	Polymers and plastics
	Metal oxide coating	Glass
Visible and near-infrared light transmission or reflectance	VO_2, FeO, SiO_2, ITO coating	Glass
	Nanoceramics coating	Solar films

Table 2.6 Continued

Enhanced properties	Nanotechnology	Building products
UV control	TiO$_2$, ZnO coating	Coating
Thermal radiation absorption or reflectance	Ceramic nanoparticles	Paints
Adhesion	SiO$_2$ or SiH$_4$ particles	Adhesives
Rheology	SiO$_2$ particles	Paints
Surface energy	SiO$_2$ particles	Paints
Electrical conductivity	CNTs	Polymers and plastics
Photoelectric effect	QDs, graphene, CNTs	Thin-film solar cells
	TiO$_2$ particles	Dye-sensitized solar cells (DSSC)
	Si particles	Silicon solar cells
Energy storage	Graphene, carbon aerogel, CNTs	Batteries, capacitors
	Zeolite particles	Wood
Friction	Fullerene	Lubricants
	Ceramic coating	Steel

its surface properties (durability, abrasion or corrosion resistance, electrical, magnetic, or optical behavior, etc.) without having to alter its entire mass. Multiple nanoscale layers can be built up to form multilayered or laminate nanostructures.

Using thin nano-based coatings is a very convenient way of exploiting the remarkable properties of nanomaterials without the high production cost associated with manufacturing large components fully out of nanocomposites. This gives them one of the greater overall market potentials for any kind of nano-based product or device.

Among nanoobjects used as a standalone end product, one of the most interesting and promising is certainly graphene, aptly nicknamed the "wonder material."

First synthesized in 2004, graphene is a 2D crystal of carbon atoms arranged in a hexagonal structure, effectively constituting a single-atom-thick sheet, with revolutionary characteristics compared to conventional materials with 3D structure. To date, graphene holds many records: it is the thinnest object ever made, the strongest material in the world, a virtually perfect electrical and thermal conductor, and constitutes a membrane impermeable to any molecule. Relatively simple manufacturing

Table 2.7 **Main new building materials properties provided through nanotechnology**

New properties	Nanotechnology	Building products
Self-healing	Nanopolymers	Concrete, gypsum
Air purifying	TiO_2, ZnO coating	Ceramic tiles, glass
	TiO_2 particles	Paints, concrete
	Zeolite particles	Gypsum drywalls, ceilings, plaster
Easy to clean	SiO_2 coating	Glass
Antigraffiti	TiO_2, SiO_2 coating	Concrete
Antistain	TiO_2, SiO_2 coating	Concrete
Scratch resistant	Al_2O_3 coating	Ceramic tiles and sanitary ware
	SiO_2 particles	Paints
Antireflection	SiO_2 coating	Glass
Antiicing	CNT	Glass
Antifogging	TiO_2 coating	Glass
Fouling resistance	TiO_2, Ag coating	Ceramic tiles, glass
	TiO_2 particles	Paint
Biocidal activity	Ag, ZrO_2 CuO coatings	Ceramic tiles and sanitary ware
	Ag particles	Paints
	Ag, Cu, Zn particles	Wood
Electromagnetic radiation block	Metal nanoparticle lattices, CNT	Sensors, paints, fabrics
Light emission	Graphene	OLED lighting
	QDs	LED lighting
Artificial photosynthesis	TiO_2 particles	Hydrogen production
Dynamic response	Semiconductive metal oxides	Dynamic glazing
Real-time structural health monitoring	CNT	Concrete

processes also make it appealing from an economic point of view in future large-scale applications.

In the biomedical field it can be employed in nanodevices capable of directly interfacing with cells for biomeasuring, administration of medicines, or reconstruction of tissues.

In electronics it allows production of transistors and integrated circuits at the nanometer scale, with huge benefits in terms of size and speed compared to traditional silicon. Its high electrical conductivity and transparency also make it an excellent candidate for integration in liquid crystal displays (LCDs), touch screens, electrochromic glass, and OLEDs. In these applications, conventional transparent conductive oxides based on indium tin oxide can be replaced by graphene sheets, which are much more flexible and resistant, as well as being applicable at low temperatures and thus simplifying production processes.

Great attention is focused on graphene integration into PVs: graphene solar cells claim conversion efficiencies of up to 60%, more than double the theoretical limit for silicon cells, and are printable on flexible sheets to allow versatile and economic applications.

Finally, perfect watertightness and low thickness make it suitable for the production of waterproof coatings for a great variety of objects and devices, and the ease of creating specific-sized holes on the surface promises a new generation of purification and filtration systems for fluids and gas.

Lately, in the form of graphene oxide, it is also finding application in the production of insulating materials, combined with nanocellulose and other nanoparticles.

2.2.5.1 Concrete

Nanotechnology application in the concrete industry creates products with unprecedented performance regarding both mechanical properties and durability over time, along with improvements in the construction process (ease of mixing, rate of setting, workability, etc.) and a reduction of carbon emissions in cement manufacturing (reduced kiln temperatures and durations during processing).[38,39] The potential environmental benefit is huge, as cement production generates over 1.6 billion tonnes of carbon worldwide, more than 8% of total carbon emissions. Concrete also accounts for more than two-thirds of construction and demolition waste, with only 5% currently recycled.[8]

Nano-enhanced concrete products may be both nanocomposites, as a cement blend with added nanoparticles to enhance its properties, and nanocoatings to modifying surface properties and interaction with the environment.

Nanocomposite concrete can become impermeable through the addition of inorganic solutions that fill its capillary void with nanocrystals during curing, and can cure much faster due to silicate hydrate nanoparticles that accelerate the hydration process without requiring energy-intensive treatments such as hot steam or hazardous additives. Addition of TiO_2 nanoparticles makes it photocatalytic, allowing for improved durability, self-cleaning action, and an air-purifying effect, with the added benefit of improved fire behavior. Regarding mechanical performance improvement, currently available UHPC with added metal and silicon oxide nanoparticles requires much less material to achieve the same structural strength compared to traditional products, thus allowing a much reduced environmental impact with a sixfold reduction in raw materials, 50% reduction in greenhouse gas emissions, and 40% reduction in manufacturing energy. CNT addition is being researched, and will allow a 90% reduction in material use.

In concrete nanocoatings, antigraffiti and antistain effects can be achieved by the application of SiO_2 nanoparticles solutions to form a nanothin layer of glass on its surface, preventing moisture or graffiti paint from penetrating the pores and dirt or algae from attaching to the surface. TiO_2 coatings impart self-healing and air-purifying effects as well.

2.2.5.2 Metals

Nanotechnology improves a number of properties in all metal applications such as structural steel, metal piping, and interior/exterior surfaces.[9]

The properties of many common metals can be greatly enhanced by the addition of relatively small amounts of nanomaterials—normally in the form of nanoparticles, nanowires, and nanotubes—as well as by the application of specific nanocoatings.

High-performance steel for the most demanding applications, such as bridge supporting cables or precast concrete, undergo specific manufacturing processes to reduce their crystalline structure to the nanoscale (nanostructured steel).

Nanocomposite steel develops improved properties concerning oxidation resistance and workability (addition of copper nanoparticles), fatigue resilience (addition of vanadium and molybdenum), and welding behavior (through calcium and magnesium particles).

Several nanocoatings give improved corrosion resistance thanks to alumina and silica nanoparticles, or employ polymer nanodispersions to promote acrylic coating adhesion.

These nanocoatings constitute more durable and environmentally friendly alternatives to traditional chromating baths. Steel used in highly hygienic applications such as healthcare facilities and medical-grade tools develops an antimicrobial effect with the addition of silver or copper nanocoatings. Metal employed in mechanical gears can greatly reduce surface friction through the use of nanoceramic coatings. Finally, architectural metal surfaces can be protected with antiscratch coatings based on the addition of aluminum oxide nanoparticles (Al_2O_3) in 1–5% concentration in aqueous coating systems, acrylate dispersions, or aliphatic polyurethane dispersions.

2.2.5.3 Plastics

Plastics and polymers (thermoplastics, thermosets, and elastomers) can take advantage of the benefits of nanoobject integration (polymer matrix nanocomposites) or nanocoating deposition on their surface.[9]

Many kinds of nanofillers (silicas, clays, CNTs, carbon blacks) are suitable for integration in polymer nanocomposites. The use of a small quantity of nanoparticles, nanotubes, or nanoplates—a few percent by weight—as a reinforcing second-phase material can dramatically alter the properties of different polymers. For instance, both strength and modulus of elasticity values can be either increased or decreased depending on the polymer host and the nanoscale reinforcing phase: materials such as polystyrene reinforced with silica nanoparticles show an elastic modulus increase, while carbon nanoparticles can increase toughness and wear resistance.

CNTs can sensibly increase compression strength if dispersed in thermoset epoxies (in the 30% range) at only 0.5–2.0% wt proportions.

Many types of widely used thermoplastics such as polycarbonates can have added CNTs, with the aim of increasing compressive and impact strength, toughness, elongation, and other properties.

Nanocoating application to plastics is challenging to carry out because of the polymers' organic nature and low heat resistance. An effective, transparent, scratch-resistant coating (+55% abrasion resistance) has, however, been obtained by chemically growing zirconia nanoparticles on top of the polymeric surface.

2.2.5.4 Ceramics

Ceramic materials in architecture include tiles both for interior and exterior application as well as sanitary ware. All these applications may benefit from nanotechnology, improving their durability or ease of maintenance characteristics or giving them entirely new properties.[9]

Through the application of nanocoatings, ceramic surfaces can repel dirt and water by becoming superhydrophobic (coatings with SiO_2 nanoparticles that fill surface pores and give an ultra-smooth surface) or superhydrophilic and photocatalytic (TiO_2 coating). SiO_2 coatings also provide antiscratch properties without affecting gloss. Nanosilver coatings provide a biocidal effect for both ceramic tiles and sanitary ware.

2.2.5.5 Glazing

Glazing is one of the fields in which nanotechnology is having the widest impact, with several nano-enhanced products already on the market with much improved performance or altogether new properties. In architecture, glass is mostly used to achieve natural lighting of interiors, and nanotechnology provides extremely thin coatings with multiple functions that do not excessively hinder the desired characteristic of visible light transparency.

In particular, nanocoatings may increase glass transparency through an antireflective effect (multiple nanosized polymer layers with different refractive angles) or by preventing dirt build-up by making its surface superhydrophobic (nanosilica coatings) or superhydrophilic and photocatalytic (TiO_2 coatings). These high transparency levels ensure maximum daylight and, in the case of building-integrated photovoltaics (BIPV), maximum efficiency in harvesting solar energy. SiO_2 nanoparticle intumescent interlayers also allow for fireproof glazing without loss of transparency. Glass fogging as well as snow or ice accumulation can be prevented by self-heating glass, obtained by metal or CNT coatings that make its surface electrically conductive.

From the energy efficiency point of view, nanocoatings are widely used to improve glazing thermal transmittance (low-e coatings using metal oxide nanoparticles) or filter and absorb UV or NIR radiation (silicon or metal oxide nanocoatings, nanoceramic solar films), avoiding heat loss in winter and excessive heat loads in summer months. In addition to static solar protection glazing, dynamic solar control coatings are now available that can vary their transparency to visible or infrared (IR) light, reacting passively to light (photochromic glass) or heat (thermochromic glass). Electrochromic glazing allows active dynamic control of light and heat transmission thanks to a thin-film coating composed of several extremely miniaturized conductive layers.

2.2.5.6 Insulating materials

Nanotechnology can be used to create new advanced high-performance insulating materials able to provide exceptional heat flow resistance thanks to their nanoporous structure achieved via sol–gel processing (silica aerogel). They can also be used in vacuum applications (VIP), or to provide additional solar radiation transparency (TIM).[40]

Due to their low thermal conductivity ($\lambda < 0.02$ W/mK), is it possible to obtain high thermal resistance values in the building envelope with extremely thin insulating layers (less than a third compared to conventional insulating materials). These products are thus particularly suitable for correction of thermal bridges and energy retrofit of existing buildings, especially historic buildings subject to architectural constraints, and in all cases in which it is necessary to increase energy efficiency and comfort with minimum space loss.

NIR reflecting coatings such as nanopaints can also be used for roof and wall applications to reduce summer overheating and the urban heat island effect (cool roofs).

Finally, nanomaterial addition can significantly alter the thermophysical properties of heat storage materials, especially in PCMs for latent thermal energy storage applications. The incorporation of specified proportions of nanomaterials can increase thermal conductivity, accelerate freezing and melting rates, improve thermal stability, and ensure thermal reliability over time.

2.2.5.7 Adhesives

Most adhesives used today are based on elastomers, thermosetting or thermoplastic polymers, and other materials, and current research on improving performance and ease of application is being spurred by the advent of nanoscience and nanomaterials.

Nanomaterials as now are mostly added to improve rheology characteristics, such as viscosity and antisagging, allowing thinner, more uniform adhesive coating applications: mostly used in this regard are nanosized silicates. The inclusion of metal or carbon nanoparticles such as iron oxide gives conductive adhesives that allow electricity transfer in electronic applications and electromagnetic-induced curing.[9]

Nanoscience is also taking inspiration from naturally occurring nanosized structures to mimic their behavior: in this sector, research has focused on the adhesive ability of mussels and geckos. Regarding the gecko, it has been observed that its excellent clinging ability stems from the nanostructured surface of its feet, which feature many small hairs, each split in even smaller hairs at the end, called satae. Each of these is covered by cuplike spatulae 200 nm in size, which radically increase the contact area between the hairs and the surface, achieving billions of contact points that take advantage of both Van der Waals forces and capillary action. Mimicking this behavior allows powerful and reversible adhesive action, and this was achieved by nanostructured surfaces with patterns akin to those of the gecko's soles, such as Poly(methyl methacrylate) (PMMA) nanopillar arrays created by electron beam lithography or CNTs. Promising results were obtained, but a decrease of adhesion in wet environments was noted. Research on wet adhesion then turned to mussels, and their ability to cling reversibly to virtually any underwater surface. It was discovered that this unique capability stems from a specialized protein secretion, rich in the amino acid L-Dopa, which is known to provide underwater adhesion and self-healing capabilities. This was mimicked in engineered polymers with an analog synthesized protein, giving a high-performance underwater glue applicable in thin films and 10 times more effective than conventional wet adhesives.[41] By coating gecko-inspired nanopatterns with this mussel-inspired adhesive, exceptional reversible performance in both dry and wet environments was finally achieved.

2.2.5.8 Paints

Nanotechnology has successfully influenced the paints sector, and the addition of nanoparticles in solution has led to the development of a whole range of innovative and high-performing nanodispersed products.[9]

In particular, nanodispersions can show improved characteristics of rheology, settling, and viscosity, improving paints' behavior during application by the addition of nanoclays and silica or aluminum nanoparticles. Photocatalytic paints include dispersed TiO_2 nanoparticles to achieve a self-cleaning, antifouling, and air-purifying effect if exposed to UV rays. Other self-cleaning paints exploit the lotus effect, caused by a bumpy surface nanopattern which is small enough to prevent water drops from attaching, causing them to roll away taking dirt particles with them. Interior environment healthiness can be raised by biocidal paints composed of nanodispersed silver particles that prevent bacteria and microorganisms from developing. Similarly, metal nanoparticles or CNTs can make surfaces antistatic or block unwanted electromagnetic radiation.

Lifetime and color fastness of organic paints exposed to UV degradation can be increased by the addition of metal oxide nanoparticles such as zinc, cerium, or titanium, while transparent protective paints (clear coats) for furniture and wood can gain increased scratch and abrasion resistance with no gloss alteration by the inclusion of silica nanoparticles. Furthermore, interesting color effects can be achieved on any surface by adding multiple polymer nanolayers with different refractive angles, thus varying the wavelength and color of reflected light according to the viewer's position (dichroic effect).

Finally, nanoparticles added to paints can improve their reflectance to solar radiation for cool roof applications.

2.2.5.9 Lighting

The energy efficiency of lighting systems is very important due to their high electrical consumption in both residential and tertiary buildings. Last-generation incandescent lamps, still widely used in existing buildings, were characterized by low efficiency, heat release, and overheating issues, especially in hot climates during summer months.

Solid-state lighting technologies such as LEDs have greatly improved the energy efficiency of lighting systems, also superseding fluorescent lamps.

However, despite their high energy efficiency, actual emission wavelengths that can be generated remain limited and not easily adjustable, and the production of full-spectrum "white light" is still problematic since it requires a final adjustment via the addition of phosphors with unsatisfactory results.

In this picture, existing LED technologies can be widely improved with the integration of a variety of nanostructures, with several beneficial effects:

- increasing light extraction or outcoupling from LED chips;
- enhancing luminescent output by increasing emitters' surface areas;
- improving the efficiency of injected electrons by using nanostructured electrodes;
- improving light control by using novel nanophosphors;
- exploiting surface fluorescence of QDs.

Among these, QD network integration in LEDs is particularly interesting, paving the way for highly efficient and innovative QLEDs (Fig. 2.7).

QDs consist of nanocrystals of semiconductor materials, such as germanium or silicon, whose electronic and luminous properties are directly related to their diminutive

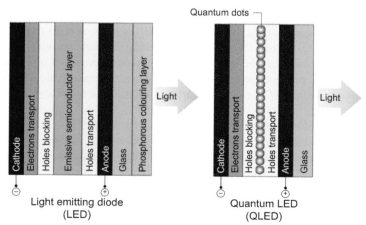

Figure 2.7 Light-emitting diodes and quantum dot light-emitting diodes cross-sections.

size. Advantages include high efficiency, better control of emitted light, and further miniaturization of form factors. Concerning light control, the actual emitted wavelength, and thus the color, does not strictly depend on the semiconductor material, as in conventional LEDs, but can be finely tuned by acting on the nanocrystals. Nanoparticle size in fact dictates energy separation levels, which in turn vary the wavelength of emitted photons. Thus precise color emittance to match specific needs, even in the case of full-spectrum white light, can be achieved by directly mixing the right kinds of constituent blue-, green-, and red-emitting QDs.

QDs also benefit from a more convenient manufacturing process: nanocrystals can be grown in large quantities with colloidal approaches, instead of being precisely shaped and cut from geranium or silicon wafers, as in macroscale semiconductors.

Further research on LEDs manufactured by using QDs regards the use of carbon-made QDs, so-called carbon dots; these can even be produced from waste bread and alcohol-free sugar beverages, giving a further reduction in raw material costs and environmental impacts.[42] Design flexibility potential is also exceptional, as QDs may be incorporated into various types of polymeric materials, rigid, flexible, and even transparent, through depositing, casting, or painting processes. Moreover, potentially longer operating life and improved stability make QLEDs effectively competitive with the other breakthrough evolution in lighting technology, OLEDs.

Recent developments in QD nanocrystal research may also lead to wireless-powered QLEDs: these would operate by absorbing and reemitting energy supplied by a nearby energy source at specific wavelengths, without the need for wires attached.

Another application of nanotechnologies concerns the use of single-walled CNTs as field emitters coupled with a phosphor screen. CNTs act as the cathode and the phosphor screen constitutes the anode of a stand alone flat emitting diode, with extremely low driving energy (0.1 Wh) and high lumen efficiency (60 lumen/W, rivaling most efficient LEDs). CNTs in fact behave like a field of nanometer-size tungsten wires, giving homogeneous and flicker-free light emission.[43]

2.2.5.10 Photovoltaics

Nanotechnology is fostering the development of third-generation PV solar cells that use nanofilms or QDs for the realization of organic (polymeric, DSSC) or multijunction-based flexible systems that offer high efficiency, transparency, integration into the building envelope, and, in time, more affordable manufacturing costs.

A further recent use of nanotechnology in the PV industry is the technology developed by the Swiss company CSEM that allows the manufacture of completely white PV modules (Fig. 2.8), without cells and connections visible, allowing better esthetic integration in buildings. This is achieved thanks to a layer of nanotechnological coating, selective and diffusing, able to diffuse the entire visible spectrum of solar radiation without affecting the transmission of the IR component, which can be converted into electricity by the PV cells beneath. This scattering selective filter consists typically of multiple transparent dielectric layers with different refractive indexes, stacked on top of each other. Any PV technology based on crystalline silicon

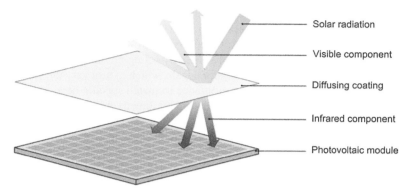

Figure 2.8 White photovoltaics.

may thus be used for the realization of white or colored modules (Fig. 2.9), although heterojunction crystalline silicon solar cells are the preferred choice as they combine higher open circuit voltages compared to standard silicon solar cells (730 mV instead of 630 mV), with an excellent response in the IR part of the solar spectrum.

Figure 2.9 Colored photovoltaics.
Courtesy of CSEM.

Current white solar prototypes performance reaches 11.4% measured on the AM 1.5 spectrum, with a 40% loss of efficiency compared to the same cell technology with the usual appearance. The size of the current module is 55 × 60 cm, and it consists of nine heterojunction solar cells interconnected in series.

This technology can be applied over existing PV modules or integrated into the manufacturing process of new panels, on flat or curved surfaces. In addition to the main application in BIPV, the technology is suitable for integration in consumer electronics (laptops and smartphones) and the automotive industry.

Furthermore, the module's white surface allows solar cells to work at temperatures 15−20°C lower than conventional panels, greatly enhancing the efficiency (PV electrical output decreases with temperature at a rate of 0.3−0.5% per degree Celsius). White solar panels may also act as cool roofs, saving energy in buildings while keeping the interior cooler and thereby reducing air-conditioning costs.

2.2.5.11 Building and environmental monitoring and control systems

NEMS sensors and actuators are currently being developed and used in buildings to monitor and control the environmental conditions and materials/structure performance. The advantage of these devices is their size, which reaches the nanoscale and allows them to be embedded in any structure or building element during the construction process itself.

One of the most promising applications of NEMS sensors is their integration within structural building materials such as steel and in particular concrete, where they can perform multiple functions.[36] The integration of sensors at the stage of concrete casting allows accurate monitoring of hydration and curing steps in cement mixture from inside the material, enabling their correction with additives where necessary and accurate prediction of the future performance as well as the achievement of needed strength thresholds to help plan the subsequent processing and thus minimize downtime on site. Once full maturity is reached, sensors can be further used for continuous monitoring of the structural elements, registering shocks, impacts, and loads, and allowing the scheduling of targeted maintenance and the safe prediction of potential collapse hazards.

Furthermore, NEMS address the issue of precisely locating detection devices in the inhabited environment without an intrusive appearance, allowing seamless integration of both sensors and actuators within the products themselves or surrounding surfaces. This way, all building components can become smarter and gather data on temperature, humidity, vibration, stress, decay, chemical or fire hazards, and many other factors such as user occupancy and behavior. Such information will prove invaluable in monitoring and improving building maintenance and safety, as well as greatly increasing energy efficiency. For instance, environmental control systems can learn and adapt to building occupancy patterns, and adjust heating and cooling or windows configuration accordingly. These embedded sensors can also interact with those worn by building users, resulting in an even smarter environment that self-adjusts to individual needs and preferences, for example, in delicate situations such as healthcare patient monitoring. These intelligent and ramified networks of smart components are a key step in enabling smart buildings and smart cities.

2.3 Smart materials

"Smart materials" are a new class of highly innovative materials able to perceive stimuli from the external environment and react to them immediately, adapting to the changed environmental conditions.[5,44–47] This reaction is usually caused by a change in the values of the surrounding energy field (change in potential, electrical, thermal, mechanical, chemical, nuclear, or kinetic energy) that triggers in the material a chemical reaction, a change in the molecular or crystallographic structure, an alteration of the voltage ranges, the absorption of a proton, and other phenomena.

The effects may show in events such as a change in color, a change in volume, a change in the distribution of stress and deformation, or a change of the refractive index.

Compared to traditional materials, smart materials are characterized by the following properties:[5]

- immediacy, as they respond to stimuli in real time;
- transiency, as they respond to multiple environmental states;
- self-actuation, as the ability to react is internal to the material, rather than external;
- selectivity, as the reaction is discrete and predictable;
- directness, as the reaction is internal to the event triggering it.

Depending on the type of reaction to the external energy field, smart materials are generally divided into property-changing materials and energy-exchanging materials (Table 2.8).

Smart materials may come at the macrometer, micrometer, or nanometer scale (smart nanomaterials).

The ability of smart materials to produce a useful effect in response to external stimuli makes them greatly interesting for their application in the fields of design and architecture, in particular, both characterized by changing and dynamic needs. Both types may also find application as sensors or actuators in smart devices or systems directly integrated into the building body (Table 2.9).

As defined earlier, "smart materials" are qualified based in their "smart" behavior, as in a dynamic response which includes all the properties mentioned. However, the term "smart" is still vague, so in general terminology it is often used more widely to describe materials or behaviors with high or unique performance and properties, such as easy-to-clean surfaces or self-healing materials, even if they do not exactly fit in the "smart materials" category.

2.3.1 Property-changing materials

This category includes materials which autonomously and reversibly change one of their mechanical, chemical, optical, electrical, magnetic, or thermal properties in response to a stimulus. They include chromogenic materials (termochromics, photochromics,

Table 2.8 Smart materials classification

Type of smart material	Environmental stimulus	Reaction/effect
Property-changing materials		
Thermochromics	Temperature difference	Color change
Photochromics	UV radiation	Color change
Mechanochromics	Deformation	Color change
Chemochromics	Chemical concentration	Color change
Electrochromics	Electrical tension	Color change
Liquid crystals	Electrical tension	Color change
Suspended particles	Electrical tension	Color change
Photocatalytics	UV radiation	Chemical reaction
Thermotropics	Temperature difference	Phase change
Electrorheological	Electrical tension	Viscosity change
Magnetorheological	Magnetic field	Viscosity change
Shape memory	Temperature difference	Crystalline phase change
Energy-exchanging materials		
Electroluminescents	Electrical tension	Visible light emission
Photoluminescents	Radiation	Visible light emission
Chemiluminescents	Chemical concentration	Visible light emission
Thermoluminescent	Temperature difference	Visible light emission
Photovoltaics	Solar radiation	Electrical tension
Bidirectional energy-exchanging materials		
Piezoelectric	Deformation	Electrical tension
Thermoelectric	Temperature difference	Electrical tension
Pyroelectric	Temperature difference	Electrical tension
Electrochemical	Chemical concentration	Electrical tension
Electrostrictive	Electrical tension	Deformation
Magnetostrictive	Magnetic field	Deformation

mechanochromics, chemochromics, electrochromics, liquid crystals, and suspended particles), PCMs, photocatalytic materials, electrorheological and magnetorheological materials, and shape-memory alloys.

To date, these materials have found application in architecture in a number of products and systems, especially in building envelope components, which can best take advantage of their properties of adaptation to different use conditions.

Table 2.9 **Main applications of smart materials in the building sector**

Requirement class	Specific need	Smart material
Controlling solar radiation transmittance through building envelope	Control of spectral absorbance and transmittance of transparent closures	Electrochromic glass Photochromic glass LCD glass Suspended particles device (SPD) glass
	Dynamic control of screening systems	External radiation sensors (PVs) Interior daylight sensors (PVs) Actuators (shape-memory alloys)
Controlling heat flows through building envelope	Control of thermal conductivity of building envelope materials according to climatic conditions	Thermotropic materials
Controlling temperature of interiors	Increase of thermal capacity of closures and internal partitions	PCMs
	Placement of heat sources outside the environment	Optical fiber systems Thermoelectric materials
	Increase lighting/heat ratio	LEDs/OLEDs Photoluminescent materials
Controlling air quality	Monitoring of CO_2 and pollutant agents	Sensors
	Air purification	Photocatalytic materials
Producing energy from renewable sources	Conversion of environmental energy into electric energy	PVs Piezoelectric materials
Optimizing lighting systems	Environment lighting monitoring	PVs
	Interior daylight monitoring	Photoelectric materials
	Interior occupancy monitoring	
	Lighting source placement	Optical fibers Electroluminescent materials

Table 2.9 Continued

Requirement class	Specific need	Smart material
Optimizing heating, ventilation, and air-conditioning systems	Temperature monitoring	Pyroelectric materials
	Humidity monitoring	Hygrometers
	Interior occupancy monitoring	Photoelectric materials
	Monitoring of CO_2 and pollutant agents	Biosensors
	Heat sources and storage placement	Thermoelectric materials PCMs
Controlling structural vibration	Eulerian buckling control	Piezoelectric materials
	Inertial damping control	Magnetorheological materials Electrorheological materials Shape-memory alloys
	Stress monitoring	Optical fibers

In particular, property-changing materials are used:

- in transparent closures for dynamic control of solar radiation (chromogenic materials);
- in closures and internal partitions to increase their thermal inertia (PCMs);
- in mortars, paints, or coatings of external and internal surfaces of the building to achieve a self-cleaning and at the same time atmospheric air-purifying effect (photocatalytic materials).

Transparent closures are certainly the most interesting application field of property-changing materials, both for new constructions and for existing buildings.

The need to combine vision and daylight through glazed closures with incoming solar radiation control via conventional fixed sun-screening systems has led to compromise solutions, skewed toward only one of these requirements and unable to guarantee the best operation in all four seasons. The integration of chromogenic materials in transparent closures allows adapting the characteristics of light transmission and solar shading to external conditions, either with passive systems, completely self-sufficient and not user-controllable (such as thermochromic and photochromic glazing), or with active systems, freely configurable by users or by an integrated management system, but with the trade-off of needing an electrical power supply (electrochromic, LCD, suspended particles devices (SPD) glazing).

Equally interesting is the application in opaque closures and partitions of PCMs, capable of reacting to changes in temperature with a state transition that can exponentially increase their thermal capacity. Available in different shapes and packages (macroencapsulated panels, beads, and bags, microencapsulated in bulk form, or integrated in various components such as bricks, gypsum, mortar, and plaster), they are used as components in low-inertia building systems such as lightweight opaque closures and interior partitions or translucent closures.

Photocatalytic materials have been finding wide use in the building industry since the early 1990s: semiconductor metal compounds such as zinc oxide (ZnO), tungsten oxide (WO_3), cadmium selenide (CdSe), and TiO_2 react to UV radiation, leading to the formation of highly reactive agents with the ability to oxidize and decompose many types of bacteria and organic and inorganic materials by photo-induced redox reactions of adsorbed substances.

Among photocatalytic construction and building materials, TiO_2 is the most widely used for its numerous advantages: it is chemically stable, safe, widely available, and relatively inexpensive; it has higher photocatalytic activity compared to other metal oxide photocatalysts; it is compatible with most traditional construction materials, such as cement, without changing any original performance; and it is also effective under weak solar irradiation in ambient atmospheric environments.[32]

Applied to the construction industry, TiO_2 has allowed the development of paints for external (Marraccini Paintek) or internal (Caparol Capasan) applications, mortars and cements (Italcementi Tx-Active and i.active Biodynamic), tiles (Ceramitex, BioTi Ecopav), and other construction products able to use solar radiation to provide a self-cleaning and depolluting action. TiO_2-induced photocatalysis is a promising technology for self-cleaning applications. Besides its photocatalytic properties, TiO_2 demonstrates an intrinsic photo-induced surface superhydrophilicity. In fact, in contrast to other photocatalytic materials, when water is adsorbed on this semiconductor material irradiated by UV light, it spreads to form a thin film instead of grouping in water droplets. When water is rinsed over the surface, contaminants like oil can be washed away.

As most photocatalytic processes are activated only by UV light waves, research is focusing on expanding these effects to a broader spectrum of sunlight to increase its efficacy and enable photocatalytic action at night or in interiors devoid of daylight.

Japan-based Showa Denko has developed a photocatalyst agent with antibacterial and antiviral functions that can be activated even by low-energy visible light, like that emitted by indoor artificial lighting devices such as fluorescent lights or LEDs. Dubbed Lumi-Resh, it has already been successfully embedded in heat insulating panels for plant factories, membrane building materials (Taiyo Hikari-Protextile), and indoor paints (Nippon Soda Co.) applicable to curtains or metallic surfaces.

Photocatalytic properties also characterize tungsten trioxide (WO_3) nanoparticles. The tungsten-based Renecat photocatalytic agent, manufactured by Toshiba, claims superior performance compared to TiO_2 regarding deodorant and antibacterial effects even with fluorescent or white LED artificial light, making it viable for interior use. Furthermore, it can also be applied on polymeric surfaces, such as PMMA or polycarbonate glazing, if provided with an inorganic binder.

Applications of electrorheological and magnetorheological fluids are currently limited to high-tech industry sectors, where their property of being able to alter viscosity as a function of an electrical or magnetic stimulus has long been exploited in active suspension systems for cars and in prosthetics manufacturing, as well as in active vibration-damping systems for machines and structures.

Finally, shape-memory alloys can have interesting applications in the construction of buildings and architectural elements, as they are able to self-adapt to ambient temperature or reassemble with only the supply of electricity or heat.

When an object made of shape-memory alloy such as nitinol (nickel titanium) is deformed from an initial shape to a new one, it will keep this last configuration. If it is heated to a certain temperature, however, either directly or by electric current, it will revert to its initial shape without any mechanical aid, as if it remembered its base configuration.

This unique shape-memory effect is caused by the internal phase changes that the material undergoes during the deformation and shape-return processes (from martensite to austenite upon heating, and vice versa upon cooling) (Fig. 2.10). This process is generally reversible, although some shape deterioration over time may occur if the bending and heating are not carefully controlled.

Shape-memory alloys are applied in several industry sectors: they are used in simple release mechanisms to reduce the need for moving parts, and shape-memory superelasticity allows metals to undergo extreme deformation (up to 8%) without consequences, and is, for instance, used in high-end eyeglass frames.

Shape memory has also been developed in the form of rigid plastics or relatively porous foams, with a wide application scope including self-unrolling surfaces and knots that tie themselves for medical applications and consumer-customizable furniture.

A notable building application is the "Translated Geometries" prototype developed by students from IAAC Barcelona in 2014, which is based on a geodetic structure of triangular panels connected by shape-memory hinges. At a standard temperature the

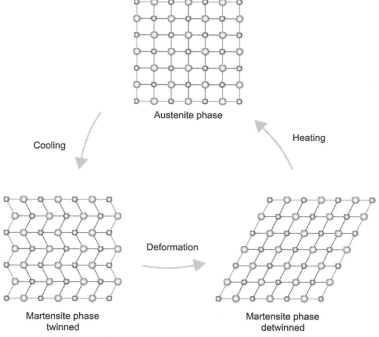

Figure 2.10 Shape-memory alloy operation.

structure is rigid and self-supporting, but at a temperature of 60—70°C connectors become flexible and the volume of construction can be increased or decreased, then maintaining the last configuration once the supply of heat stops.

2.3.2 Energy-exchanging materials

The second category of smart materials includes those capable of converting more or less reversibly an input energy into another form of energy in accordance with the first law of thermodynamics. They include PV, luminescent (LED, OLED), piezoelectric, thermal, pyroelectric, electrostrictive, and magnetostrictive materials.

The main applications of energy-exchanging materials in construction concern PVs and electro and photoluminescent materials.

PV materials enable the production of electrical energy by transforming the incident solar radiation (photoelectric effect).

PV technology has reached the third generation of solar cells and boasts decades of application. Research trends today are focused on the development of processing techniques that would allow the production of high-efficiency, low-cost, and high-yield solar cells made of copper indium gallium selenide (CIGS), cadmium Telluride, and organic materials (polymer composites and DSSC). Further promising smart technologies for solar energy capture regard the direct conversion of light waves into DC current. Using multiwall CNTs and tiny rectifiers fabricated on to them, researchers have recently demonstrated the first "optical rectenna," a device that combines the functions of an antenna and a rectifier diode to convert light directly into electricity; for now it has a conversion efficiency of 1%, but there is good growth potential.[48]

Luminescent materials are commonly classified as chemiluminescent, electroluminescent, or photoluminescent according to the excitation energy source, albeit light emission can depend upon other energy mechanisms (such as friction).

Electroluminescent materials are the basis of the solid-state lighting technology used in LED systems, characterized by high efficiency, small size of light sources, and application versatility. In LEDs light emission is obtained by driving electrical current through a p-n semiconductor junction (conventionally made of silicon or germanium), which causes fluorescence to occur. Other materials used in LEDs include p- and n-type GaN, AlGaN, InGaN, InGaAlN, GaAs, GaAsP, and various heterostructures of such compounds.

Promising hybrid single-layer LEDs using organometal halide perovskite and polyethylene oxide sandwiched between indium tin oxide and indium—gallium eutectic alloy have also been studied, showing higher light emittance with lower tension, and having a much simplified manufacturing process.[49]

The wavelength, and thus the color, of light emitted is quite narrow and dependent on the nature of the semiconductor used. To obtain full-spectrum white light, current approaches mix red, green, and blue phosphors applied to the semiconductor material.

A new generation of solid-state lighting is constituted by QLEDs made of QD networks. These work similarly to conventional LEDs, but show greatly improved functionality and even wider applications.

Photoluminescent materials are capable of absorbing solar light and reemitting it at a different wavelength in the visible spectrum. When the time delay between absorption and emission is in fractions of seconds, the photoluminous effect is called fluorescence, whereas when the time delay is longer and there is still a decayed afterglow after the excitation source is removed, the effect is instead called phosphorescence. By varying the nature of the photoluminous material, and particularly its embedded impurities, emitted light in various colors can be obtained.

The fluorescence effect is commonly used in fluorescent lighting, where a fluorescent agent (a compound of three or more halophosphates) absorbs UV radiation emitted by the electroluminescent gases (neon, argon, xenon, krypton) and reemits it in the visible range.

Photoluminescent materials are widely employed in the fields of interior decoration (phosphorescent or fluorescent paints) and building materials, such as glass bricks containing phosphorescent crystals able to glow in the dark (TGP Veluna, Seves Luminescent Glass Block). A phosphorescent material of particular interest is Starpath Spray-On developed by British-based Pro-Teq, which can be applied on the surface of any concrete, asphalt, or wood pavement to make it glow without the need of electric energy. This special coating absorbs light radiation during the day and returns it at night after sundown. The coating, which is nonslip and water resistant, is currently being tested in the central park of Cambridge in the United Kingdom, where it was sprayed on a total area of 150 square meters (the application time is only 30 min, allowing the surface to be ready for use approximately four hours after application). The coating is also antireflective, and is available in 11 different colors.

The concept was further developed with the launch of Starpath Pro, which forgoes the spray-on application for the safer, more environmentally friendly solution of a ready-to-use photoluminous blended stone aggregate. Installation simply requires mixing the aggregates with a specific two-part polyurethane resin binder for three minutes, and once spread and leveled the surface is ready for foot traffic in four hours. The product can only be applied to a hard substrate such as concrete or tarmac. The result is a natural-looking pathway in the daytime which transforms into a soft, blue luminescent glow at night (Fig. 2.11).

Figure 2.11 Photoluminous pavement.
Courtesy of Pro-Teq Surfacing Ltd.

Figure 2.12 Van Gogh cycle track.
Courtesy of Studio Roosegaarde ©.

Similarly, the Smart Highway project in the Netherlands applied road surface markings on a 500 m stretch of highway N329 at Oss with paint enriched with "glow-in-the-dark" luminescent powder to provide lighting in place of traditional street lamps. The phosphorescent paint is loaded by sunrays during the day and provides up to 10 h of night illumination. The same team turned 600 m of the Van Gogh cycle track in Nuenen into a visual art piece, inspired by the "Starry Night" painting by the same artist, using a photoluminescent pavement (Fig. 2.12). The effect was obtained by covering individual paving granules with a photosensitive coating in shades of green and blue, and depositing them on a bed of concrete. Solar energy stored during the day is sufficient for up to eight hours of night lighting, and on cloudy days is aided by LEDs integrated into the paving and powered by solar panels.

A recent application of luminous materials is photo-transparent solar collectors, which are PV cells constituted by organic salts that absorb invisible UV and IR wavelengths and then reemit them as invisible IR radiations toward conventional PV cells, to be converted into electrical energy.

More recently, a new type of nanoparticle has been created which is able to convert invisible NIR light to higher-energy blue and UV light with record high efficiency. The potential applications of this multilayered nanoparticle are in solar energy harvesting, bioimaging, and light-based security techniques.[50]

Finally, from the automotive industry comes a special energy paint that could find future applications in the construction industry. Developed by researchers at the Mercedes-Benz Advanced Design Studio in Beijing, this special paint, dubbed "multi-voltaic silver," allows production of electricity from both solar and wind power. The paint behaves as a PV paint and an accumulator of electrostatic charge when, during driving, the wind strikes the surface of the car. Electricity obtained is bound to the electrolysis of water to produce hydrogen, which in turn powers the vehicle.

Thermoelectric materials are capable of transferring heat from one pole to another if powered by an electric potential difference, and are already used in heat dissipation devices in electronic equipment such as laptops and small appliances.

Pyroelectric materials are based on crystals which are given permanent polarization due to aligned molecules with fixed dipole moment. When the temperature changes, the polarization varies accordingly, triggering surface electric charges or, if the surfaces are connected, a pyroelectric current. The pyroelectric effect is commonly used in intruder-detection systems and thermal imaging.

Finally, the piezoelectric effect is the property of some crystalline materials to polarize when subjected to a mechanical deformation (direct piezoelectric effect), generating a potential difference, and at the same time to deform in an elastic manner when traversed by electrical current (inverse piezoelectric effect). The piezoelectric effect occurs only in one direction, and deformations associated with it are the order of nanometers. Many common devices, including doorbells, speakers, microphones, and quartz watches, are based on piezoelectric technologies. Piezoelectric elements are also the key of recent proposals for harvesting energy from pavements (Fig. 2.13) and roads by exploiting kinetic stress, such as by using Pavegen tiles (Fig. 2.14).

A single 45 × 60 cm Pavegen tile, made of recovered polymers with an upper surface of recycled truck tires, is able to produce up to 7 W when compressed by 5 mm, thanks to a hybrid system that takes advantage of both piezoelectric effect and electromagnetic induction. The tile can be equipped with self-powered Wi-Fi transmission systems to provide real-time data on pedestrian traffic and electricity production without the need for cabling. Finally, the use of piezoelectric systems on roads for motor vehicles is particularly promising, as demonstrated by the pilot project Innowatech in Israel, constituting a network of piezoelectric generators embedded 5 cm in the asphalt. Tested on a 10 m stretch of road, this technology promises 200 kWh/km energy yield.

Another building application of piezoelectric systems is that proposed by SMIT in New York, which drew inspiration from ivy plants growing on buildings to develop its Solar Ivy hybrid leaves. The system comprises a distributed energy generator constituted by small leaves, customizable in shape and color, fixed to a stainless steel mesh support structure. The leaves are able to produce electricity by a thin-film photovoltaic

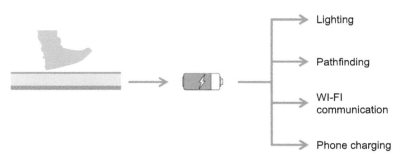

Figure 2.13 Piezoelectric pavement.

Figure 2.14 Pavegen tiles.
Courtesy of Pavegen Systems Ltd.

module (available in organic, a-Si, or CIGS variants with power yields up to 4 Wp per leaf) and by a piezoelectric generator that converts wind swaying. This leaf skin can vary in density to adapt to opaque or transparent supports, create visual patterns, and regulate solar radiation reaching the building.

Much more striking is the system devised by Swedish firm Belatchew Lab Arkitektur, which intends to achieve an urban wind farm by fixing flexible electricity-generating straws on the outside of the uncompleted Soder Torn tower in Stockholm. Dubbed Strawscraper, the resulting building would bring unprecedented dynamism to the urban landscape with its swaying skin. Each straw integrates piezoelectric crystals capable of producing electricity from deformation stresses caused by wind. Straw generators are able to operate even with lower wind speeds, with advantages over traditional building-installed wind turbines since they are quiet and pose no hazard to wildlife.

2.4 3D printing for architecture

3D printing, also known as additive manufacturing, refers to the various processes of rapid prototyping used to synthesize a 3D object with the support of 3D computer-aided design (CAD) data. In 3D printing, successive layers of material are deposited under computer control to create the desired object. These objects can be of almost any shape, geometry, material, and dimension, and are produced from a 3D model custom designed or obtained by a 3D scan.

Numerous experiments have been conducted in the architecture and design fields showing the enormous potential of 3D printing technology in all phases of the construction process, from design to production of components and up to the construction of entire buildings both in new realizations and in renovation and restoration interventions.

Advantages concern the reduction of construction time and waste production, the possibility of creating products with superior mechanical properties, of greater lightness, or with unique forms, the reduction of transportation costs, the improvement of management, maintenance, and restoration of buildings by overcoming the concept of obsolescence of equipment, and the possibility of having at any time *ad hoc* printed substitutional elements at hand. There is also the possibility of using recovered or recycled materials with a consequent reduction of impacts on the environment, and the ability to print objects able to reach their final form independently at a later time (4D printing). Scientists at the University of Boulder, Colorado have recently developed a "4D printing" process in which shape-memory polymer fibers are deposited in key areas of a composite material item as it is being printed.[51] By carefully controlling factors such as the location and orientation of these fibers, designated areas of the item will fold, stretch, curl, or twist in a predictable fashion when exposed to an appropriate stimulus such as water, heat, or mechanical pressure. One 4D printed object could change shape in different ways, depending on the type of stimulus to which it is exposed. This functionality would allow, for instance, printing a PV panel in a flat shape, exposing it to water to cause it to fold up for shipping, and then exposing it to heat to make it fold out to yet another shape that is optimal for catching sunlight.

A number of 3D printer models are commercially available, using different additive manufacturing techniques. The main differences concern the type of additive process used and the possible need to provide support structures, the materials that can be used for printing (including multimaterial printing), the shapes that can be printed and colors that can be obtained, the size of the generation chamber and thus of printed objects, print times, and possible applications of printed objects (conceptual model, functional prototype, finished product, etc.).

The most widely used technologies today include:

- stereolithography;
- selective laser sintering;
- fused deposition modeling;
- laminated object manufacturing;
- powder binder printing.

As concerns in particular 3D printing of metal objects, main techniques include selective laser melting, direct metal laser sintering, and direct metal deposition, which allow production of objects with complex shapes and superior mechanical properties compared to those obtained via conventional processes.

Among experimental techniques, very promising is the continuous liquid interface production technology developed by Carbon3D, a new approach to 3D printing designed to speed up the 3D manufacturing process drastically by "growing" objects out of a pool of resin rather than printing them layer by layer. The innovative process is claimed to produce commercial-quality objects from a range of polymer-based material at speeds between 25 and 100 times faster than conventional 3D printing.

Possible applications of 3D printing in architecture are the realization of building envelope components and structural elements, up to the entire building.

As for the envelope, California-based Bot Laboratory, in its research on 3D printing processes applied to the construction industry to make them faster and cheaper, presented for the first time at Maker Fair 2014 the m_Wall prototype, a 3D printed wall made of plastic material with an incredible load resistance despite the reduced thickness (6.35 mm). In a few hours, Euclid Robot 3D, the printer used to realize m_Wall, is able to erect a wall of about 2 m in height in a single sweep. This system could be extremely useful for the realization of the structural elements of a building, reducing construction time and also taking advantage of material recycled (here, plastic pellets) or recovered from previous demolitions.

London-based architecture firm Foster + Partners is instead focusing on the prefabrication of individual building elements, partnering with Swedish construction company Skanska and Loughborough University to develop a robotic computer numerical control machine for printing concrete that is fully marketable and capable of producing finished products characterized by a very high degree of complexity. The project aims to implement a complete supply chain, involving manufacturers such as Buchan Concrete, ABB, and Lafarge Tarmac.

Among the applications in the restoration of existing buildings, particularly interesting is the 2014 intervention at the Annie Pfeiffer Chapel in Florida, built by Frank Lloyd Wright between the 1940s and 1950s, in which the very special concrete blocks that make up the outer shell, unique elements individually designed by Wright and built by hand, were restored thanks to the 3D reproduction of the original molds, which had been lost.

As regards structural elements, a team led by the Arup firm has developed a method of designing and 3D printing steel joints that promises to reduce greatly the time and cost of implementing the complex nodes in tensile structures (Fig. 2.15). The procedure employs a laser to heat and melt a thin layer of steel powder, which then solidifies

Figure 2.15 3D printed structural steel joints.
Courtesy of Davidfotografie/Arup ©.

to form part of the structure. A new layer of powder is then added, and the structure is gradually built up. This technique allows building the structure in a very precise manner with great savings on material used.

Combining Inca construction traditions with the technological potential of 3D printing, architectural firm Emerging Objects developed a structural column, dubbed QuakeColumn, which is incredibly resistant and light without needing any binder, mortar, glue, or reinforcement (Figs. 2.16 and 2.17). In this structure, bricks have been

Figure 2.16 QuakeColumn.
Courtesy of Emerging Objects.

Figure 2.17 Detail of QuakeColumn.
Courtesy of Emerging Objects.

CAD designed to fit perfectly together in a sort of 3D puzzle, and then printed with cement. Once assembled, the characteristics of the joints make the column earthquake resistant, preventing horizontal displacement. Bricks are hollow, thereby achieving an excellent weight-resistance ratio, and during printing each code-marked element is provided with special handles to make on-site assembly easy and fast, as if using Lego blocks. The same architects also designed Cool Bricks, interlocking modular blocks whose porous ceramic structure, 3D printed, is able to retain rainwater like a sponge while remaining permeable to air, and to cool interiors by evaporating it (Fig. 2.18).

In the construction of entire 3D printed buildings (3D house printing technology), the most interesting projects definitely include the 3D Print Canal House, a real printed condominium designed by DUS Architects, to be completed within a few years along the banks of one of the canals of Amsterdam (Fig. 2.19). An indispensable tool for the project is the maxi KamerMaker printer that will print the entire building on site.

In just three weeks after the works started, the corner of the 3D Print Canal House had already been completed. Once finished, the house will include 13 rooms, fully featured. The KamerMaker 3D printer is 3.5 m high, and located inside a container for easy movement. Before moving on to the final version, all 3D Print Canal House components are printed in 1:20 scale, to be thoroughly tested and approved before printing the structural elements in 1:1 scale (Fig. 2.20). The first section of the house has been built with plastic material filled with concrete, but the DUS Architects team aims at early replacement with more natural or recycled materials, such as bioplastics, vegetable oil, and microfibers, including recovery of discarded materials from the

Figure 2.18 Cool Bricks.
Courtesy of Emerging Objects.

Advanced materials for architecture

Figure 2.19 3D Print Canal House printed sections.
Courtesy of Olivier Middendorp/Hollandse Hoogte ©.

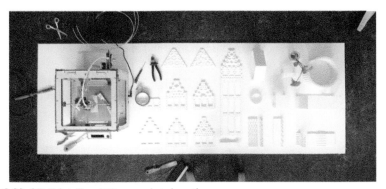

Figure 2.20 3D Print Canal House printed sections.
Courtesy of DUS Architects ©.

construction industry. Being a fully digitized process, once tested and established the 3D Print Canal House production system could put architectural projects directly online and make them available to future builders, who would thus have the ability to adapt them to their every need.

In parallel to the Dutch project, a Chinese firm has managed to build 10 rudimentary printed prototype houses in just 24 h. The project was led by the Chinese firm Winsun

Decoration Design Engineering, which started by printing individual structural components and assembling them on site. Although extremely basic, with no special features, these 10 structures completed within the experiment confirmed the potential use of 3D printers in the building sector, by succeeding in just one day in completing a small village of printed houses.

The system used by Winsun is based on a 3D printing process, used for over 12 years, that employs an "ink" made from a mixture of glass fiber, steel, cement, hardening agents, and recycled construction waste. The 3D printer, which is 6.6 m tall, 10 m wide, and about 150 m long, works by printing, layer by layer, large sections of buildings (such as wall panels) which are then assembled on site, much like prefabricated concrete, to create the final structure, enabling the realization of houses of about 200 m^2 surface, with 10 and 20 m sides, and about 6.6 m high. By reusing waste materials, the system patented by Winsun can also partially mitigate the environmental impact of construction.

Winsun also realized in January 2015 in Suzhou Industrial Park, Jiangsu Province (China) the world's tallest 3D printed building—a five-story apartment block—and an 1100 m^2 mansion, further demonstrating the potential of 3D house printing technology.

The Italian company WASProject, active for years in the field of 3D printers, has released its Big Delta model, 12 m high and intended for building applications (Fig. 2.21). Easy to transport and with an estimated cost of €40—50,000, it allows construction of sustainable buildings in less than 24 h by using local materials such as clay and straw, thus being extremely promising for disaster relief operations and small contractors.

Within the smart construction field, parallel to 3D printing, rapid prototyping techniques are becoming more and more widespread, allowing on-site manufacturing by remotely controlled robots of special pieces and components directly translated from 3D models made with advanced CAD programs, and able to build even the most complex projects.

Finally, there is growing use of drones for maintenance and monitoring of buildings and 3D mapping of construction sites, transmitting data directly to intelligent, automated earth-moving machines.

Figure 2.21 WASP Big Delta 3D printer.
Courtesy of WASProject/CSP s.r.l.

2.5 Conclusions and future trends

The introduction of advanced construction materials is contributing substantially to the improvement of buildings' energy and environmental efficiency throughout their entire life cycle. By introducing new functionalities and improved properties, advanced materials add value to existing products and processes in a sustainable integrated approach from cradle to cradle.

Some of these advanced materials and technologies are already commercially available and have proven cost-effective, with payback periods of less than five years. Others are more costly and will require government intervention if they are to achieve a wider market uptake.

Nanotechnology is certainly offering essential scope for development and formulation of innovative building products with increased mechanical, physical, and chemical properties, with application potential in all technical elements, from structures to opaque and transparent closures, internal partitions, systems, and appliances.

3D printing technology is also showing enormous potential in terms of creation of products with superior mechanical properties and unique forms, including the construction of entire 3D printed buildings (3D house printing technology).

A new class of highly innovative materials able to perceive stimuli from the external environment and to react immediately to such stimuli, adapting to the changed environmental conditions, the so-called "smart materials" are finding application in architecture in a large number of products and systems, especially in building envelope components which can best take advantage of their properties of adaptation to different use conditions.

The development of a new generation of multifunctional coatings, synthesized by nanotechnology, such as nanocomposite coatings, nanoscale multilayer coatings, superlattice coatings, and nanograded coatings with passive functionality, but the capacity to respond actively to changes in their surroundings, is currently a hot technological topic. These will provide a key technology for the fabrication of future high-tech "smart" coatings.

Future approaches will focus on tailor-made coatings for specific functions, with particular attention to self-healing coatings designed with the ability to repair themselves after chemical or mechanical degradation, based on nanocontainers or nanocapsules that can be filled with inhibitors to protect the substrate from corrosion after damage or scratching in the coating layer.

To foster diffusion of advanced materials, and nanomaterials in particular, a number of barriers still need to be addressed, not only related to costs but also mostly to health and environmental concerns based on reliable and sufficient scientific data.

For this purpose, short, medium and long-term effects of nanomaterials on health and the environment are to be investigated, taking into account the transport mechanism between boundaries as well as levels and type of exposure (direct ingestion, inhalation, dermal intake). The issue of nanomaterials eventually entering the human body is closely related to the form in which they are used in products, with particular concern about their ease of becoming unbound: for instance, nanoparticles covalently bound between themselves are less likely to become free, as are those firmly embedded in a metal matrix in nanometallic composites.[8–10,19] Once detached, the hazard of free nanoparticles would depend on their mobility, inherent toxicity, or potential to carry other pollutants.

Environmental concerns also tend to merge nanoparticles indistinctly with the ill-famed ultrafine particles and particulates; while the former are the specific result of an engineering process and serve specific purposes, and thus come in a great variety of sizes, compositions, and coatings, the latter are an uncontrolled and unwanted by-product of other activities and processes, and are as such more difficult to assess.[8–10,19]

In conclusion, most important nanotechnology challenges relate primarily to establishing univocal methods and tools for detecting, characterizing, and analyzing nanomaterials, to investigating their potential hazard, and to assessing the effective exposure. In this picture, regulations should guarantee high levels of health, safety, and environmental protection, while at the same time allowing market uptake of advanced products and fostering innovation and competitiveness in this crucial sector of the industry.

References

1. Smith WF. *Foundations of materials science and engineering*. New York: McGraw-Hill; 2004.
2. Lyons A. *Materials for architects and builders*. Oxford: Elsevier; 2007.
3. Brownell B. *Material strategies*. New York: Princeton Architectural Press; 2012.
4. Torricelli MC, Del Nord R, Felli P. *Materiali e tecnologie dell'architettura*. Bari: Editori Laterza; 2007.
5. Addington DM, Schodek D. *Smart materials and technologies*. Oxford: Architectural Press; 2005.
6. Featherston C, O'sullivan E. *A review of international public sector strategies and roadmaps: a case study in advanced materials*. Cambridge: Centre for Science Technology and Innovation, Institute for Manufacturing, University of Cambridge; 2014.
7. Pacheco-Torgal F, Labrincha JA. The future of construction materials research and the seventh UN millennium development goal: a few insights. *Constr Build Mater* 2013;**40**:729–37.
8. Pacheco-Torgal F, Diamanti MV, Nazari A, Goran-Granqvist C, editors. *Nanotechnology in eco-efficient construction: materials, processes and applications*. Cambridge: Woodhead Publishing; 2013.
9. Michael F, Ashby MF, Ferreira PJ, Schodek DL. *Nanomaterials, nanotechnologies and design an introduction for engineers and architects*. Burlington: Elsevier Butterworth-Heinemann; 2009.
10. Smith GB, Granqvist CG. *Green nanotechnology solutions for sustainability and energy in the built environment*. Boca Raton: CRC Press; 2011.
11. Rifkin J. *The third industrial revolution; how lateral power is transforming energy, the economy, and the world*. London: Palgrave MacMillan; 2011.
12. Feynman R. There's plenty of room at the bottom (reprint from the speech given at the annual meeting of the West Coast section of the American Physical Society). *Eng Sci* 1960;**23**:22–36.

13. Taniguchi N. On the basic concept of 'nano-technology'. *Proc Int Conf Prod Eng Tokyo Part II Jpn Soc Precis Eng* 1974:18−23.
14. Drexler KE. *Engines of creation the coming era of nanotechnology.* Anchor Books; 1986.
15. Horizon 2020 Work Programme 2016−2017 5.ii. *Nanotechnologies, advanced materials, biotechnology and advanced manufacturing and processing.* European Commission Decision C; 2015. 6776 of 13 October 2015.
16. RNCOS. *Global nanotechnology market outlook 2022.* December 2015.
17. Germany. European Patent Office (EPO). *Nanotechnology and patents.* Munich: EPO; 2013.
18. EU. European Commission. *Nanotechnologies, advanced materials, biotechnology, and advanced manufacturing and processing. HORIZON 2020 work programme 2016−17.* Bruxelles: European Commission; 2015.
19. Finland Finnish Institute of Occupational Health. *Nanosafety in Europe 2015−2025: towards safe and sustainable nanomaterials and nanotechnology innovations.* Helsinki: Finnish Institute of Occupational Health; 2013.
20. Edwards SA. *The nanotech pioneers: where are they taking us?.* Weinheim: Wiley-VCH; 2006.
21. ISO/TS 80004-1. *Nanotechnologies − vocabulary − Part 1: core terms.* 2015.
22. ISO/TS 80004-4. *Nanotechnologies − vocabulary − Part 4: nanostructured materials.* 2011.
23. ISO/TS 80004-8. *Nanotechnologies − vocabulary − Part 8: nanomanufacturing processes.* 2013.
24. Javey A, et al. Near-unity photoluminescence quantum yield in MoS_2. *Science* 2015; **350**(6264):1065−8.
25 Kheiri F. Material follows function: nanotechnology and sustainability in steel building constructions. *Int J Sci Res* 2013;**2**(12):1−5EU.
26. European Commission. *Nanotechnology: Research and Innovation the invisible giant tackling Europe's future challenges.* Luxembourg: European Commission; 2013.
27. Serrano E, Rus G, Garcia-Martinez J. Nanotechnology for sustainable energy. *Renewable Sustainable Energy Rev* 2009;**13**:2373−84.
28. Germany Hessian Ministry of Economy, Transport, Urban and Regional Development. *Application of nanotechnology in energy sector.* Wiesbaden: HA Hessen Agentur; 2008.
29. Heiranian M, Farimani AB, Aluru NR. Water desalination with a single-layer MoS_2 nanopore. *Nat Commun* 2015;**6**(8616):1−6.
30. Brandl F, Bertrand N, Martins Lima E, Langer R. Nanoparticles with photoinduced precipitation for the extraction of pollutants from water and soil. *Nat Commun* 2015;**6**(7765): 1−10.
31. Lei W, Mochalin VN, Liu D, Qin S, Gogotsi Y, Chen Y. Boron nitride colloidal solutions, ultralight aerogels and freestanding membranes through one-step exfoliation and functionalization. *Nat Commun* 2015;**6**(8849):1−8.
32. Chen J, Poon C. Photocatalytic construction and building materials: from fundamentals to applications. *Build Environ* 2009;**44**:1899−906.
33. Sev A, Ezel M. Nanotechnology innovations for the sustainable buildings of the future. *Int J Civ Environ Struct Constr Archit Eng* 2014;**8**(8):867−77.
34. Sobolev K. Nanotechnology and nanoengineering of construction materials. In: Sobolev K, Shah SP, editors. *Nanotechnology in construction. Proceedings of NICOM5*; 2015. p. 3−13.
35. Niroumanda H, Zain MFM, Jamilc M. The role of nanotechnology in architecture and built environment. *Procedia Soc Behav Sci* 2013;**89**:10−5.
36. Atwa HM, Al-Kattan A, Elwan A. Towards nano architecture: nanomaterial in architecture − a review of functions and applications. *Int J Recent Sci Res* 2015;**6**(4):3551−64.

37. Ibrahim Anous IH. Nanomaterials and their applications in interior design. *Am Int J Res Humanit Arts Soc Sci* 2014:16−27, 14−512.
38. Hossain K, Rameeja S. Importance of nanotechnology in civil engineering. *Eur J Sustain Dev* 2015;**4**(1):161−6.
39. Leone MF. Nanotechnology for architecture. innovation and eco-efficiency of nanostructured cement-based materials. *Archit Eng Technol* 2011;**1**(1):1−9.
40. Gao T, Jelle BP, Sandberg LC, Ng S, Tilset BG, Grandcolas M, et al. Development of nano insulation materials for building constructions. In: Sobolev K, Shah SP, editors. *Nanotechnology in construction. Proceedings of NICOM5*; 2015. p. 429−34.
41. Kollbe Ahn B, Das S, Linstadt R. High-performance mussel-inspired adhesives of reduced complexity. *Nat Commun* 2015;**6**(8663):1−7.
42. Sarswat PK, Freea ML. Light emitting diodes based on carbon dots derived from food, beverage, and combustion wastes. *Chem Phys* 2015;**17**:27642−52.
43. Bahena-Garrido S, Shimoi N, Abe D, Hojo T, Tanaka Y, Tohji K. Plannar light source using a phosphor screen with single-walled carbon nanotubes as field emitters − nanotubes as field emitters. *Rev Sci Instrum* 2014;**85**(104704):1−6.
44. Schwartz M. *Smart materials*. Boca Raton: CRC Press; 2009.
45. Ritter A. *Smart materials in architecture, interior architecture and design*. Switzerland: Springer Science; 2007.
46. Schwartz M. *Encyclopedia of smart materials*. New York: John Wiley and Sons; 2002.
47. Shahinpoor M, Schneider HJ. *Intelligent materials*. Cambridge: Royal Society of Chemistry; 2007.
48. Sharma A, Virendra Singh V, Bougher TL, Cola BA. A carbon nanotube optical rectenna. *Nat Nanotechnol* 2015;**10**:1027−32.
49. Li J, Bade SGR, Shan X, Yu Z. Single-layer light-emitting diodes using organometal halide perovskite/poly(ethylene oxide) composite thin films. *Adv Mater* 2015;**27**(35):5196−202.
50. Chen G, Damasco J, Qiu H, Shao W, Ohulchanskyy TY, Valiev RR, et al. Energy-cascaded upconversion in an organic dye-sensitized core/shell fluoride nanocrystal. *Nano Lett* 2015;**15**(11):7400−7.
51. Ge Q, Qi HJ, Dunn ML. Active materials by four-dimension printing. *Appl Phys Lett* 2013;**103**(13):131901.

Part Two

Smart insulation

Building insulating materials

3.1 Heat transfer physics

The transmission of heat through the opaque components of the building envelope takes place according to three different modes[1–4] (Fig. 3.1):

- Convection between the inner and outer surfaces of the building envelope and the air that skims them. Air that touches the envelope exchanges heat, and varies in temperature and consequently in density, triggering upward or descending movements that contribute to heat transfer (convective motions).
- Irradiation between the inner and outer surfaces of the building envelope and the bodies present in the external and indoor environment, through the emission and absorption of infrared (IR) radiation.
- Conduction, or diffusion, inside the continuous media (solids or fluids) that constitute the building envelope as a result of exchanges of energy at the atomic level between adjacent particles.

Since heat transmissions by convection and radiation coexist, and their effects are difficult to separate, they are usually considered together as adduction transmission (or external conduction).

The heat flux (Φ) which in stationary conditions passes by adduction and diffusion through the building envelope is directly proportional to the temperature

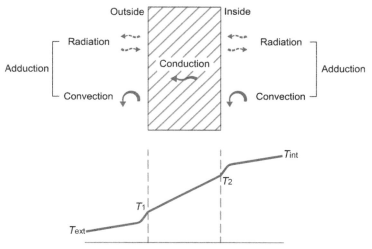

Figure 3.1 Heat transmission through a single wall.

difference (ΔT) between the internal and external environments, to the area (A) affected by the flow, and to the thermal transmittance (U) of the envelope according to the formula:[5]

$$\Phi = \frac{Q}{t} = U \cdot A \cdot \Delta T \text{ (W)} \qquad [3.1]$$

Thermal transmittance, conductance, or global transmission factor

$$U = \frac{\Phi}{A \cdot \Delta T} \text{ (W/m}^2\text{K)} \qquad [3.2]$$

is defined as the amount of heat that in steady state is transmitted in the time unit (heat flow) through a single-area unit of wall, when the temperature difference between the two environments separated by the wall itself is equal to 1°C. Its inverse ($1/U$) expresses the thermal resistance R to the passage of heat (m²K/W).

Thermal transmittance takes account of all phenomena of heat transfer affecting the building envelope, and is defined as the inverse of the sum of the thermal resistances of the surface adduction resistances and the internal resistance of the individual layers constituting the envelope itself:

$$U = \frac{1}{R} = \frac{1}{R_{is} + R_{int} + R_{es}} \qquad [3.3]$$

The value of the surface thermal resistance varies in relation to the direction of the heat flow (ascending, descending, or horizontal), the characteristics and speed of the fluid in contact with the building envelope, and the emissivity of the surface. It is higher for the inner surface ($R_{is} = 0.10-0.17$ m²K/W) compared to the outer surface ($R_{es} = 0.02-0.08$ m²K/W according to wind speed).[6]

As regards internal diffusion, the heat flow transmitted through a layer of material of thickness d is directly proportional to the temperature difference ΔT at the ends of the layer and to the area A of the surface perpendicular to the heat transfer direction, and is inversely proportional the thickness d of the layer itself:

$$\Phi = \frac{\lambda}{d} \cdot A \cdot \Delta T [W] \qquad [3.4]$$

The proportionality between the heat flow and the temperature difference is expressed by the parameter λ, which is termed thermal conductivity. It depends on the nature and physical state of the material, and is defined as the amount of heat that is transmitted in a unit of time, under steady state, through a pair of elements with isothermal surfaces with a unit area and unit distance apart, when the difference between the temperatures of the two elements is also equal to 1°C (W/mK). Its value depends on the porosity of the material, its density, and the degree of humidity.

Building insulating materials

For a given thermal conductivity of a material, the thermal resistance of a homogeneous layer of thickness d is given by the expression:

$$R = \frac{d}{\lambda}\left[\frac{m^2 \cdot K}{W}\right] \qquad [3.5]$$

In case of a nonhomogeneous material, such as a hollow brick, the resistance R is given by the expression

$$R = \frac{1}{C}\left[\frac{m^2 \cdot K}{W}\right] \qquad [3.6]$$

where C is the thermal conductance of the construction element.

At ordinary temperatures, λ values range between a few tens and a few hundreds of units for metals and a few tenths and a few units for building materials; they decrease to a few cents for thermal insulation materials. In liquids, λ has values between a few hundredths and a few tenths ($\lambda = 0.6$ for water). Thermal conductivity of gases is particularly low: air has a conductivity of $\lambda = 0.026$ W/mK.[7]

Table 3.1 presents the values of thermal conductivity and volume mass of the main construction materials.

Table 3.1 **Thermal conductivity and volume mass of main construction materials**[7]

Materials	Thermal conductivity (W/mK)	Volume mass (kg/m³)
Metals	420/17	11,300/2700
Natural stones	4.1/0.63	3000/1500
Plaster and mortars	1.40/0.29	2000/600
Close-structure concrete	1.16/0.75	2000/1700
Open-structure concrete	1.06/0.13	1900/250
Glass	1	2500
Porcelain	1	2300
Bricks	0.90/0.25	2000/600
Waterproofing materials	0.70/0.23	2100/1100
Compact plastic materials	0.50/0.28	2000/1050
Water (calm at 293K temperature)	0.60	1000
Wood	0.22/0.12	850/450
Cellular plastic materials	0.059/0.032	80/10
Still air	0.026	1.025

Table 3.1 shows that the ratio between the thermal conductivity values of the best insulators and that of one of the best conductors is just 10^4. Moreover, closer examination shows that the lowest λ values are found in porous materials (glass wool, polystyrene foam, etc.), while, considering only compact materials, said ratio does not exceed 10^3. If one compares this value with the corresponding ratio related to electrical conductivity, which is higher than 10^{20}, it seems that it is incorrect to speak of thermal insulation and insulating materials; rather, it would be more accurate to speak of poor conductors.

Thermal resistance to the passage of heat provided by cavities is the sum of two different resistances: one related to transmission by radiation between the two surfaces (independent of the interspace thickness), and one related to transmission by conduction–convection through a fluid. If the gap is devoid of any fluid, the thermal flow is transmitted only by radiation between the two faces (as is the case in double-glazed windows with a vacuum cavity). In the presence of fluid, the thermal flow is certainly greater than that which occurs in an empty cavity, since the transmission by internal conductivity through the fluid is added to that by irradiation.

However, convection motions usually also develop in the fluid (the magnitude depends on characteristics such as width and orientation of the cavity relative to the vertical and the direction of heat flow), so transmitted heat is greater than it would be with a completely still fluid. In particular, thermal resistance at first grows with the cavity thickness as long as this remains small enough for convection to be almost absent, but then decreases as the extent of convection increases, until reaching, for sufficiently large thickness, the value provided for free convection.[6] Highest values of global thermal resistance are therefore obtained for intermediate thicknesses, in the order of a few centimeters.

For interspaces of such thickness, at ordinary temperatures thermal resistance per area unit can vary between 0.11 and 0.23 m^2K/W when the emissivity of the surfaces which delimit the interspace is high ($\varepsilon = 0.8-0.9$), and much higher values if the emissivity is very low.[6]

Finally, regarding the orientation of the cavity with respect to uprightness and the direction of heat flow, a horizontal cavity possesses a higher thermal resistance when the flow is directed downward (the temperature of the upper surface is higher than that of the lower face), because the convection is particularly hampered, and vice versa if the flow is directed upward.[6]

Envelopes of existing buildings without thermal insulation materials show on average thermal transmittance values between 3.5 and 1.0 W/m^2K in relation to thickness, constituting materials, and the presence of any air gap. Current European standards oriented to the achievement of zero-energy buildings require, depending on the climatic area, minimum values of 0.45–0.20 W/m^2K, making the use of special insulating materials with low thermal conductivity necessary.[8-10]

3.2 Classification and thermal properties

Building envelope insulation is normally achieved through the use of materials (so-called insulating materials) with specific thermal and physics characteristics (thermal conductivity $\lambda \leq 0.065$ W/mK), to give the various components to which they are applied high levels of thermal resistance, or R-value (m^2/KW), thus reducing heat flows for the same environmental conditions (the inverse of R-value is the overall heat transfer coefficient, or U-value W/m^{2}K) (Table 3.2).

Table 3.2 Thermal insulation materials[12]

Insulating material	Thickness (mm) for achieving $U = 0.25$ W/m^{2}K	Thermal conductivity λ (mW/mK)
Innovative high-performance products		
Vacuum insulating panels	30	8
Aerogel	50–55	13–14
Polyurethane		
Polyurethane with pentane up to 32 kg/m^3	105–115	27–30
Soy-based polyurethane	100–145	26–38
Coated polyurethane with pentane up to 32 kg/m^3	75	20
Coated polyurethane with CO_2	130	35
On-site-applied polyurethane (sprayed/injected)	80–100	23–28
Polyisocyanurate		
Polyisocyanurate up to 32 kg/m^3	95–105	25–28
Coated polyisocyanurate up to 32 kg/m^3	80–85	22–23
On-site-applied polyisocyanurate (sprayed/injected)	80–100	23–28
Polyisocyanurate up to 32 kg/m^3	95–105	25–28
Coated polyisocyanurate up to 32 kg/m^3	80–85	22–23
On-site-applied polyisocyanurate (sprayed/injected)	80–100	23–28
Phenolic foam		
Phenolic foam	80–95	20–25
Coated phenolic foam	75–85	20–23
Expanded polystyrene foam (EPS)		
EPS up to 30 kg/m^3	115–165	30–45
EPS with graphite (gray)	115–120	30–32
Extruded polystyrene foam		
Extruded polystyrene foam with CO_2	95–140	25–37
Extruded polystyrene foam with hydrofluorocarbons 35 kg/m^3	110–120	29–31

Continued

Table 3.2 Continued

Insulating material	Thickness (mm) for achieving $U = 0.25$ W/m²K	Thermal conductivity λ (mW/mK)
Mineral insulating materials		
Glass wool	135–180	30–44
Rock wool up to 160 kg/m³	150–170	34–40
Vermiculite	235	39–60
Cellular glass	140–185	38–50
Expanded perlite panels	190	51
Biobased insulating materials		
Cotton	165–170	39–40
Cork 120 kg/m³	155–200	41–55
Wood fiber	145–225	39–61
Sheep wool 25 kg/m³	150–215	34–54
Injected cellulose fiber 24 kg/m³	150–190	35–46
Hemp fiber	165	39

The behavior of insulating materials is explained by the fact that the base material occupies only a small part of their entire volume—as is shown clearly by the comparison between their apparent density and the density of the base material itself—while the rest is occupied by air. This is distributed in pores or capillaries, arranged in many cavities so small as to exclude convective motions. The transmission of the pores is therefore to be attributed to the internal conductivity of the air, which, like all gases, is modest at ordinary temperatures (0.026 W/mK).

Based on their origin, insulating materials are classified into vegetal (cork, wood fiber, hemp fiber, coconut fiber, etc.), animal (sheep wool), mineral (glass fiber, mineral wool, expanded vermiculite, pumice, etc.), or synthetic (polyurethane, polystyrene, polyethylene foam, etc.) insulators.[11]

The structure of these materials can be fibrous (mineral wool, wood fiber), porous (cork, expanded vermiculite, pumice), or cellular (polyurethane, polystyrene, polyethylene foams).

The cells or pores may be intercommunicating (open-cell insulators), as in the case of fibrous materials, or noncommunicating (closed-cell insulators), as in the case of synthetic insulators with cellular structure (expanded polystyrene or polyurethane and sintered polystyrene). With materials of the first type, it is essential to avoid the risk of moisture condensation inside the cavities, which would drastically reduce heat insulation.

In terms of the method of installation, some insulation materials are manufactured in the factory (blocks, panels, plates, mats, bulk) and some materials are applied on site by injection, printing, or spraying.

Recent advances in the development of insulating materials are due to the progress in nanotechnology and material sciences, and have allowed production of high-performance thermal insulators with a thermal conductivity below 0.02 W/mK, compared to an average value of conventional insulating materials in the range of 0.025−0.040 W/mK.

Such performance is achieved thanks to the higher rarefaction of air inside the insulation materials due to nanoporous solid structures (aerogel), application of a partial vacuum (vacuum insulating panels—VIPs), or by a combination of both methods (VIPs with nanoporous core).

Thus overall thermal conductivity of porous materials depends on convection within the pores, conduction within the pores, conduction in the solid matrix, and radiation. In particular:

- convective heat transfer decreases with the decrease of air motion inside the pores;
- conduction decreases with the decrease of pressure around and inside the pores;
- radiation decreases significantly with decreases in temperature and pore size;
- conduction in the solid part of the porous matrix is defined by the type and amount of material used.

In nanoporous materials, high porosity reduces heat conduction through the solid part, while the small size of the pores reduces radiation and conduction in the gas. Reducing the pore size to nanoscale level increases the collisions between the gas molecules and the pore walls, leading to decreased gas conduction (the Knudsen effect).

Conduction in the gas also drops with any decreases of pressure that can be obtained. Ideally, a vacuum inside the pores produces the best insulating properties. In practice, the smaller the pores become, the less vacuum needs to be achieved for the same insulating properties.

In addition to outstanding thermal properties, nanoporous insulating materials show interesting sound insulation characteristics. Inside these materials, which are over 97% constituted by air, the direct vibration through the solid phase conducted via the solid matrix is greatly reduced, with evident sound insulation effects mostly at lower frequencies (<400 mHz).[13,14]

The choice of the most suitable insulating material or installation technology depends on the type of application and the operating conditions. In addition to thermal and thermohygrometric behavior, insulation materials must be evaluated for aspects such as limited impact on living space, acoustic behavior, fire behavior, compressive resistance, and behavior in the presence of water. Resistance to damaging agents such moisture and rodents should be considered when relevant for the application. The choice should also consider aspects like lightweight construction and ease of installation, competitive price, full compatibility with all relevant material combinations, enhanced durability for increased operative life, reduced maintenance, and reduced costs.

Particular attention must be paid to the life-cycle sustainability of selected materials in terms of energy incorporated in the product, emission of substances harmful to health or the environment, ease of deconstruction, and reusability/recyclability at the end of life, evaluated via specific life-cycle assessment studies.

Advanced insulation materials are beginning to enter the market in various niche applications. The cost, at roughly 10 times higher than traditional insulating materials, is a primary barrier to wider application, and in some cases there are concerns about long-term performance. There also is a lack of knowledge about innovative applications, and detailed design guidelines are limited. Greater effort is needed to highlight applications that are viable in market terms, such as locations in buildings with space limitations that will usually require a combination of high thermal performance insulation with lower material cost. Furthermore, in a systems perspective high-performance insulation can reduce labor costs, especially for building renovations (eg, interior wall insulation in historic buildings), so consideration of cost-effectiveness does not have to be limited just to the material cost of a system. High-performance insulation should offer the greatest value in applications with space constraints and in the existing building stock.[15–18]

3.3 Functional model and building facade applications

It is a fundamental principle to insulate to the greatest level that is justified, based on life-cycle costs, when constructing a building or retrofitting an existing building. The marginal cost of installing additional insulation is generally low.[17,18]

Higher levels of insulation can be justified during new construction or deep renovation by considering full-system impacts that allow for downsizing mechanical equipment in accordance with life-cycle cost assessment. The primary drivers that determine optimal levels of insulation are climate, the cost of energy, the heating system type and efficiency, and the installed cost of the insulation.

Regardless of the type of material used, the insulation for building facades can be placed on the inside, outside, or in an intermediate layer (Fig. 3.2). With the same insulating material and thickness, the thermal resistance is obviously the same in all three cases, but the insulation position has a strong influence on the dynamic behavior of the envelope because it determines a physical separation of the thermal mass of the inner layers from the external ones. Placing the insulating layer on the outside protects the envelope from solar radiation during summer, slowing the build-up of heat inside walls and accumulating part of the heat supplied by the heating plants or picked up through passive solar systems in the winter, thus contributing to the building heating.[11,19]

Compared to internal and cavity insulation, external insulation does not present the risk of undesirable interstitial vapor condensation, and for interventions on existing buildings allows the maximum reduction of thermal bridges within the structure, which often are responsible for over 30% of building energy consumption (Fig. 3.3).

The main factors in choosing between the three insulating methods are best assessed at the local or regional level, and relate to the type of building intervention

Building insulating materials 115

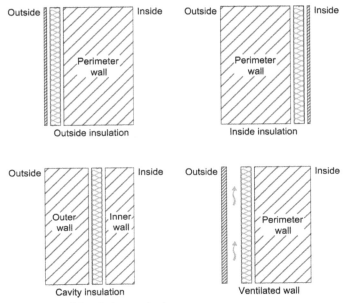

Figure 3.2 Perimeter wall insulation methods.

(new construction or renovation), conditioning system operation regime (continuous or discontinuous attenuation), the presence of passive solar systems, outdoor climate (humidity and solar irradiation levels), safety standards, and material and labor costs (time and estimated construction cost).[17,18]

3.3.1 Insulation on the outside

Insulation placed on the outer face of the wall uses an external thermal insulation composite system (ETICS). With these solutions, the internal thermal mass is entirely enclosed in the inner spaces, with the result that fluctuations in surface and air

Figure 3.3 Thermal bridge correction and insulated cavity wall.

temperature are more moderate and rooms take longer to heat up and cool down. The transitional phase of air-conditioning systems will then be considerably longer, as all the thermal inertia of the walls is borne by heating, ventilation, and air-conditioning (HVAC) plants. This solution is therefore most useful when predominantly continuous operation of air conditioning or heating is planned. In this case, in fact, transient shutdowns of conditioning systems do not generate rapid changes of surface temperatures in the environment.

Being applied from the outside, the insulation can greatly reduce any thermal bridges and water vapor condensation and protects the outer walls against weather, as well as decreasing internal thermal shocks with a remarkable reduction of thermoplastic stress. Technical solutions require the use of insulators capable of withstanding the stresses induced by atmospheric agents. It is preferable to employ high-density insulating materials, such as cork or glass foam, which show better response to solar radiation due to the higher specific heat capacity. In climatic conditions with frequent phenomena of external condensation, it is appropriate to use damp-resistant materials or provide a vapor barrier on the outside to protect the insulating material.

External insulation systems have been widely used for a long time in the Northern European countries, where there is greater need for thermal insulation, and until a few years ago were considered, at least in Southern Europe, ideal for interventions on existing buildings. When normal maintenance works of the exterior facade such as plaster reconstruction are planned, it is convenient to consider adding envelope insulation in view of its reasonable impact on overall spending. Moreover, such works would not require temporary removal of occupants. In recent years, however, the construction sector is moving toward ETICS installation for new constructions. This is because such systems allow on the one hand maximum correction of thermal bridges, the incidence of which becomes increasingly important with the increase of the thermal resistance of the building envelope, as well as avoiding possible interstitial vapor condensation, and on the other hand reduce construction time and cost.

3.3.2 Insulation on the inside

When insulation is arranged on the inner face of the wall, it effectively separates all its thermal mass from the interior spaces, resulting in reduced overall thermal inertia of the wall itself. Rooms and walls will then be brought up to the desired temperature more quickly by air-conditioning systems, but will move away from this condition just as fast once the plant shuts off; the influence of external conditions on the dynamics of the internal temperatures will therefore be minimal.

With this insulation mode, periods in which the temperature of the walls generates discomfort, even if the temperature values of the ambient air are deemed acceptable, will be significantly reduced. This type of application is therefore appropriate in environments with discontinuous use, such as holiday homes or buildings intended for tertiary activities, in which HVAC plants are frequently turned on and off.

Furthermore, since the thermal jump occurs in a very small space, it is easy to get condensation behind the insulating layer, saturating it with water with consequent swelling and rot in the inner lining. In this case, it is appropriate to provide a vapor barrier on the interior side of the insulating material. Finally, it is not technically easy to achieve perfect continuity of the insulating layer, for example in proximity to floor slabs, so several thermal bridges can occur at discontinuity points.

Inner insulation is mainly suitable in the case of energy retrofit of existing buildings of particular architectural or historical value, in which it is not possible to intervene on the external facade. Moreover, its application is quicker, requires no external scaffolding, and is more economic.

3.3.3 Insulation inside cavity walls

As an alternative to the two solutions proposed earlier, it is possible to position the insulating layer within the wall itself (Fig. 3.3). In some cases the thermal resistance of the air present in the cavity is alone sufficient to ensure the insulation desired. In other cases the gap can be partially or entirely filled with bulk insulating material (cellulose fibers, perlite granules and expanded vermiculite, cork granules) or panels. The greater protection of the insulating layer from atmospheric agents and mechanical action compared to ETICS allows using insulation with inferior mechanical resistance and rigidity.

As regards thermal inertia, the inner layer of the wall behaves as part of the interior thermal mass; effectiveness in terms of thermal inertia of interiors therefore depends on the available thickness. For these reasons, the most effective technical solutions normally provide for the interior resistant layer to have a greater mass, and consequently a larger thermal capacity, than the outer layer. The inner wall is generally made with bricks or concrete blocks, or with on-site cast or prefabricated concrete walls; the outer wall is made with lightweight brick or boards, thin elements of cellular concrete, or a masonry veneer.

Cavity insulation can be very effective for limiting thermal bridges, provided that the insulating layer also covers the structural nodes (beams, columns, connections between windows and masonry). However, double walls with cavity insulation can in winter have a worse response to vapor transfer from the heated interiors to the colder exterior. Water vapor permeating the wall toward the outside may suddenly meet cold surfaces beyond the insulating layer and condense if the layer sequence has not been carefully studied. In particular, in the case of an insulating layer made with rigid or semirigid panels, if an air chamber for vapor diffusion is not provided, it is appropriate to place a vapor barrier on the inside of the insulation. In particularly damp and cold environments, the interposition of an air chamber between the external resistant layer and the insulation is recommended to allow the diffusion of vapor coming from the outside and keep the outer wall dry. In this case, the air chamber may include evacuation devices placed on the floor to allow the removal of precipitated water.

3.3.4 Ventilated walls

A special type of external insulation is the ventilated wall (Fig. 3.4) or rain-screen system, defined as a "type of advanced screen facade in which the cavity between the cladding and the wall is designed so that the air present therein can flow by stack effect in a

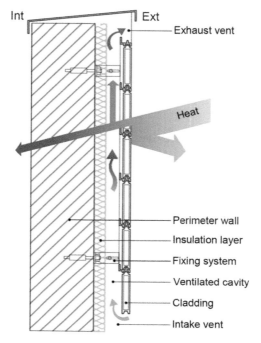

Figure 3.4 Ventilated wall operation (summer—daytime).

natural way and/or in artificially controlled way, depending on seasonal and/or daily needs, in order to improve overall thermal energy performance."[20]

This technology to control the effects of solar radiation developed traditionally in hot, dry climates, and is based on the consistent decrease of the heat flow transmitted from the outside to the inside by convective dissipation through the ventilated cavity exposed to the radiation.

The system is composed essentially of three parts: the vertical perimeter wall (load-bearing or nonload-bearing); the thermal insulation placed on the outside; and cladding panels (prefabricated slabs, metal planks, stone, wood, or others) fixed to the structural elements via a metal scaffold and spaced from the insulation by a gap of 20—40 mm to create an air cavity which, through ventilation grids placed at the base and the top, allows a flow of ascending air determined by the temperature difference due to solar radiation (stack effect). The combined effects of ventilation and insulation reduce incoming thermal flow and therefore the internal temperature of the wall.

This technique gives good performance where enough temperature difference to induce the ventilation occurs, for example, between the upper and lower parts of elements very stressed by solar radiation, such as east and west elevations.

Ventilated walls are characterized by two different behaviors in summer and winter. In the summer configuration, incident radiation is partly reflected and partly transmitted, triggering air movement in the cavity to cool the inner layer. In winter operation the temperature of the ventilation layer is close to the external one; therefore air movement decreases and determines a thermal conductance similar to an isolated but not ventilated wall. To maximize the wall insulation, reduce the convection, and exploit the incident radiant heat, closing the air vents is recommended during the winter.

Ventilated walls can be used both as an efficiency upgrade of an existing building and in new constructions. In conclusion, the main advantages resulting from the application of ventilated walls are:

- protection of the load-bearing layer of the building from the elements;
- reduction of solar radiation heat load;
- control of thermal bridges and surface condensation;
- good sound insulation;
- control of external layer stress;
- ease of maintenance (cladding elements can be replaced).

The main disadvantages are represented by:

- higher costs compared to a simple ETICS;
- increase of heat loss in winter (if the air vents remain open);
- low incident solar energy gain;
- flame spread in case of fire;
- poor exploitation of natural lighting;
- difficulty in workmanlike execution of joints between the elements.

Among cladding materials, terracotta elements are particularly interesting, thanks to modern industrial technologies that overcome traditional limits of form (from a classic rectangular to a wing shape), finish (rustic, sandblasted, lapped, glazed), color (from more traditional and warm hues like red, ocher, and tobacco to colder ones such as blue and gray), and size (even more than 1 m in length), offering the designer a wide range of solutions that combine tradition and innovation (Fig. 3.5).

Thanks to the extrusion process it is also possible to obtain slabs with different profiles, provided with holes along the direction of the extruder, which help to lighten the product and to obtain special milling used for fixing the elements to the substructure.

Figure 3.5 Ventilated wall in terracotta elements.
Photo by the author.

Ventilated walls can be further developed to integrate air filtration or purification functions within their components. In addition to photocatalytic surfaces, which chemical technology requires UV light to operate and only affects the air that comes into contact with them, physical purification can be obtained by forcing air through filters or employing other separation techniques. An example is the Breathe Brick, developed at California Polytecnic State University - School of Architecture, which is designed to integrate with a building's regular ventilation system, with a double-layered ventilated facade of the specialist bricks on the outside complemented by a standard internal layer providing insulation. Breathe Bricks contain a cyclone chamber cast directly into the concrete form which causes the incoming air to spin, winnowing out particulates in a way akin to modern vacuum cleaners. Heavy pollutant particles are separated from the air, dropped in a vertical shaft, and collected into a removable hopper at the base of the wall, while filtrated air is extracted from the top of the brick into the ventilated wall plenum behind. This air can then be delivered to the building interiors through mechanical equipment or trickle vents driven by passive systems such as stack ventilation.

The system is composed of two key parts: concrete bricks with a faceted surface to direct airflow into the cyclone system, and a recycled plastic coupler which helps to align bricks, creates a route from the outside into the brick's hollow center, and collects extracted dust (Fig. 3.6).

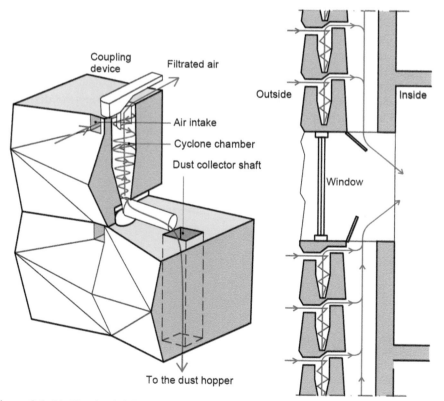

Figure 3.6 Air filtration brick operation and ventilated wall integration.

Wind-tunnel testing displayed 30% filtration of fine particles (such as airborne pollutants with a diameter of 2.5 µm or smaller) and 100% of coarse particles (10 µm and larger), such as dust. The entire system is relatively inexpensive and requires no power to operate, and may represent a suitable way to lower pollution levels in developing countries where rapid expansion of industry and less stringent environmental regulations often cause problems. Regular emptying of the collector hoppers is required to prevent the system from seizing up.

3.3.5 Advanced pitched roofs

The configuration of pitched or sloped roofs allows for promising integration of several technological solutions for thermal insulation, solar radiation control, ventilation, and renewable energy integration; the same effects can be achieved in existing buildings by intervening above the roof surface. Oak Ridge National Laboratory's researchers devised an integrated vented, insulated, and radiant barrier cavity with excellent performance, building-integrated photovoltaics, and energy renovation potential.[21] This prototype roof deck consists of profiled and double foil-faced expanded polystyrene insulation to be attached on top of an existing roof or fitted over rafters in new constructions (Fig. 3.7). The panels are profiled to create a 2.5 cm air gap between the rafters, triggering convective airflow to extract the heat penetrating the deck and expel it through a ridge vent. Both the top and bottom sides of the panels are foil faced: the top one acts as a radiation barrier toward the inclined air cavity, while the bottom side performs as a low-e radiant barrier toward the interior of the attic. Air slots near the eaves, just above the soffit, allow air from the attic and soffit vents to enter the cavity and be driven out, creating a negative pressure at the eaves that pulls fresh air in and creating a semiconditioned attic space. EPS insulation prevents convection heat from penetrating and ensures a cooler radiant barrier temperature, further reducing radiation heat exchange between the deck and the attic floor. External finishing is composed of traditional oriented strand boards (OSB) and asphalt shingles.

Summer testing of the vented radiant-insulated deck against traditional roof assemblies with the same outer finishing showed a drop in indoor temperature of almost 18°C compared to a simple shingle roof system and almost 28°C compared to a foil-faced OSB roof assembly. Winter operation is not as efficient, since heat gains during the day are lower than those of noninsulated decks, but this is counterbalanced by lower heat losses at night.

3.3.6 Thermal reflective surfaces

Thermal reflective surfaces are meant to improve a building's thermal insulation performance by reflecting some of the heat exchanged with the outside by radiation (Fig. 3.8). Building walls and roofs may become thermally reflective by simply applying one or more coats of specific paints or paint additives. These take advantage of the inclusion of hollow microspheres (Fig. 3.9) of ceramic (Insuladd, Nanoceramix, Thermoshield, Thermodry) or glass (3M Glass Bubbles), and claim a significant reflection of near-infrared (NIR) and thermal IR radiation. When applied on inner surfaces, they help reduce thermal losses during winter by reflecting back inside the heat radiated by warm bodies and heating systems (thermal radiation) that would otherwise

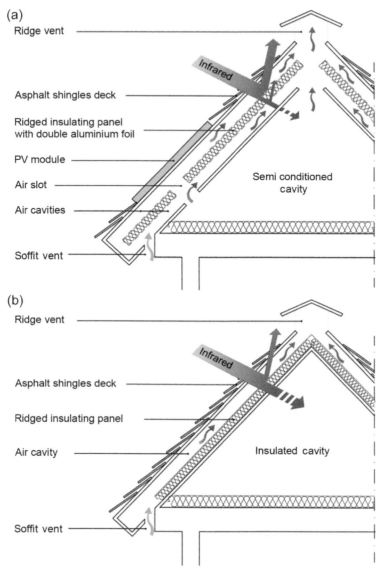

Figure 3.7 (a) Advanced and (b) traditional pitched roofs.

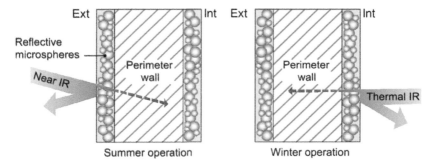

Figure 3.8 Thermal reflective coatings.

leave the building. Exterior applications, on the other hand, reflect part of incoming solar heat (NIR radiation), keeping the building cooler during summer. An application of two coats of IR reflective paint on both internal and external walls of a building claims an efficacy comparable to that of a 3 cm thick ETICS.

The macroporous structure of these paints also favors the dispersion of water vapor, preventing moisture formation and mold proliferation.

Other products employ nanomolecules of Hydro-NM-Oxide (Nansulate), and promise to improve the thermal insulation performance of the support up to 30% with a three-coat application with a total thickness of 190 μm, along with giving protection against corrosion, rust, and mold. Their nanoscale, maze-like internal architecture impedes the free flow of air molecules by the Knudsen effect, reducing heat flow through the painted surface. Nansulate's microengineered structure also makes it suitable for lead encapsulation in old paint removal or nuclear plant decommissioning.

Heat-reflective paints can also be very effective if applied on flat or inclined roofs, acting as high-performance cool roof coatings and thus achieving savings in energy consumption for cooling while reducing the heat island effect in urban areas.

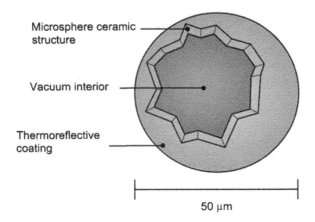

Figure 3.9 Reflective microsphere structure.

3.4 Conclusions and future trends

The role of thermal insulation is fundamental to achieve the objectives of zero-energy building in both new constructions and retrofit interventions and the consequent reduction of global CO_2 emissions. More effective insulation of buildings will also bring significant environmental, economic, and social benefits for both public administrations and citizens.

Many different insulating materials are currently available on the market, but there is still a great need to improve their performance, ensuring both economic and environmental sustainability measured by accurate life-cycle costing and environmental impact analyses.

Future scope concerns the development and characterization of improved insulation materials and solutions based on advanced sustainable materials and/or nanotechnologies to provide enhanced insulation properties and environmental performance, addressing issues such as thermal bridge correction and ensuring the best architectural quality.

In particular, there is room for improvement regarding enhanced durability to increase operating life and reduce maintenance cost, reduced embodied energy and environmental impact, ease of installation, noise reduction, adaptability to new constructions as well as renovation interventions, and reduction of hazardous materials.

There is promising scope of research into multifunctional insulating materials, such as those characterized by high load-bearing capacity, ability to perform structural as well as insulating functions (structural insulation materials), and reduced need for often-hazardous products such as glues or adhesives. These include, for instance, insulating concrete forms, which use interlocking polystyrene concrete formwork to create a seamless wall through which air cannot penetrate, and structural insulated panels that work by sandwiching insulation into interlocking sheets of building material to create uniform coverage and can be integrated into a number of materials, including particle and gypsum board, sheet metal, plastics, and foams.

Among established insulating techniques, the ventilated wall is still one of the most promising technologies for energy upgrading of existing building stock, since this solution can often be applied on top of existing building shells without too many issues. At present research is focusing on novel ways to integrate ventilated cladding with intelligent HVAC and energy production management systems, creating versatile retrofitting systems to help bring the existing building stock to the near-zero-energy building standards required by the latest regulations across the world.

Further development regards advanced foam insulations with high performance, lower cost, less petroleum demand, and greater applications and adhesion in cold climates.

Finally, more effective thermal reflective surfaces may allow walls and roofs of existing buildings to be made thermally reflective by simply applying one or more coats of specific paints or paint additives without the need to add an excessively thick thermal insulation layer.

Important core activities related to performance research are also needed to promote high-performance buildings globally: first, the promulgation at a global level of scalable, affordable, and repeatable test mechanisms allowing designers, builders, and building code officials to ensure that the most appropriate insulation materials are being specified and installed; and second, the harmonization of test mechanisms across the world to foster commerce and reduce market barriers.[15]

References

1. Carlslow HS, Jaeger JC. *Conduction of heat in solids*. Oxford: Oxford University Press; 1959.
2. Chapman AJ. *Heat transfer*. London: McMillan; 1960.
3. Grober H, Erk S, Grigull U. *Foundamental of heat transfer*. New York: McGraw Hill; 1961.
4. Rohsenow WM, Harnett JP. *Handbook of heat transfer*. New York: McGraw Hill; 1973.
5. ISO 7345. *Thermal insulation — physical quantities and definitions*. 1987.
6. ISO 6946:2012. Building components and building elements - Thermal resistance and thermal transmittance - Calculation method http://www.iso.org/iso/rss.xml?csnumber=40968&rss=detail%20Building%20components%20and%20building%20elements%20-%20Thermal%20resistance%20and%20thermal%20transmittance%20-%20Calculation%20method
7. UNI 10351. Materiali da costruzione — Conduttività termica e permeabilità al vapore.
8. EN 1745:2012. *Masonry and masonry products — methods for determining thermal properties*.
9. UNI 10355:2012. Murature e solai — Valori di resistenza termica e metodo di calcolo.
10. UNI/TR 11552:2012. Abaco delle strutture costituenti l'involucro opaco degli edifici. Parametri termofisici.
11. Casini M. *Costruire l'ambiente. Gli Strumenti e i metodi della progettazione ambientale*. Milano: Edizioni Ambiente; 2009.
12. Energy Saving Trust. *CE71 — insulation material chart — thermal properties and environmental ratings*. London: Energy Saving Trust; 2012.
13. Michael F, Ashby MF, Ferreira PJ, Schodek DL. *Nanomaterials, nanotechnologies and design an introduction for engineers and architects*. Burlington: Elsevier Butterworth-Heinemann; 2009.
14. Pacheco-Torgal F, et al., editors. *Nanotechnology in eco-efficient construction: materials, processes and applications*. Cambridge: Woodhead Publishing Limited; 2013.
15. Jelle BP. Traditional, state-of-the-art and future thermal building insulation materials and solutions — properties, requirements and possibilities. *Energy Build* 2011;**43**:2549—63.
16. Jelle BP, Gustavsen A, Baetens R. The path to the high performance thermal building insulation materials and solutions of tomorrow. *J Build Phys* 2010;**34**(2):99—123.
17. IEA. *Technology roadmap: energy efficient building envelopes*. Paris: OECD/IEA; 2013a.
18. IEA. *Transition to sustainable buildings: strategies and opportunities to 2050*. Paris: OECD/IEA; 2013b.
19. Arbizzani E. *Tecnologia e tecnica dei sistemi edilizi. Progetto e costruzione*. Rimini: Maggioli Editore; 2015.
20. UNI 11018. *Rivestimenti e sistemi di ancoraggio per facciate ventilate a montaggio meccanico — Istruzioni per la progettazione, l'esecuzione e la manutenzione — Rivestimenti lapidei e ceramici*. 2003.
21. US DOE (United States Department of Energy). *R&D roadmap for emerging window and building envelope technologies*. Washington: DOE; 2014.

Advanced insulating materials

4.1 Nanoporous insulating materials: aerogels

4.1.1 Origin and properties

As an alternative to traditional thermal insulators, the market now offers innovative products in the form of panels, rolls, or loose granulates that exploit the latest advances of nanotechnology in the fields of materials science, and are thus able to provide high levels of thermal protection with very low thickness, allowing architectural quality to be combined with energy efficiency.

These nanotechnological insulating products are made of aerogel (air + gel), a solid nanoporous material with ultra-low density obtained through the dehydration of a colloidal gel by replacing the liquid component with a gaseous one.

Invented in 1931 by Samuel Stephens Kistler at the University of the Pacific in Stockton in California, with important innovations introduced in the manufacturing process at the end of the 1960s, aerogel has since the 1990s found significant applications in the highest-tech aerospace, chemical, and pharmaceutical sectors. In the last 10 years it has been employed in the sports sector, and more recently in the building industry.[1]

Aerogel can be obtained from gels made of different inorganic or organic materials[2] for applications such as insulators, sensors, actuators, electrodes, and thermoelectric devices, or to trap space-dust particles. Among different aerogels, those based on silica (dubbed silica aerogels) are the most widely used for all applications, and in particular for thermal insulation, due to their properties and relatively simple and reliable preparation method.[3]

Also known as frozen smoke, solid smoke, or blue smoke due to its transparency, silica aerogel is an amorphous material that appears as a solid foam with a tactile feeling akin to foam rubber (Figs. 4.1–4.3). Consisting of more than 90% air, aerogels are the solid substance with the lowest weight per volume unit known today (the current record is held by graphene aerogel created by researchers at China's Zhejiang University, with a weight of only 0.16 kg/m^3, while silica aerogels can reach 3 kg/m^3).[4,5]

Its extraordinary low thermal conductivity values, typically 0.015 W/m K up to 0.004 W/m K in modest vacuum, are due to its high porosity, with interconnected pore sizes typically ranging from 5 to 100 nm and an average pore diameter between 20 and 40 nm.[1,4]

This nanoporous structure best exploits the Knudsen effect (Fig. 4.4), which comes into play when the scale length of a system is comparable to the mean free path of gas particles involved (70 nm for air molecules). In this condition, instead of freely moving through the pores of the medium keeping most of their energy intact, air molecules collide with the pore walls more often than with each other, dispersing most energy

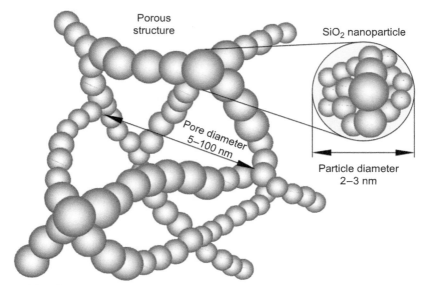

Figure 4.1 Silica aerogel structure.

Figure 4.2 Monolithic silica aerogel.
Courtesy of Aspen Aerogels Inc.

within the porous structure. Moreover, the aerogel structural network is heavily cross-linked and full of dead ends, helping to trap air molecules inside. This makes an effective block to heat transfer by convection.

Aerogel's peculiar nanoporous structure ensures thermal performance conservation regardless of operating temperature, whereas traditional insulating materials usually become more conductive as air temperature rises and air molecules excite and move

Advanced insulating materials

Figure 4.3 Monolithic silica aerogel.
Courtesy of Aspen Aerogels Inc.

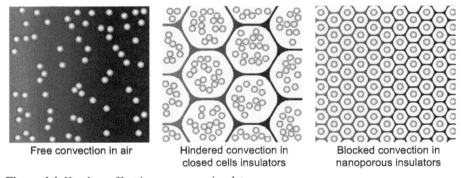

Free convection in air   Hindered convection in closed cells insulators   Blocked convection in nanoporous insulators

Figure 4.4 Knudsen effect in nanoporous insulators.

more freely within the porous network (Table 4.1). These characteristics make aerogel the best material for thermal insulation in the world, capable of operating in a temperature range between −200 and +650°C.

The open structure of the pores makes aerogel permeable to water vapor, while its composition based on hydrophobic silica makes it waterproof and fireproof.

In addition, thanks to its high porosity and high surface area, aerogel shows excellent acoustic performance: particle sound velocity in monolithic aerogel is as low as 40 m/s, while commercial nonmonolithic products reach 100 m/s.[6] In its granular form, aerogel has high sound reflection and absorption characteristics: by combining different granular sizes in multiple layers, it is possible to achieve a 60 dB sound attenuation in just 7 cm thickness.[7]

Table 4.1 **Technical properties of silica aerogel and glass**[1,4,9]

Properties	Silica aerogel	Glass
Bulk density (kg/m^3)	5–200	2300
Internal surface area (m^2/g)	500–800	0.1
Refractive index at 632.8 nm	1.002–1.046	1.514–1.644
Light transmission at 632.8 nm	90%	99%
Thermal expansion coefficient at 20–80°C (1/C)	2×10^{-6}	10×10^{-6}
Thermal conductivity at 25°C (W/m K)	0.010–0.030	1
Sound speed in the medium (m/s)	40–1300	5000–6000
Acoustic impedance (kg/m^2/s)	10^4	10^7
Electrical resistivity (Ω cm)	1×10^{18}	1×10^{15}
Dielectric constant 3–40 GHz	1.008–2.27	4.0–6.75

High luminous transmission values (90%) make aerogel suitable as a transparent insulator inside glazing in both monolithic and granular form (see Section 4.4 and Section 7.3 of Chapter 7).[6]

Despite being extremely friable, its dendritic microstructure, with a specific surface area up to 800 m^2/g, gives aerogel the compression resistance to bear a load of up to 4000 times its own weight.[1] To improve its mechanical properties, reinforcing nanofibers or special additives can be added to the gel matrix, resulting in "composite aerogels" for more varied applications in addition to thermal insulation.[8]

4.1.2 Preparation method

The rapid development of sol—gel techniques over the past two decades has led to fast progress in the deliberate synthesis of porous materials. These techniques complement conventional procedures used for the preparation of amorphous solids or glasses, such as precipitation or impregnation methods followed by high-temperature treatments.

Aerogels are synthesized via a sol—gel process consisting of three main steps: sol synthesis, in which a precursor solution develops nanoscale sol particles; sol—gel transition (gelation), as the sol particles are cross-linked and assembled into a wet gel; and finally gel—aerogel transition (drying), when the solvent inside the pores is extracted without compromising the gel structure (Fig. 4.5).[1,3,6] The choice of precursors and the optimization of sol—gel parameters determine the physical properties of the final aerogel product.

In sol—gel preparation, the most commonly used precursors are alcohol-soluble alkoxides, as they are easily available. In particular, silica aerogel requires cheap silicon alkoxides such as TEOS (tetraethyl orthosilicate) or TMOS (tetramethyl orthosilicate); water-soluble precursors such as sodium silicate (Na_2SiO_3) are also viable.[2]

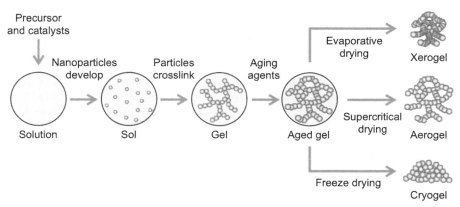

Figure 4.5 Aerogel preparation process.

During the sol synthesis step in silica aerogel preparation, the solid silicon alkoxide precursor is dispersed in a liquid medium and silicon dioxide nanoparticles develop, gradually agglomerate, and form a reticulate network spanning the entire solution, transforming the sol into a gel (gelation process). Particles aggregate by bonding or electrostatic forces, or by the addition of other chemical precursors (such as acidic or basic catalysts) that trigger molecular bonding between the atoms. At the end of this sol−gel transition, a solid three-dimensional open network structure that takes the shape of the container is achieved, characterized as hydrogel or alcogel according to the liquid medium (water or alcohol).[10]

The tenuous solid skeleton of this gel may not withstand stresses induced by subsequent dehydration and collapse, so the solution is often subjected to an aging process to reinforce it by developing additional cross-links, filling holes in the structure caused by unreacted alkoxide groups, and reducing average pore size. The aging process may involve additives to enhance hydrolysis and condensation reactions. To obtain a transparent aerogel, water must be completely removed during the aging process.

Gel dehydration is usually achieved by supercritical drying, since alternative routes such as ambient evaporation lead to xerogel, with high density (>0.25 g/cm^3) and low porosity (50%) because of shrinkage phenomena and capillary tensions in the pores. These cause the collapse of the gel structure, and may also result in cracked or even powder samples. Freezing the solvent and sublimating it (freeze drying) leads to cryogel, but it is always obtained as a powder because the crystallization of the solvent in the pores destroys the gel network.[1] Supercritical drying occurs at a pressure greater than the vapor pressure, allowing the liquid to turn into gas without the two phases coexisting at any time (supercritical fluid). Moreover, in supercritical conditions superficial tension is null, as there are no solid−liquid bonds, thus compressive forces triggered by capillary tension are avoided. Supercritical drying may be carried out either at high temperature or, more usually, at low temperature. The latter method requires placing the aged gel sample in an autoclave at 100 bar and 4–10°C, and filling it with CO_2 to replace the liquid solvent within the pores. The gel is then heated up to 40°C at 100 bar pressure to evaporate CO_2 at a supercritical stage. The aerogel sample is then isothermally depressurized and finally cooled at room temperature.[9]

4.1.3 Insulating products

Silica aerogel is mainly available on the market in monolithic blocks or slabs, loose granular form, or as mats, rolls, or fiber-reinforced panels (rigid or semirigid).

Aerogel in monolithic blocks or slabs is still difficult to manufacture and market because of its brittleness and the difficult of obtaining defect-free specimens from the production process, with applications still at the prototype stage. Nevertheless, monolithic aerogel shows promising characteristics in terms of thermal conductivity ($\lambda = 0.01$ W/m K in moderate vacuum), visible light transmission, and solar heat gain (respectively 90% and 75% for a 10 mm slab), making it particularly interesting as an insulator inside double-glazing units (see Section 7.4 of Chapter 7).[11] To address the brittleness issue, opaque composite monolithic aerogels have recently been introduced on the market. These self-supporting products can be directly applied for insulating the building envelope (Airloy).

Aerogel in loose granular form with hydrophobic processing (Cabot P100–P400) can be used for filling wall cavities or mixed in cement for the construction of internal or external insulating plaster. Plasters are also available premixed with lime and hydraulic cement with aggregates in aerogel and mineral particles (Fixit 222), with $\lambda = 0.028$ W/m K and an application thickness of 30–150 mm.

Small aerogel particles of a size ranging from 2–40 µm to 1.2 mm can be added to paints and coatings for specialized thin-film insulation coating applications, such as reducing burn risks in hot piping or preventing condensation in cold ones; the coatings are effective even if only a few millimeters in thickness (Cabot Enova, Tnemec Aerolon).

Due its optical properties, aerogel in granular form is also used in manufacturing transparent insulating materials (TIMs). By placing aerogel particles in the 0.7–3.5 mm size range in the cavity between glass or polycarbonate panes, light-diffusing panels of unmatched thermal performance are achieved (see Section 4.4).

Silica aerogel light transmission is less desirable for thermal insulation of opaque building enclosures, since it allows infrared radiation to pass through and escape heated spaces. For this reason, carbon black can be added as an opacifier to absorb infrared radiation, and sometimes even add to the material's mechanical strength. A 9% carbon black addition to silica aerogel lowers thermal conductivity from 0.0170 to 0.0135 W/m K at ambient pressure.[9]

To improve its mechanical characteristics and ease its use in insulating opaque enclosures, aerogel is usually integrated in a polyethylene terephthalate (PET) fibrous support structure to give fiber-reinforced aerogel blankets (FRABs). The manufacturing process (Fig. 4.6) first impregnates a fibrous felt with silica sol–gel solution (sol–gel casting); when impregnation is complete, felts are rolled and placed in a controlled environment for chemical aging of the gel. Finally, supercritical extraction of the

Advanced insulating materials 133

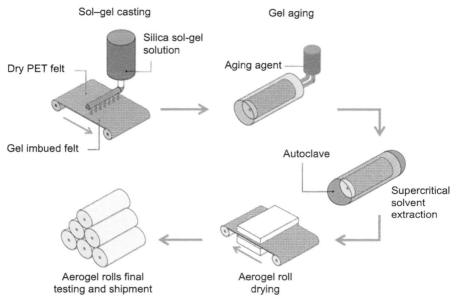

Figure 4.6 Fiber-reinforced aerogel blankets manufacturing process.

solvent is carried out in large autoclaves with CO_2 recovered from other external industrial processes. As the gel dries, aerogel particles are firmly embedded into the mat fibers. After a final drying step, an aerogel saturated mat, easy to transport and install, is achieved (Aspen Aerogels Spaceloft, Cabot Thermal Wrap) (Table 4.2). FRABs can then be coupled with support materials such as rigid polypropylene panels (AMA Composites Aeropan) (Figs. 4.7 and 4.8).

Final FRAB composition is 40—55% silica, 20—45% PET/glass fiber, and 0—15% additives. Specific fire-resistant FRABs forgo PET in favor of only glass fiber to achieve a Euroclass A2 fire rating (Spaceloft A2).

In addition to building insulation, most FRAB production is employed in the industrial and petrochemical sectors, with specific products for applications where extreme cold (Aspen Aerogels Cryogel) or heat (Aspen Aerogel Pyrogel) make it irreplaceable (see Tables 4.3 and 4.4).

Compared to conventional insulation materials (Table 4.5; Figs. 4.9 and 4.10), such as vegetal (cork, wood fiber, hemp fiber, coconut fiber, etc.), animal (sheep wool), mineral (fiberglass, rock wool, expanded vermiculite, pumice, etc.), or synthetic materials (polyurethane (PUR), polystyrene, polyethylene foams, etc.), aerogels have important advantages as well as offering greater thermal insulation values ($\lambda = 0.015$ W/m K):

- constant thermal performance regardless of operating temperature;
- high hydrophobicity values while maintaining high water vapor permeability (Fig. 4.11);
- low flammability (Euroclass C to A2: fire-resistant materials);
- mold proofing and ultraviolet (UV) and element resistance;

Table 4.2 **Main aerogel insulating product specifications**

Aerogel product	Aspen Aerogels Spaceloft
Thermal conductivity (W/m K)	0.015
Maximum use temperature (°C)	200
Water vapor permeability (μ)	5
Compressive strength (10% deformation) (kPa)	80
Density (kg/m^3)	150
Specific heat (J/kg)	1000
Fire reaction (Euroclass)	C, s1, d0
Visible light transmission (thickness)	0.203 (5 mm) 0.055 (10 mm)
Solar transmittance	0.196 (5 mm) 0.052 (10 mm)
Site adaptability	Cut to size, flexible
Appearance	Translucent white/gray
Use	Underfloor insulation, thermal bridge correction in buildings

Figure 4.7 Aeropan Basic and Spaceloft fiber-reinforced aerogel blankets. Courtesy of AMA Composites and Aspen Aerogels Inc.

- no performance drop over time due to perforation or material decay;
- ease of installation thanks to lightness and ease of adaptation;
- ease of handling and storage;
- high environmental sustainability and no toxicity.

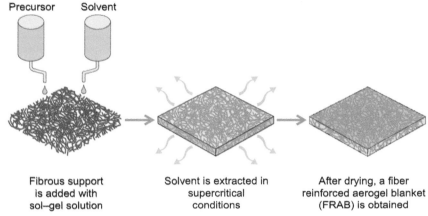

Figure 4.8 Fiber-reinforced aerogel blankets (FRABs) aerogel batting.

Table 4.3 Silica aerogel operating temperatures and applications (Aspen Aerogels)

−200 to +200°C	−100 to +200°C	+650°C
Cryogenic	Building and construction	Industrial
Liquified natural gas	Clothing	Hot process
Petrochemical	Appliances	Fire protection
Industrial	Services	District heating
Cold storage	Vehicles	Appliances and transport

Table 4.4 Comparison of main Aspen Aerogels FRAB products

Aerogel product	Spaceloft	Cryogel Z	Pyrogel XT-E
Thermal conductivity (W/m K)	0.015	0.017	0.021
Maximum use temperature (°C)	200	125	650
Density (kg/m^3)	150	160	200
Appearance	Translucent white/gray	Opaque white	Maroon
Use	Underfloor insulation, thermal bridge correction in buildings	Subambient and cryogenic pipelines, vessels, and equipment	Wall and ceiling insulation, thermal bridge correction

Table 4.5 **Comparison between aerogel and traditional insulating materials**

Insulating material	Thermal conductivity λ (W/m² K)	Equivalent thickness (cm)	Reaction to fire Euroclass	Vapor resistance μ	Specific heat (J/kg K)
Fiber-reinforced aerogel blanket	0.013	1.00	C	5	1000
Phenolic resins	0.023	1.77	C	35	1400
Polyisocyanurate foam	0.026	2.00	E	60	1464
Expanded PUR	0.027	2.08	E	60	1400
Extruded expanded polystyrene (EPS)	0.034	2.62	E	150	1450
Molded EPS	0.035	2.69	E	60	1450
Stone wool	0.036	2.77	A1	1	1030
Cork	0.042	3.23	F	10	1560
Calcium silicate	0.044	3.38	A1	3	1300
Mineralized spruce wood–wool	0.068	5.23	B	5	1810

Reproduced from Casini M. Smart materials and nanotechnology for energy retrofit of historic buildings. *Int J Civ Struct Eng* 2014;1(3):88–97.

Despite these advantages, the cost per square meter of aerogel insulating material is still high, about 8–10 times more than the traditional insulation. Supercritical drying is the most expensive and risky aspect of aerogel production. A highly desirable goal in aerogel preparation is the elimination of the supercritical drying process in favor of ambient pressure drying techniques to make industrial preparation much cheaper and thus make aerogels more competitive.

From an environmental point of view, FRABs require a relatively modest 53.9 MJ/kg for manufacturing (embodied energy), compared to 88.6 MJ/kg for EPS and 101.5 for PUR foam. Most energy is needed for obtaining and mixing the silica precursor and other raw materials for the sol–gel solution. Embodied CO_2 amounts to 4.23 kg per kg of finished product, of which 3.2 kg are required for solution preparation alone (see Table 4.6).

Moreover, FRABs are completely harmless to the ozone layer (ozone depletion potential = 0), employ recycled materials and are completely recyclable, are lead

Advanced insulating materials

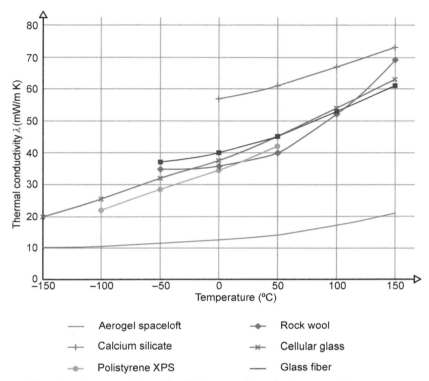

— Aerogel spaceloft
♦ Rock wool
+ Calcium silicate
× Cellular glass
● Polistyrene XPS
— Glass fiber

Figure 4.9 Thermal conductivity of insulating materials related to operating temperature. Reproduced from Casini M. Smart materials and nanotechnology for energy retrofit of historic buildings. *Int J Civ Struct Eng* 2014;**1**(3):88–97.

Figure 4.10 Thickness comparison between rock wool, polyurethane, and fiber-reinforced aerogel blanket.
Photo by the author.

Figure 4.11 Fiber-reinforced aerogel blanket hydrophobic behavior. Photo by the author.

Table 4.6 Aspen Aerogels Spaceloft Fiber-reinforced aerogel blanket environmental characteristics[13]

Component	Embodied energy (MJ/kg)	Embodied CO_2 (kg/kg)
Silica precursor and other raw materials	35.5	3.2
Fibrous reinforcement	12.1	0.6
Production process	3.5	0.4
Supercritical extraction	2.1	Recovered from other industrial processes
Pollution control equipment	0.7	0.03
Total	**53.9**	**4.23**
International transport (from United States to Europe)	2.2	0.14

free and Restriction of Hazardous Substances Directive (ROHS) compliant, and release zero volatile organic compounds (VOCs).[13]

Finally, their excellent aging resistance, measured by accelerating aging techniques (12 weeks at 90°C, equal to 60 years in real-world conditions, showed no decay in insulating performance) ensures maintenance-free operating life and the possibility of further reuse of the products.[14]

In the installation phase FRABs tend to produce dust during movement, but this is not considered harmful to health as it contains silica in amorphous and not crystalline form (Fig. 4.12), posing no carcinogenicity concerns according to the International Agency for Research on Cancer.[7]

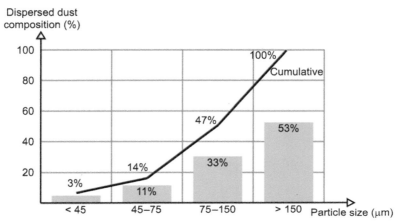

Figure 4.12 Fiber-reinforced aerogel blanket dust composition.

However, dust coming from aerogel blanket manipulation may cause mechanical irritation to the upper respiratory tract, and its high hydrophobicity dries skin and may damage the lubricated parts of machines used for processing and installation. Mat encapsulation with rolled flame-retardant polymers was proved a viable solution to this issue, causing little increase of thermal conductivity (from 0.018 to 0.020 W/m K).[15]

As an alternative to FRABs, research is under way to address aerogel's main issues, namely the brittleness and fragility that hamper its integration in manufacturing processes in its monolithic form. New alloys are being developed, such as Airloy, which claims to bridge aerogel with conventional plastics by combining the resistance of the former with the lightness and thermal insulation of the latter (Table 4.7). Airloy belongs to the category of so-called X-aerogels, namely aerogels in which the nanoporous structure is cross-linked with a polymer to increase its mechanical resistance and density. This result is achieved during the sol–gel process by soaking the wet gel in a cross-linking solution containing the monomer precursor to a polymer that can develop chemical bonding with the surface of the nanoparticles. During the supercritical drying process, the polymer evenly coats the skeletal nanoparticle framework that cross-links and fuses adjacent nanoparticles together, rendering the interparticle necks wider and more robust. A threefold increase in density keeps the material in the lightweight category, but is paired by an increase in mechanical strength up to 300 times. Polymers such as PURs/polyureas, epoxies, and polyolefins are suitable for X-aerogel preparation, while cross-linking of around 35 different oxide aerogels yields a wide variety of dimensionally stable, porous, lightweight materials with interesting structural, magnetic, and optical properties.[8]

Airloy may be mechanically processed in a number of ways and is self-supporting, thus rivaling structural plastics in weight-conscious applications such as aeronautics. Airloy can be manufactured from ceramics, polymers, carbon, metals, or carbides, in densities ranging from 0.1 to 0.9 g/cc. Compared to traditional aerogel, mechanical

Table 4.7 **Airloy specifications**

Product	X50	X60	X100	X110	X400
Density range (g/cc)	0.1–0.6	0.2–0.6	0.01–0.6	0.05–0.7	0.4–1
Ultimate compressive strength (MPa)	260	950	640	100	900
Yield compressive strength (MPa)	7	14	12	2	10
Compressive modulus (MPa)	150–670	200–400	300	30	500
Maximum specific energy absorption (J/g)	10–80	200	105	50	100
Thermal conductivity (mW/m K)	18–30	20–35	18–30	18–30	–
Maximum operating temperature (°C)	180	180	160–200	290	1000
Internal surface area (m^2/g)	150–720	70–150	150–320	250–400	1000–3000
Electrical conductivity	No	No	No	No	Yes
Appearance	Transparent blue to white	Opaque green	Opaque white or pink	Opaque white or yellow	Opaque black

resistance is greatly increased (up to 14 MPa yield strength), with the trade-off of higher thermal conductivity (18–35 mW/m K). Internal surface area ranges from 70 to 400 m^2/g, compared to 800 m^2/g for standard aerogel; specific product lines for use in electronics are electricity conductive and reach 3000 m^2/g. Cost is still prohibitive, however, with 30 × 30 cm panels totaling $490.

4.1.4 Building applications

FRAB products are suitable for any type of building, new or under renovation, but are particularly convenient for external and/or internal interventions for building refurbishment or retrofit, especially of historic buildings subject to architectural constraints, and in all cases when it is necessary to increase energy efficiency and living comfort in as little space as possible.

FRABs can be applied in their original mat form (Spaceloft) or precoupled with support materials such as fiberglass mesh, a polypropylene-reinforced layer, or gypsum wallboard for different applications (Figs. 4.13 and 4.14).

In addition to insulation of building envelopes, eventually combined with radiant heating systems, nanoporous insulating materials can be used for hot and cold piping and tank insulation.

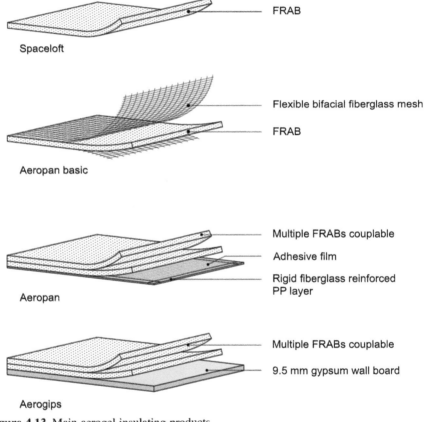

Figure 4.13 Main aerogel insulating products.

Figure 4.14 Aeropan (left) and Aerogips (right) insulating products.
Courtesy of AMA Composites s.r.l.

4.1.4.1 Floors and roofs

For floor insulation, FRABs can be applied as simple insulating mats (Spaceloft), as they are flexible, resistant to compression, breathable, and hydrophobic. This product can be easily cut to size by conventional tools and adapted into any shape, greatly reducing the installation time. Thanks to its low thermal conductivity, it gives a high degree of isolation with low thickness (5–10 mm). Due to its characteristics the product is suitable for use in all cases in which the use of traditional underfloor insulation would result in an excessive reduction in the height of indoor spaces, not compliant with statutory minimum values, or for external areas where the added thickness would result in a rise of the walkable surface incompatible with the height of access thresholds.

For roof insulation, FRABs are best used precoupled with a plaster-supporting fiberglass mesh (Aeropan Basic, Spaceloft SW) or with gypsum wallboard (Aerogips, Aerowall) for easier finishing.

4.1.4.2 Perimeter walls

Due to aerogel's hydrophobicity, to be installed on walls, FRABs must be coupled with materials that allow plaster to attach to the surface.

Products are available on the market in which a FRAB is precoupled with a rigid glass fiber-reinforced polypropylene panel (Aeropan), plaster fixing mesh (Aeropan Basic, Spaceloft SW), cork panels (Spacecork), or gypsum wallboards (Aerogips, Aerowall, Aerorock).

For external cladding insulation, products such as Spaceloft SW and Aeropan (Figs. 4.15–4.18; Table 4.8) are available in the form of panels (700 × 1400 mm or 720 × 1440 mm) that are easily applied to existing walls by using glue and dowels, then laying the reinforcement mesh, leveling the reinforcing coat, and finishing with

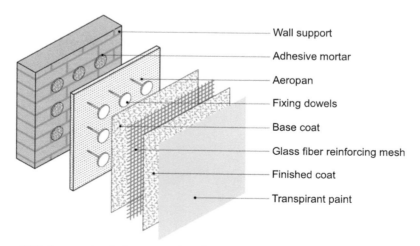

Figure 4.15 Aeropan insulation system installation.

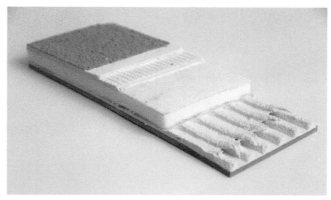

Figure 4.16 Aeropan insulation system.
Photo by the author.

Figure 4.17 Aeropan installation.
Photo by the author.

colored paste or painting. Resistant to moisture and stresses induced by elements, proof against mold and mildew, and totally breathable, aerogel products applied externally give high levels of thermal insulation with extremely low thickness (16–36 mm) thanks to a thermal conductivity λ of only 0.015 W/m K, ensuring consistent performance over a range of temperatures between -200 and $+200°$C.

For internal insulation of walls, products are available in the form of rolls (Spaceloft), semirigid panels (Aeropan), or precoupled gypsum panels (Aerogips, Aerowall). The latter are particularly interesting for their ease of installation, mechanical strength

Figure 4.18 Installed Aeropan.
Photo by the author.

Table 4.8 AMA Composites Aeropan specification and performance

Aerogel product	Aeropan
Thermal conductivity at 10°C (W/m K)	0.015
Water vapor permeability (μ)	5
Compressive strength (10% deformation) (kPa)	80
Density (kg/m^3)	230
Specific heat (J/kg)	1000
Fire reaction (Euroclass)	C, s1, d0
Visible light transmission (thickness)	n.d.
Solar transmittance	n.d.
Site adaptability	Cut to size
Appearance	Opaque white
Use	Wall and ceiling insulation, thermal bridge correction

(Figs. 4.19—4.22), and, in case of Aerowall, the possibility of combining thermal insulation and air purification.

Aerowall (1200 × 1440 mm, 1200 × 2880 mm, thickness 13 mm) is made from an insulating aerogel-based felt coupled with an Activ-air gypsum slab (also available reinforced), and can be glued and doweled to the wall or mounted on a suitable

Figure 4.19 Installation of Aerogips panels.
Courtesy of AMA Composites s.r.l.

Figure 4.20 Installation of Aerogips panels.
Courtesy of AMA Composites s.r.l.

Figure 4.21 Installed Aerogips panels.
Courtesy of AMA Composites s.r.l.

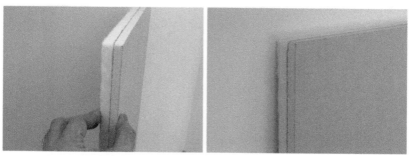

Figure 4.22 AMA Composites Aerogips panel.
Photo by the author.

structure as a normal plasterboard. The installation is completed with joint sealing. There is no need for plasters or finishes: once the panel is installed and the joints are leveled, the wall is simply whitewashed.

The presence in the gypsum slab of Activ-air (1/1000 of the mass) captures up to 70% of VOCs present in indoor air, such as aldehydes and formaldehyde, and transforms them into inert compounds, with absorption values of 60 $\mu g/m^2$ h.

4.1.4.3 Thermal bridge correction

On-site flexibility and low thickness make aerogel blankets a recommended choice for the correction of thermal bridges in new constructions, and even more so in existing buildings (Fig. 4.23). Structure thermal bridges alone may account for more than 30% of a building's energy consumption, and often lead to local condensation, decay, and mildew growth.

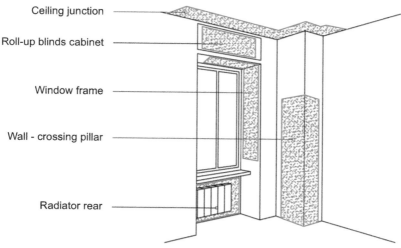

Figure 4.23 Aerogel thermal bridge correction.

An aerogel blanket is so thin as to be able to improve local conductivity effectively when only a couple of centimeters thick, a size compatible for installation behind radiators, around the edges of windows and openings (Fig. 4.24), or at ceiling and floor junctions between walls and floor slabs without unsightly and unpractical bulges, since it easily flushes within the plaster or underfloor screed. Concrete pillars and beams are also easily corrected with an inside or outside application, as are wood or steel mullions for light framing constructions.

4.1.4.4 Low-temperature heating

Aerogel's excellent insulating performance in a reduced thickness paves the way to high-performance radiating devices able to minimize heat dispersed backward outside the heated environment. An example is AktivePan, a radiating panel developed by Italian AMA Composites and German Frenzelit, which integrates a hicoTHERM PET/carbon film with an aerogel-based Aeropan mat (Fig. 4.25). When supplied with electrical current (130 W power per supply line), the 60 μm thick carbon film heats up by Joule effect, with operating temperatures of 5–70°C. It forms a radiative surface to be installed on walls and ceilings or under floors, while the aerogel support prevents heat from being dispersed through the back of the panel. Installation is completed with a thin lime plastering, giving an invisible radiating surface that prevents mold formation and enhances indoor comfort.

4.1.4.5 Tensile membranes

Aerogel blankets can also be integrated in polytetrafluoroethylene (PTFE) tensile membranes to enhance their thermal insulation performance greatly without adding excessive weight. Commercially available products such as Birdair Tensotherm

Figure 4.24 Window thermal bridge correction with Aerogips panels.
Photo by the author.

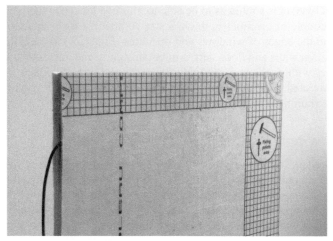

Figure 4.25 AMA Composites Aktivepan.
Photo by the author.

employ fiber-reinforced aerogel rolls sandwiched between a PTFE- or polyvinyl chloride (PVC)-coated fabric exterior skin and an acoustic or vapor barrier internal membrane. The resulting tensioned membrane system is light (>5 kg/m^2) and flexible, able to cover long spans, and can be adapted to complex anticlastic shapes. A 24 mm aerogel blanket ensures good thermal transmittance ($U = 0.56$ W/m^2 K) while still diffusing daylight (visible light transmission $= 2.2\%$, solar heat-gain coefficient (SHGC) $= 2.3\%$), with good noise absorption (70% of broadband noise), especially at low frequencies, and sound transmission reduction up to 21 dB.

Aerogel blanket inclusion also overcomes one of the drawbacks of traditional insulators such as fiberglass or rock wool when placed inside tensile membranes: while the steel and cable support structure strongly compresses the traditional material and squeezes out the air, undermining open-cell insulator efficiency, this effect in fact further enhances an aerogel's performance by packing the aerogel crystals closer together.

4.1.4.6 Appliances

For the insulation of pipes, pressurized containers, tanks, or mechanical systems, highly flexible insulating mats are available, resistant to temperatures up to 650°C and composed of silica gel reinforced with a nonwoven felt of carbon and glass fiber (Pyrogel). Some products even have an integrated vapor barrier for low-temperature applications such as water-cooling piping (Cryogel Z, Pyrogel XTZ).

For high-temperature solar applications up to 200°C a solar stainless steel twin pipe (Nanosun2) is available, with a high-efficiency ($\lambda = 0.014$ W/m K), 5 mm thick nanotechnological aerogel coating, already prepared for connection of the hot-water storage tank to the solar panel.

4.2 Vacuum insulating panels

Recently introduced in the market, vacuum insulating panels (VIPs) today hold the record for thermal insulation, with thermal resistance values up to 20 times higher than conventional insulating materials such as polystyrene foam or mineral wool (Fig. 4.26).[16]

VIPs, available as panels 10–50 mm thick, are constituted by an inner core of open pore structure material enclosed in a multilayer airtight plastic or aluminum envelope (Figs. 4.27–4.29), inside which air is extracted to create a vacuum, with internal pressure values of 0.1–5 mbar.

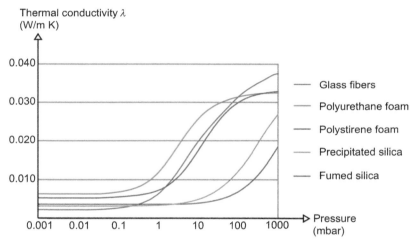

Figure 4.26 Thermal conductivity of vacuum insulating panels related to operating pressure and core material.

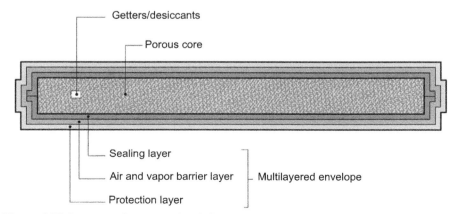

Figure 4.27 Structure of a vacuum insulating panel.

Figure 4.28 Vacuum insulating panel core and envelope.
Courtesy of va-Q-tec AG.

Figure 4.29 Vacuum insulating panel thickness.
Courtesy of va-Q-tec AG.

The extreme rarefaction of the air inside the panel eliminates the transmission of heat by conduction and convection, by lengthening the mean free path of gas molecules and thereby limiting impacts and mutual exchange of energy (the Knudsen effect). It is thus possible to ensure thermal conductivity values of the order of 0.005–0.002 W/m K.

Concerning the core, the first products used organic or inorganic materials, such as PUR or EPS foams or glass fibers, all characterized by rapid performance decay in the case of air infiltration and thus an expected service life of below 10–15 years, too short for building application.[16,17]

Panels on the market today instead employ microporous nanostructured materials such as pyrogenic or fumed silica or aerogel. The very small pore size (300 nm) and high specific surface (100–400 m^2/g in the products on the market) of these materials physically hinder the motion of air particles, limiting impacts between them and dispersing energy, making them effective even in the presence of relatively high pressure values (20–100 mbar). Their thermal conductivity at pressures up to 50 mbar is 0.003 W/m K, giving a good value of 0.020 W/m K at atmospheric pressure and allowing a good service life (0.008 W/m K thermal conductivity after 30 years) and acceptable performance even in case of increased internal pressure due to panel failure or puncture. To impede the transfer of heat by radiation, the core materials also include opacifying agents to absorb or scatter the infrared radiation. Mostly used in pyrogenic silica cores is silicon carbide (SiC), but iron oxide (Fe_3O_4), titanium dioxide (TiO_2), and carbon black are also employed.[18]

Finally, the core must be robust enough to withstand the pressure exerted on the external walls, whose difference with the internal pressure values can be up to 100 kPa.

Although lower performing than silica powder in terms of thermal insulation, requiring 0.01 mbar pressure to achieve 0.04 W/m K thermal conductivity at the center of the panel, fiberglass-core panels are suitable for high-temperature insulation, such as boilers and hot-water tanks and pipes, due to their excellent thermal stability.[18]

The function of the barrier film is to protect the panel during handling and installation, and especially to slow down as much as possible the entry of air and vapor within the core, to ensure optimum performance over time (Fig. 4.30).

The envelope is normally composed of three layers.[19] A first outer protective layer of PET, which is cheap and has good resistance to scratches and punctures, or nylon six polyamide with a higher melting point (225°C) for high-temperature applications, serves as support to a second intermediate layer which functions as a barrier against the entry of air into the core. This layer can be made from a single sheet of aluminum laminated to plastic films, which shows outstanding impermeability to the passage of water vapor and oxygen (oxygen and water vapor transfer rates equal to 0.001 g/m^2 day), or PET films metallized with a vapor deposition of aluminum 20–100 nm thick, varying in number from one to three, combining elements with good resistance to the passage of oxygen (transfer rate of 0.01–0.5 cm^3/m^2 day) with others to block the entry of water vapor (transfer rate of 0.01–0.5 g/m^2 day).[20] The third layer, the innermost, has the function of sealing the core and consists of two high- or low-density polyethylene films heat sealed together between two hot bars under pressure.

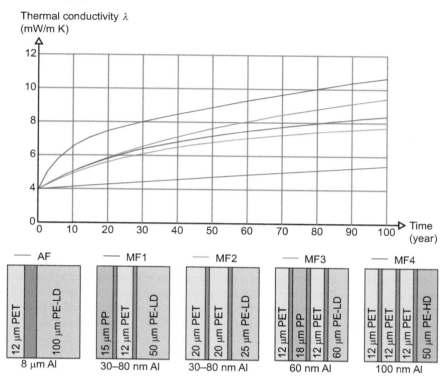

Figure 4.30 Aging of different 100 × 100 × 2 cm vacuum insulating panel envelopes. Drawn from Wegger E, Jelle BP, Sveipe E, Grynning S, Gustavsen A, Thue JV. Accelerated ageing of vacuum insulation panels (VIPs). *Rehab 2014 — proceedings of renewable energy conference 2010, Trondheim, Norway*. Barcelos: Green Lines Institute; 2010.

Although they ensure the best performance in terms of airtightness in the long term, it is preferable to avoid the use of metal sheets in the constitution of the envelope since they increase the heat transfer at the perimeter of the panel, triggering thermal bridges that limit its effective performance.

Thermal conductivity of aluminum is in fact several orders of magnitude greater than polymer layers and evacuated core material (210 W/m K compared to 0.25−0.36 W/m K and 0.004 W/m K). With reference to a 25 mm thick vacuum panel, a 7.5 μm aluminum foil, corresponding to the minimum thickness achievable before rupturing the surface, can double the passage of heat for a 50 × 50 cm panel or even triple it for a 30 × 30 cm one (Fig. 4.31). A cheaper 25 μm aluminum foil can nullify the benefits of the vacuum insulation even for larger panels.[20]

Silicon oxides (SiO_x) and silicon nitride (SiN_x) coatings are currently under research to replace metallized films, and promise effective barrier properties along with reduced thermal bridging.[18]

In addition to increasing thermal bridge phenomena, envelope-to-volume and edge-to-volume ratios are a key factor in long-term performance, as larger and thicker panels

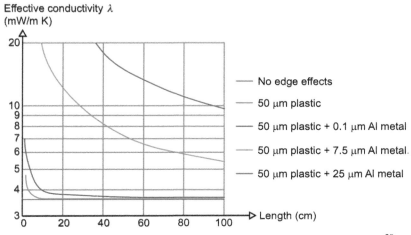

Figure 4.31 Envelope edge effect on vacuum insulating panel thermal conductivity.[20]

tend to leak pressure more slowly than smaller ones. Service life is also influenced by external relative humidity and temperature, which depend on both climate and the panel position within the building envelope: in so-called vapor-open application the relative humidity around the panel closely follows internal or external air values, which are typically higher than 80% in climates suitable for VIP use; in vapor-closed application, humidity fluctuations are primarily related to temperature. Higher water vapor pressure values require lower permeability barriers to avoid performance degradation and ensure an acceptable service life.[21]

To extend the performance of the panel further over time, it is possible to insert inside the core drying agents and getters, able to cope with the possible entry of small quantities of air or vapor, or the spontaneous emission of gas by the core under low-pressure conditions. These substances, however, are generally expensive and their heavy-metals composition makes them harmful to health and difficult to dispose of, so combinations of materials for casing and core that do not rely on their utilization are preferable. Such is the case of fumed silica, one of the most suitable core materials for VIPs, which thanks to its high porosity and sorption capability acts as a natural desiccant. In any case, the internal pressure of the panels should never exceed 100 mbar after 30–50 years of use.[17]

The VIP manufacturing process involves several steps, requiring molds and facilities which drive up prices and limit versatility, in particular relating to size and shape of the panels. First, the core material is prepared, such as a board of fumed silica cut to size from prestressed panels or pressed in a specific-size mold. The multilayer envelope is then prepared by joining the desired layers and sealing the sides—three-sided seals are preferred to four-sided seals because of reduced permeability—leaving one open for subsequent core insertion. The panel is then moved to a vacuum chamber, whose dimensions dictate maximum panel size.[18] Sealing is performed by heat welding the polyethylene (PE) layers at the open side of the bag while the desired vacuum is maintained. To end up with a mostly rectangular board to ensure best adhesion to neighboring panels, parts sticking out of the envelope have to be wrapped and fixed on one face of the panel. While prewelding, filling, evacuation, and sealing of the bag are processed semiautomatically, wrapping and other finishing steps are done manually and contribute to driving up the final price.

4.2.1 Specifications and performance

VIP panels currently available on the market perform significantly better than all other insulation products (including aerogel), with thermal conductivity values (λ) of 0.004−0.005 W/m K at manufactured pressure of 0.5 mbar for panels with a silica core and 0.002−0.004 W/m K for those with an aerogel core. The inevitable increase in internal pressure over time (in the order of 1 mbar/year), water vapor penetration, and the thermal bridge constituted by the thickness of the envelope and the gaps between panels lead to a rated installed conductivity value of 0.007−0.008 W/m K according to various rating standards (in Switzerland a safety increment of 0.002 W/m K accounts for moisture accumulation of 4% total mass, and another 0.002 W/m K accounts for a dry air pressure increase of 50 mbar, both over a 25-year time span).[22]

Long-term performance is still of concern, since the relative novelty of VIPs has prevented extensive testing on site and accelerated aging tests are not standardized. In addition, many manufacturers do not declare effective service life expectancy or provide long-term guarantees for their products. Scientific testing such as that conducted by NRC-IRC in Canada over seven years on 20 VIP products showed an average loss of thermal resistance of 2% per year, enough to ensure a service life of at least 25 years.[23]

Commercial products (Fig. 4.32; Table 4.9) are suitable for use in horizontal and vertical closures in standard encased format (Vakum Vaku VIP, Bitbau Vacupor NT,

Figure 4.32 Vacuum insulating panel products.
Courtesy of va-Q-tec AG.

Table 4.9 Performance and specifications of main VIPs on the market

Manufacturer	Model	Thickness (mm)	Factory λ (W/m K)	Rated λ (W/m K)	Core material	Envelope material	Pressure (mbar)
Vakum	Vaku VIP	10–40	0.0043	0.008	Silicic acid	Aluminum	1–5
va-Q-tec	va-Q-vip	10–50	<0.005	n.d.	Fumed silica	Aluminum	<5
va-Q-tec	va-Q-pur	10–40	0.007–0.009	n.d.	PUR foam	n.d.	<1
va-Q-tec	va-Q-mic	14–20	0.0028–0.0035	n.d.	PET microfleece	n.d.	<1
Bitbau	Vacupor NT	10–30	0.005	0.008	Silicic acid	Multilayer film	n.d.
Bifire	Vacunanex	13–36	0.0045	n.d.	Amorphous silica	n.d.	0.1
Kingspan	Optim-R	20–60	0.0042	0.007	Fumed silica	Metallized polyester	n.d.
Promat	Slimvac	10–40	0.0042	0.007	Reinforced silica	Multilayer film	<5
Porextherm	Vacupor NT-B2-S	10–50	<0.005	0.007	Fumed silica	Metallized polyester	<5
American aerogel	Aerocore	12–25	0.00186	0.00418	Organic aerogel	Metallized polyester	<1.3

Bifire Vacunanex, Kingspan Optim-R, va-Q-tec va-Q-vip, Porextherm Vacupor NT, American Aerogel Aerocore), or strengthened with additional rubber coverings (Vaku VIP Gum, Vacupor RP, va-Q-vip F GGM), polystyrene (Vaku VIP SP, Vacupor PS, va-Q-vip F EPS), glass-fiber textile (va-Q-tec va-Q-vip B), or plaster support materials (Vaku VIP Blauplatte, Vacupor TS), with dimensions of individual panels up to 3000 × 1250 mm and thicknesses up to 50 mm.

As with aerogel, VIPs are suitable for any type of building, but are ideal for external and/or internal restructuring or building restoration interventions, especially of historic buildings subject to architectural constraints, and in all cases where it is necessary to increase energy efficiency and living comfort while using as little space as possible.

In the case of energy retrofits, the minimum thickness required using conventional insulation (6−10 cm) can often be incompatible both with the esthetic features of the building and the needs of users (reduced net area or minimum height), whereas VIPs allow sizes 5−10 times lower (a 20 mm vacuum panel offers the same insulating performance as 10−20 cm of traditional insulation).

Acoustic properties of VIPs are still being researched. Early experiments have shown that a single VIP panel underperforms, as its critical frequency corresponds to traffic and speech noise (1 kHz). Loss of pressure in the panel following failure or perforation further reduces acoustic insulation. More meaningful testing on VIPs integrated into sandwich elements and massive walls is still lacking.[23]

Fire reaction classification is still questionable, since silica cores are certainly nonflammable (class A1), but polymer barrier materials are combustible, which could be an issue for facade applications. Classification and testing requirements are currently under discussion in countries where VIPs are most used, such as Germany and Switzerland.[22]

Nowadays, however, large-scale use of VIPs is still hindered by both cost (100−300 €/m^2) and lack of versatility, due to the fact that they must be sized in the factory and cannot be adapted during installation. Furthermore, any cut or puncture of the envelope leads to the loss of internal pressure and resulting irreparable decay of thermal performance.

To overcome this last drawback partially, at least for new constructions, it is advisable to integrate VIP insulation in prefabricated elements such as cladding panels of curtain walls so that the most delicate building phases are carried out in a controlled environment and careful sandwiching of the VIP may ensure panel protection and slow aging. These integrated elements are sometimes referred as structural vacuum panels.[22]

Compared to aerogel insulating products such as FRABs, VIPs show even better thermal conductivity performance (rated $\lambda = 0.007-0.008$ W/m K compared to 0.015 W/m K) and better core fire behavior (although fire reaction of the whole VIP assembly is still an issue). On the other hand, FRABs are much less delicate during both installation and operating life, can be easily adapted on site, and show no degradation of performance over time, with accelerated aging test reaching 90 years. VIPs' inherent vapor impermeability can also be limiting if a degree of breathability of the building envelope is desired (Table 4.10).

Current research and development is focused on increasing the performance, containing costs, and overcoming the constraints associated with maintaining internal

Table 4.10 Performance and real-world application comparison between traditional and innovative insulating materials

Insulating materials and solutions	Thermal conductivity	Performance preservation over time	Perforation resistance	On-site adaptation	Vapor permeability	Fire behavior
Mineral wool	High	Good	Good	Good	Good	Excellent
PUR/polystyrene foam	Medium	Good	Good	Good	Poor	Poor
Aerogel mat	Low	Good	Good	Good	Good	Medium
VIP	Lowest	Uncertain	None	Poor	None	Excellent

pressure. Materials with higher porosity would work effectively even at higher pressure, increasing the expected service life of VIPs and dispensing with the need to integrate desiccants and getters.

To make these products economically viable, alternative materials to aerogels and precipitated silica are being tested, derived from natural and renewable raw materials such as some byproducts of wood, or easy-to-supply minerals such as perlite, together with cheaper manufacturing processes. The goal is to halve the cost of the panels compared to traditional silica-based VIPs, and at the same time improve their environmental sustainability using renewable or low-incorporated-energy materials.

4.2.1.1 Modified atmosphere insulation panels

A most interesting development of VIP technology is modified atmosphere insulation (MAI) panels.[24] As with conventional VIPs, MAI panels achieve high insulation values through partial vacuum and subsequent sealing of the envelope barrier, but instead of manufacturing the panel in extremely reduced pressure conditions, a complex and cumbersome operation that drives manufacturing costs up, the innovative process devised by Nanopore in collaboration with Oak Ridge National Laboratory (ORNL) researchers blows hot steam inside the nanoporous silica core, which then cools and condenses to liquid, occupying a fraction of its original volume and leaving the remaining space as vacuum. This means the barrier-sealing phase can be carried out at atmospheric pressure with standard equipment, and the subsequent evacuation phase happens at ~ 50 mbar pressure and takes only a few seconds, compared to conventional VIP evacuation which takes place at the final panel pressure (~ 5 mbar) and lasts several minutes.

This more streamlined and cheaper procedure greatly reduces the processing and overhead stages costs, which alone account for 75% of VIPs' final price (with the silica core taking 20% and barrier film 5%). The research goal is achieving the same thermal performance as VIPs at 40% lower cost, thus greatly enhancing vacuum insulating product penetration in the market.

MAI panels share the same durability issues as conventional VIPs, however, namely delicacy and puncture vulnerability. Testing is being carried out by ORNL to assess MAI performance over time, and Firestone Building Products partnered in this endeavor to develop a composite that uses polyiso foam to encapsulate the MAI cores, allowing MAIs to be more easily integrated into buildings.

4.2.1.2 Future developments

Innovative research is focusing on GFPs (gas-filled panels), where the porous core is replaced by a metal matrix and air is replaced by heavier, less conductive gases such as argon, krypton, and xenon; the use of a filling at atmospheric pressure does not put the stress of large loads on the internal structure, which can be constituted by a honeycomb lattice of infrared radiation reflective metallized films called baffles. Theoretical thermal conductivity values λ range from 0.012 (krypton filling) to 0.020 (argon filling), but experiments on GFP prototypes gave results up to 0.040 W/m K.[25]

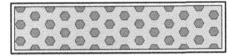 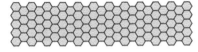

| VIP - vacuum insulating panel | VIM - vacuum insulating material |

Figure 4.33 Vacuum insulating panels and vacuum insulating materials.

The new generation of vacuum insulation will be marked by the passage from VIP panels to vacuum insulating materials based on a homogeneous closed-pore structure where any loss of pressure remains localized instead of compromising entire regions of the panel (Fig. 4.33). The difficulty in manufacturing these materials derives from the impossibility of creating a vacuum in the absence of communication between the pores: it is therefore necessary to obtain the porous structure in vacuum conditions, for example, by using a rapid drying sol—gel process, or a grid structure material able to absorb completely the gas molecules contained within the pores via a chemical reaction process.[26]

4.2.2 Building applications
4.2.2.1 Floors and roofs

VIP panels are suitable for installation over internal or external floors in cases where there is a need for insulation with extremely low thickness, due to insufficient interior heights or exterior terraces with in-line entrances.

On a sufficiently clean and smooth area, it is possible to use standard panels with aluminum or polyester casing, adequately protected with a screed pouring, but otherwise products with additional rubber protection should be used (Vaku VIP GUM, Vacupor RP1) as these are more suitable for irregular surfaces. For flooring use, panels with integrated acoustic cladding in recycled plastic are also available (Vacupor TS).

4.2.2.2 Walls

VIPs are particularly suited for internal insulation of existing buildings thanks to their reduced thickness, which allows optimum adaptation to arches, frames, and alcoves and limits the reduction of floor area. Their delicacy requires the use of adhesives in installation of panels, linings, and skirtings, to prevent puncturing the core under vacuum and compromising its performance. The inability to cut or adjust the panels in place requires careful study of the positioning of equipment (cables and electrical outlets, switches, water and thermal piping) that could result in ad hoc sizing of the panels.

VIPs' high insulation performance triggers important temperature drops within walls, thus extra care must be taken regarding interstitial condensation to prevent damage and mold growth.

It is also possible to use VIPs in high-performance external thermal insulation composite systems (ETICS) with dedicated products such as VakuVIP SP, va-Q-safe, or Vacupor PS/XPS, which have a thick polystyrene layer on one or both sides

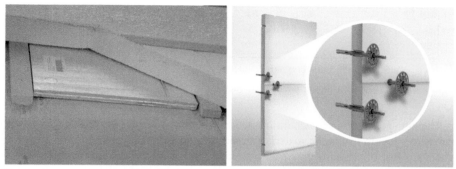

Figure 4.34 External thermal insulation composite systems installation of va-Q-vip (left) and va-Q-safe (rights) panels.
Courtesy of va-Q-tec AG.

(Fig. 4.34), or Vaku VIP Bauplatte, already prepared with a plaster support trim in recycled material. All have a 2-cm polystyrene perimeter frame to allow safe mechanical fastening with dowels.

4.2.2.3 Other uses

VIPs have widespread use in specific technological sectors where their performance is not affected by their delicacy and poor versatility, as in manufacturing of cold-storage cells, electrical appliances, and in particular containers for the transport of low-temperature material, in which the high performance with low thickness and weight give advantages for reduced payload users such as carriers of air cargo.

VIP insulation is also viable for high-temperature applications such as domestic ovens, furnaces, hot-water tanks, and concentrated solar plants, especially with heat-resistant glass-fiber core panels.

These technological and logistics applications are not affected by VIP performance degradation over time, since they seldom require more than 10 years of operating service.

4.3 Biobased insulating materials

Along with the development of high-tech insulation products, the use of natural materials, in the vein of so-called green architecture, has been growing over time. These materials are rapidly renewable, of vegetable or animal origin (biobased), and possibly recycled or recovered and reused following processes of selective demolition (Table 4.11).

This trend responds to demands for the use of environmentally friendly building materials and products that ensure the reuse or recyclability of construction works, their materials, and their parts after demolition. The use of bioecological insulation is also encouraged by the main environmental building certification systems (LEED,

Table 4.11 **Main renewable insulating materials of vegetable or animal origin on the market**

Origin	Material	Thermal conductivity λ (W/m K)	Available products
Vegetable	Wood fiber	0.038–0.052	Panels
	Mineralized wood fiber	0.086–0.107	Panels
	Cellulose fiber	0.040	Panels, blow-in products
	Kenaf fiber	0.039	Panels
	Hemp fiber	0.039–0.043	Panels
	Linen fiber	0.037–0.040	Panels, mats, felts, mending wool
	Corn fiber	0.040	Blow-in products, filling
	Coconut fiber	0.043–0.047	Mats, felts, mending wool
	Jute fiber	0.050–0.055	Panels
	Common reed	0.045–0.056	Mats
	Expanded cork	0.036–0.040	Panels, filling material
Animal	Sheep wool	0.037–0.044	Mats, felts, mending wool

BREEAM, ITACA), which award points to projects for use of renewable, recyclable, or recycled materials or recycled content.

Use of these products aims to reduce the environmental impact of the whole life cycle of materials through greater resource conservation and a reduction of energy consumption and pollutant emissions during production and disposal (Table 4.12).

Compared to insulating materials of synthetic (EPS, extruded polystyrene foam (XPS), PUR foam) and mineral (rock wool, glass wool, expanded clay) origin, insulating materials of organic origin (animal or vegetable) have, with equal performance, lower environmental impact at every stage of the life cycle of the product and a lower embodied energy value.

Besides the more traditional renewable insulating materials of natural origin, such as wood fiber, cork, or sheep wool, innovative products made with fungi roots, wood foam, marine algae, and the fibers of recycled jeans are currently being researched.

Mushrooms root (mycelium) insulating materials are grown from agricultural and food waste, giving a material with good thermal performance, inherently flame retardant, and stable over time that does not need chemical additives that emit VOCs. First developed by the New York firm Ecovative and applied in the packaging industry, products for the building industry in the form of panels and ceiling slabs, or in bulk form, are now close to entering the market (Fig. 4.35). An innovative application allows controlled cultivation of mycelium directly within the walls: in the course of

Table 4.12 Environmental characterization of insulating materials

Information	Definition
Renewable resource (UNI 11277)	Resource for which, in a given period, the time of depletion of reserves is equal to or greater than the time required to maintain reserves themselves continuously available.
Reduced use of resources (ISO 14021)	A reduction in the amount of material, energy, or water used to produce or distribute a product or packaging or specified associated component.
Recovered energy (ISO 14021)	Characteristic of a product manufactured using energy recovered from other materials or energy that would have been otherwise wasted but was collected through managed processes. In this context, the recovered energy itself can be the product.
Embodied energy	Nonrenewable energy used to transform raw materials into building products (MJ/kg). It refers to overall energy consumed during raw materials acquisition, transportation from the quarry to the factory, and transformation into finished products. In the building's life-cycle analysis, it is also extended to the stages of transportation to the construction yard and installation of the products.

Recycled content (ISO 14021)	1. Recycled content
	Proportion, by mass, of recycled material in a product or packaging. Only "preconsumer" and "postconsumer" materials should be considered as recycled content, consistent with the following terms.
	  a. "Preconsumer" material
	  Material subtracted from the waste chain during a manufacturing process. Reutilization of reworked or reground materials or scrap generated in a process and able to be recovered in the same process that generated them is excluded.
	  b. "Postconsumer" material
	  Material generated by households or commercial, industrial, and institutional facilities in their role as end users of the product, which can no longer be used for its intended purpose. This includes the return of material from the distribution chain.
	2. Recycled material
	Material that has been reprocessed from reclaimed material (regenerated) by means of a reworking process and transformed into a final product or a component to be incorporated in a product.
	3. Material recovered (reclaimed)
	Material that would otherwise have been disposed of as waste or used for the recovery of energy, but that was instead collected and recovered (regenerated) as supply material, instead of a new raw material, for a recycling or production process.

Table 4.12 Continued

Information	Definition
Waste reduction (ISO 14021)	Reduction in the quantity (mass) of material entering the waste chain as a result of a product, process, or packaging variation. Waste may include air and water outlets, as well as solid waste from production or treatment processes.
Compostable (ISO 14021)	Characteristic of a product, packaging, or associated component that allows it to biodegrade, producing a relatively homogeneous and stable humus-like substance.
Degradable (ISO 14021)	Characteristic of a product or packaging that, with respect to specific conditions, allows it to decompose up to a specific degree in a given time.
Designed for disassembly (ISO 14021)	Design feature of a product that allows disassembly of the product at the end of its service life in such a way that parts and components may be reused, recycled, recovered for energy, or in some other way subtracted from the waste chain.
Product with extended service life (ISO 14021)	Product designed to provide prolonged service, based on improved durability or improvability features, which causes lower resource utilization or waste reduction.
Recyclable (ISO 14021)	Characteristic of a product, packaging, or associated component that can be rescued from the waste chain through available processes and programs and can be collected, processed, and returned to use in the form of raw materials or products.
Reusable (ISO 14021)	Characteristic of a product or packaging which has been conceived and designed to achieve in its life cycle a certain number of runs, rotations, or actuations for the same purpose for which it was conceived.

Figure 4.35 Mycelium insulating materials and preparation.
Courtesy of Ecovative.

about a month the mycelium grows, completely seals the wall cavity, stabilizes, and dries up to form an airtight insulation package without the toxicity and ecological effects normally associated with foam fillings.

Wood foam was developed by researchers at the Fraunhofer Institute in Munich looking for environmentally friendly alternatives to traditional fossil-derived PUR foams employed in construction. Byproducts of various sawmill processes of any size and origin (both coniferous and deciduous trees) are combined with hemp, hay, or other cellulose-based materials, ground finely to obtain a viscous paste, and subsequently expanded with the addition of gas. The resulting foam is solidified in an oven thanks to substances naturally present in the wood and can be used in panels or mats. Wood foam promises improved performance in dimensional stability and resistance to moisture and defibration compared to traditional wood fibers or wood−wool insulation. Insulating performance is on par with polystyrene, with beech foam thermal conductivity in the 0.32−0.40 W/m K range for available densities of 45−280 kg/m^3, and a 0.02−0.82 N/mm^2 compression resistance that makes it suitable for further machining processes. Fire behavior is similar to other wood-fiber products: wood foam burns and smolders, often showing self-extinguishing properties. Flame resistance may be easily improved by mixing specific ecological additives in the manufacturing process. This innovative product was deemed worthy of the GreenTec Award 2015 in the "Construction and Living" category. Further research is now focusing on identifying most suitable tree species and streamlining the production process.

Another innovative application of ecological materials for building insulation uses algae deposited on the shore by waves. During the spring, summer, and autumn tons of *Posidonia oceanica* leaves clump on Mediterranean Sea beaches in the form of piles or balls called egagropili (Fig. 4.36), generally disposed of in landfills or used for composting. Their incombustibility, mold resistance, and ability to accumulate and dispose of water vapor make them interesting for application in the building industry. A process developed by German-based Neptutherm separates the algae from the sand and shreds them into fibers 1.5−2.0 cm long; these are then used to make insulating panels or fill cavities (Fig. 4.37). The material has excellent thermal capacity (2502 J/kg K, 20% more than wood), which makes it suitable for use in hot environments for heat

Figure 4.36 Shored egragopili
Courtesy of Neptutherm GmbH.

Figure 4.37 Algae-based insulation.
Courtesy of Neptutherm GmbH.

protection. Thermal conductivity values ($\lambda = 0.039$ W/m K) are aligned with other ecological insulation materials such as cork or wood fiber.

Insulators made of seagrass also lend themselves well to synergy with other bioecological products, dramatically improving the characteristics of thermal inertia. Sardinia-based Edilana has devised the Edimare insulating panel, which combines a layer of fibers of beached *Posidonia*, handpicked in coordination with coastal municipalities, with a layer of pure sheep wool obtained from shearing leftovers. The finished product combines good thermal conductivity ($\lambda = 0.039$ W/m K) with high values of heat capacity (2529 J/kg K) and phase shift (+60% compared to insulating minerals or fossils). Its exceptional environmental sustainability won the product the "Living Green for EXPO2015" prize.

Hemp fibers are also enjoying good diffusion on the market. The hemp plant is quick and easy to grow, alternates well with other cultivation species such as barley or rye, and does not need herbicides or pesticides. While hemp outer fibers are used in the textile industry, its inner core lends itself well as an additive to lime binders to achieve building materials with good mechanical and insulation characteristics. Hemp core has a naturally high silica content, allowing it to bind perfectly with water and lime to manufacture nonload-bearing blocks and insulating panels, or to spray as cavity infilling or an insulating layer between walls and external plaster. It is ideal as a light screed for floor heating systems. Compared to concrete, so-called hempcrete has about 5% of compressive resistance and 15% of density, with a much better thermal transmittance of 0.08 W/m K, enough to reach good thermal resistance values without requiring additional insulation layers. Finally, hempcrete is inherently fire resistant and does not need added flame-retardant substances, does not attract infestations, and has excellent hygroscopic behavior by combining vapor permeability of lime with vapor storage capacity of hemp shiv. Environmental impact is obviously extremely reduced, with each ton of grown hemp absorbing 1.7—1.9 tonnes of CO_2, and each ton of hempcrete absorbing an additional 249 kg of CO_2 in its 100-year life cycle.

Beyond wood, among insulators of vegetal origin deriving from recycled raw materials or processing wastes, those that use textile fibers and recycled cellulose are of particular interest.

Recycled cellulose comes from waste, byproducts of paper processing, and returned newspapers. Cellulose fiber comes in flakes or wadding, and is mainly applied via blowing inside cavity walls, roofs, and attics. This insulating material is composed of 85% pure cellulose and 15% safe additives such as lime and verdigris, which do not release hazardous substances into the atmosphere (boron or ammonium salts, print inks, or glues). Thermal conductivity is on par with synthetic insulation materials ($\lambda = 0.036$ W/m K), with good thermal capacity (2.15 kJ/kg K) for easily improving thermal mass at the same time—a 180 mm cellulose layer yields a thermal phasing of 10 h. Furthermore, cellulose is mold, rodent, and insect proof, does not set or lose volume over time, has good fire behavior (Euroclass B, s1-d0), and has a net negative environmental impact, as carbon emitted in manufacturing is 220 times lower than that saved over operating life. The main cellulose manufacturers on the market are Turin Polytechnic spinoff Nesocell and German-based Climacell.

Recycled textile insulation materials are also gaining interest, in particular the fibers of recycled jeans because of the excellent acoustic and thermal insulation properties of the jeans textiles (Fig. 4.38). The waste can be preconsumer (waste from production) or postconsumer (denim insulation).

Among the most recent and renowned examples of the use of denim insulating materials are the Academy of Sciences by Renzo Piano in San Francisco, where 68% of thermal insulation is made from waste jeans, and the renovation of the historic building at 110 Embarcadero in San Francisco, where about 7800 pairs of Levi's jeans

Figure 4.38 Recycled jeans insulation.
Photo by the author.

otherwise destined for landfill have been used in the insulation as a tribute to the history of the city homeland of Levi's. Bonded Logic, Applegate Insulation, and Le Relais are among companies that supply insulation products in recycled textile fibers. Bonded Logic was the first to patent a specific product, UltraTouch Denim insulation, containing 90% postconsumer recycled natural fibers free of chemicals or irritating substances and without VOC emissions.

Research in the field of biobased insulating materials also relates to nanotechnology applications: a research team led by Turin Polytechnic and financed by the Swedish Strategic Foundation has developed an innovative, environmental friendly insulating material made by combining a graphene oxide suspension with nanocellulose and sepiolite nanoparticles (renewable and abundant materials).[27] This new insulator has been produced through a process called "freezecasting," which allows the production of ultralight foams whose thermal insulation and fire behavior performance is considerably higher than those of traditional materials, even showing self-extinguishing properties without the need for toxic additives.

4.4 Transparent insulating materials

The need to satisfy both requirements of thermal insulation and those of maximization of daylight in interior spaces has led to the development of light-diffusing insulating materials capable of combining high direct transmission of visible solar radiation with low heat transmission (Fig. 4.39).[28–30]

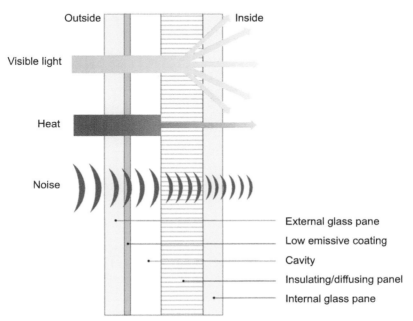

Figure 4.39 Transparent insulating materials.

These materials, called transparent insulating materials (TIMs) or, more properly, translucent insulating materials, with thickness between 30 and 100 mm, have values of light transmission (20−67%) and solar heat gain (SHGC = 0.69−0.07) comparable to those of a double-glazing unit and values of thermal transmittance that can reach those of a well-insulated opaque wall ($U = 1.8-0.28$ W/m^2 K).

TIMs are typically made of polycarbonates, polymethylmethacrylate (PMMA), PVC with a honeycomb or capillary structure, monolithic or granular aerogel, or even phase-change materials (PCMs), and are available as self-supporting panels or slabs inserted inside the double-glazing glass (Figs. 4.40 and 4.41; Table 4.13).

All TIM types give high visual comfort, free of glare phenomena, due to their light-diffusing behavior that transmits daylight inside diffusely in all directions. Since they do not allow vision through them, however, TIM cannot be substitutes for glass where external visibility is required.

As regards the geometrical configuration of the cellular insert, TIMs can be classified as having:

- parallel structure geometry, in which transparent layers are laid parallel to the panel;
- perpendicular structure geometry, in which transparent or opaque slats are laid perpendicular to the panel in a honeycomb pattern;
- mixed geometry or cavity structure constituted by a combination of the two previous geometries, such as multiple transparent ducts or even transparent bubble foams;
- homogeneous material, such as loose glass fibers or aerogel granules.[28]

Figure 4.40 Okalux panel (left) and detail of capillary slab (right). Courtesy of Okalux GmbH.

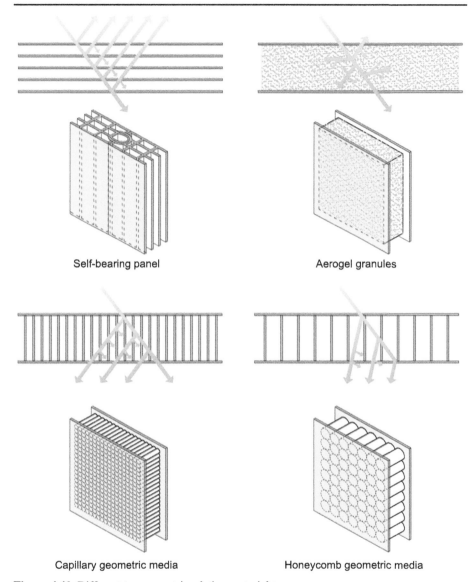

Figure 4.41 Different transparent insulating material types.

For horizontal roof windows, through which vision is seldom required, the use of TIMs is even more advantageous for reducing thermal loss. Due to convection, the performance of a gas-filled standard insulating glass unit (IGU) drops significantly (up to 50%) when used in a roof application compared to rated values, which refer to traditional vertical installations (Fig. 4.42).

In addition to their light transmission and thermal insulation properties, TIMs ensure high protection against noise (up to more than 50 dB) and UV radiation

Table 4.13 **Classification of transparent insulating materials**

Self-bearing panels	Non self-bearing panels			
	Aerogel	PCM	Geometric media	
			Capillary	Honeycomb
Polycarbonates PMMA PVC	Monolithic Granular	Granular	Polycarbonates PMMA Glass	Polycarbonates

(<1%), allow realistic rendering of colors, and can assume different hues. These properties make them particularly suitable for use in school and university buildings, hospitals, sports centers, museums, libraries, shopping malls, railway stations, airports, and exhibition halls to replace vertical or horizontal opaque closures or transparent roof windows (Figs. 4.43 and 4.44).

TIMs are also integrated in passive solar systems (Thermische Wärme Dämmerung), especially in Northern Europe. In this application perpendicular geometry has proven the most effective, since solar radiation is bounced forward toward the absorber surface instead of being reflected back, as in parallel geometry structures. Honeycomb or capillary structures can provide enough radiation transmittance while also blocking upward thermal convection and ensuring thermal insulation. Care shall be taken in choosing the right TIM material, as high temperatures reached in thermal

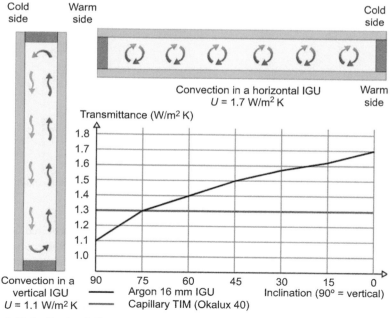

Figure 4.42 Inclination influence on transparent insulating material and IGU transmittance.

Figure 4.43 Aragon Convention Center, Nieto Sobejano Arquitectos, Zaragoza. Courtesy of Roland Halbe Photography©.

Figure 4.44 British Museum Great Court, Foster + Partners, London. Photo by the author.

collectors may be incompatible with polymer TIMs and require glass honeycomb inserts or even aerogel/glass fiber homogeneous panels.[31]

Self-supporting panels can be of various types (compact flat panels, hollow core slabs, corrugated panels) and different materials. Generally, they are manufactured from plastics, such as polycarbonate, PMMA, or PVC, with a structure in parallel rather than intercommunicating chambers to ensure thermal insulation and mechanical strength.

Used in the past for mundane buildings, such as warehouses and canopies, today, thanks to the improved resistance to sunlight (especially UV rays), weather, and fire, they are also used in more valuable buildings such as sport structures, greenhouses, schools, industrial premises, and libraries (Table 4.14).

The extreme lightness of the panels, combined with airtight and watertight tongue-and-groove mounting systems, ensures ease and speed of installation. Panels can be

Table 4.14 **Thermal and daylight performance of main transparent insulating materials on the market**

Product	Thickness (mm)	Light transmission (%)	UV transmission (%)	SHGC	U (W/m² K)	Weighted sound reduction index (Rw) (dB)
Glass						
Double-glazed glass SGG climaplus silence	28	76	1	0.58	1.30	40
Self-bearing panels						
Rodeca Pc 2530-4	30	67	<1	0.69	1.40	20
Rodeca Pc 2540-7	40	53	<1	0.56	1.00	24
Rodeca Pc 2560-12	60	41	<1	0.43	0.71	27
Capillary geometric media						
Kapilux T	20	67	n.d.	0.62	1.80	n.d.
Okalux 40	40	24	n.d.	0.28	1.30	45
Kalwall 70	70	30	n.d.	0.33	1.30	31
Kalwall 100	100	5	n.d.	0.07	0.45	35
Honeycomb geometric media						
Clearshade CS TC7 (sun angle 40–75°)	24	68–46	1	0.33–0.14	1.90	n.d.
Clearshade CS TTW7 (sun angle 40–75°)	24	44–24	1	0.24–0.11	1.90	n.d.
Granular aerogel filling						
Okagel 60	60	45	n.d.	0.54	0.30	52
Kalwall Lumira Aerogel	70	21	n.d.	0.26	0.30	35
TGP K25	25	38	n.d.	0.31	1.20	44
Duo-Gard	40	40	n.d.	0.46	0.51	n.d.

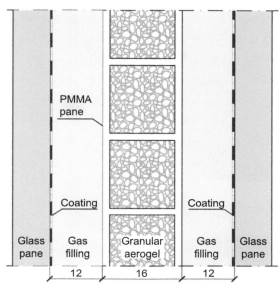

Figure 4.45 Aerogel transparent insulating material section (Zae Bayern prototype). Drawn from Reim M, Koerner W, Manara J, Korder S, Arduini-Shuster M, Ebert HP, et al. Silica aerogel granulate material for thermal insulation and daylighting. *Sol Energy* 2005;**79**(2): 131−9.

custom made, with length up to 25 m, and lend themselves easily to staining or printing processes on one or both faces while still maintaining a high degree of transparency. Unlike glass, for roof applications they require a minimum slope of just 5 degrees.

TIMs of the nonself-supporting type are inserted inside a double-glazing unit and are made of polycarbonates, PMMA, PVC, glass with capillary or cellular honeycomb structure (geometric media), loose glass fibers, and PCMs in granular form or aerogel granules.[32]

Currently available aerogel TIMs (Fig. 4.45) are made by placing aerogel particles (0.7−3.5 mm size) in the cavity (10−70 mm thick) between two sheets of fiber-reinforced polymer with an aluminum support structure (Kalwall Lumira) or two glass panes (Okagel; Fig. 4.46). While not allowing see-through vision because of its granular form, a thickness of only 16 mm is able to ensure a coefficient of light transmission of 70% and a thermal transmittance U equal to 1 W/m^2 K.

Performance can be further increased by coupling the aerogel-packed bed with the gas-filled cavities found in traditional IGUs. Prototypes investigated by Zae Bayern employed glazing composed of two glass panes with low-e coating on the inside, two 12 mm gas gaps, and a 16 mm aerogel-granulate-filled PMMA double-skin sheet. By combining different lowe coatings (more or less optimized toward solar control or thermal transmittance), semitranslucent or translucent aerogel, and argon or krypton filling, it was possible to achieve visible light transmission values ranging from 19% to 54% and SHGC values from 0.17 to 0.45. Thermal transmittance values

Figure 4.46 Okagel panel.
Courtesy of Okalux GmbH.

followed the same trend, with dark samples achieving U_g as low as 0.37 W/m² K and clear ones up to 0.56 W/m² K.[33]

Commercially available granular aerogel TIMs can reach thermal and acoustic insulation values comparable with opaque insulated walls that comply with building codes ($U = 0.30$ W/m² K, Rw = 52 dB), while they have the obvious advantage of diffusing daylight (91% light transmission per cm), and their good SHGC (up to 0.54) allows exploitation of infrared radiation in colder climates to reduce winter heating energy consumption. This can understandably be detrimental in more moderate climates, where it can lead to interiors overheating during summer. In addition, aerogel TIMs can be integrated with phosphorescent smart materials to give night illumination without the use of electrical energy.

A very recent development of TIMs regards the integration of PCMs inside glazing systems to improve thermal inertia characteristics (see Section 5.5.6 of Chapter 5).

4.5 Conclusions and future trends

Advanced insulating products are suitable for any type of building, but are ideal for external and/or internal interventions of renovation or building restoration, especially of historic buildings subjected to architectural constraints, and in all cases in which it is necessary to increase energy efficiency and comfort with minimum space loss.

In renovation projects using conventional insulating materials ($\lambda = 0.02-0.04$ W/m K), even if it is possible to increase insulation thickness at all, significant esthetic and functional compromises are often required to fit more insulation on the inside or outside of the building envelope. The use of advanced insulation materials ($\lambda < 0.02$ W/m K) in refurbishment interventions instead allows significant reduction of the energy requirements of buildings while maintaining their integrity and architectural quality, to achieve the goal of reducing primary energy consumption in the building sector.

Advanced insulating materials such as aerogel and VIPs are beginning to enter the market in various niche applications. Cost, roughly 10 times higher than traditional insulating materials, is a primary barrier to greater application, and in the case of VIPs there are concerns about long-term performance. There also is a lack of knowledge about innovative applications, and detailed design guidelines are limited. Greater effort is needed to highlight applications that are viable in market terms, such as locations in buildings with space limitations that usually require a combination of high thermal performance insulation with lower material cost.

Regarding aerogel, the most promising insulating material, two main development scopes are on the one hand the elimination of the supercritical drying process from its manufacturing, to make industrial preparation much cheaper and thus make aerogels more competitive, and on the other hand substituting the fiber-reinforced blanket with other substrates (such as a melaminic substrate) to avoid dust dispersion in both manufacturing and application.

Regarding VIPs, the goal is to develop high-performing vacuum panels with long life and lower cost that can be used in building systems to achieve very high system performance, such as when embedded in ETICS systems. The research goal is achieving a cost reduction of at least 40%, thus greatly enhancing penetration of vacuum insulation products in the market. In this direction, MAI panels could be an interesting development of VIP technology.[24]

Along with the development of high-tech insulation products, the use of natural materials with low environmental impact has been growing over time. An interesting future direction regards the production of environmental friendly insulating materials by combining biobased materials with nanotechnology using a low-tech/high-tech approach.

References

1. Aegerter MA, Leventis N, Koebel MM, editors. *Aerogels handbook*. New York: Springer-Verlag; 2011.
2. Acharya A, Joshi D, Gokhale VA. AEROGEL — a promising building material for sustainable buildings. *Chem Process Eng Res* 2013;**9**:1—6.
3. Dorchech AS, Abbasi H. Silica aerogel: synthesis, properties and characterization. *J Mater Process Technol* 2008;**199**:10—26.

4. Gao T, Jelle BP. Silica aerogels: a multifunctional building material. In: Sobolev K, Shah SP, editors. *Nanotechnology in construction. Proceedings of NICOM5*; 2015. p. 35–41.
5. Sun H, Xu Z, Gao C. Multifunctional, ultra-flyweight, synergistically assembled carbon aerogels. *Adv Mater* 2013;**25**(18):2554–60.
6. Baetens R, Jelle BP, Gustavsen A. Aerogel insulation for building applications: a state-of-the-art review. *Energy Build* 2011;**43**(3):761–9.
7. Pacheco-Torgal F, et al., editors. *Nanotechnology in eco-efficient construction: materials, processes and applications*. Cambridge: Woodhead Publishing Limited; 2013.
8. Leventis N. Mechanically strong lightweight materials for aerospace applications (x-aerogels). In: *Proceedings of 56th international astronautical congress. Fukuoka, Japan, October 17–21, 2005*. Paris: International Astronautical Federation/AIAA; 2005.
9. Thapliyal PC, Singh K. Aerogels as promising thermal insulating materials: an overview. *J Mater* 2014;**14**:1–10.
10. Vansant EF, Van Der Voort P, Vrancken KC, editors. *Characterization and chemical modification of the silica surface*. Amsterdam: Elsevier Science; 1995.
11. Buratti C, Moretti E. Transparent insulating materials for buildings energy saving: experimental results and performance evaluation. In: *Third international conference on applied energy. Perugia, May 16–18*; 2011.
12. Casini M. Smart materials and nanotechnology for energy retrofit of historic buildings. *Int J Civ Struct Eng* 2014;**1**(3):88–97.
13. Aspen Aerogels. *Spaceloft® aerogel insulation environmental product declaration according to ISO 14025 and EN 15804. Declaration number S-P-00725*. UK, Anglesey: Renuables Ltd; 2015.
14. O'Connor M. Aerogel life cycle assessment and life cycle costs. In: *NanoItaly conference 1st ed. Technical session (Ts) I.4. Aerogel*. Rome: Sapienza University of Rome; September 21–24, 2015.
15. Carty L, Garnier C, Williamson JB, Currie J. New thin aerogel for high performance internal wall insulation of existing solid wall buildings. In: *Rehab 2014 – proceedings of the international conference on preservation, maintenance and rehabilitation of historic buildings and structures*. Barcelos: Green Lines Institute; 2014.
16. Wang X, Walliman N, Ogden R, Kendrick C. VIP and their applications in buildings: a review. *Constr Mater* 2007;**160**(4):145–53.
17. Baetens R, Jelle BP, Gustavsen A, Roels S. Long-term thermal performance of vacuum insulation panels by dynamic climate simulations. In: Gawin D, Kisilewicz T, editors. *Research on building physics: proceedings of the 1st central European symp. on building physics*. Cracow: Technical University of Lodz; 2010.
18. Alam M, Singh H, Limbachiya MC. Vacuum insulation panels (VIPs) for building construction industry – a review of the contemporary developments and future directions. *Appl Energy* 2011;**88**(11):3592–602.
19. Wegger E, Jelle BP, Sveipe E, Grynning S, Gustavsen A, Thue JV. Accelerated ageing of vacuum insulation panels (VIPs). In: *Rehab 2014 – proceedings of renewable energy conference 2010, Trondheim, Norway*. Barcelos: Green Lines Institute; 2010.
20. Porextherm Dämmstoffe GmbH. *VACUPOR® vacuum insulation panel technical information*. Kempten, Germany: Porextherm Dammstoffe GmbH; 2014.
21. Tenpierik M, Cauberg H. Vacuum insulation panel: friend or foe? In: *Proceedings of PLEA 2006 – the 23rd conference on passive and low energy architecture. Geneva*; September 6–8, 2006.

22. Simmler H, Brunner S, et al. *Vacuum insulation panels. Study on VIP components and panels for service life prediction of VIP in building applications IEA/ECBCS Annex 39.* 2005.
23. Kalnæs SE, Jelle BP. Vacuum insulation panel products: a state-of-the-art review and future research pathways. *Appl Energy* 2014;**116**:255–375.
24. Oak Ridge National Laboratory. *Modified atmosphere insulation — next generation insulation material.* Oak Ridge, Tennesse: UT-Battelle; 2015.
25. Baetens R, Jelle BP, Thue JV, Tenpierik MJ, Grynning S, Uvsløkk S, et al. Vacuum insulation panels for building applications: a review and beyond. *Energy Build* 2010;**42**(2): 147–72.
26. Baetens R, Jelle BP, Gustavsen A, Grynning S. Gas-filled panels for building applications: a state-of-the-art review. *Energy Build* 2010;**42**(11):1969–75.
27. Wicklein B, Kocjan A, Salazar-Alvarez G, Carosio F, Camino G, Antonietti M, et al. Thermally insulating and fire-retardant lightweight anisotropic foams based on nanocellulose and graphene oxide. *Nat Nanotechnol* 2015;**10**:277–83.
28. Kaushika ND, Sumathy K. Solar transparent insulation materials: a review. *Renew Sustain Energy Rev* 2003;**7**(4):317–51.
29. Huang Y, Niu J. Application of super-insulating translucent silica aerogel glazing system on commercial building envelope of humid subtropical climates e Impact on space cooling load. *Energy* 2015;**83**:316–25.
30. Dowson M, Harrison D, Craig S, Gill Z. Improving the thermal performance of single-glazed windows using translucent granular aerogel. *Int J Sustain Eng* 2011;**4**(3):266–80.
31. Granqvist CG. *Materials science for solar energy conversion systems.* Oxford: Pergamon Press; 1991.
32. Jensen KI, Schultz JM, Kristiansen FH. Development of windows based on highly insulating aerogel glazings. *J Non-Cryst Solids* 2004;**350**:351–7.
33. Reim M, Koerner W, Manara J, Korder S, Arduini-Shuster M, Ebert HP, et al. Silica aerogel granulate material for thermal insulation and daylighting. *Sol Energy* 2005;**79**(2):131–9.

Phase-change materials

5.1 Thermal mass and latent heat storage

Each element of the building envelope and structure, as well as any other object inside the building with an appreciable mass, is characterized by its thermal capacity C (or thermal mass, J/K in SI units), which measures the quantity of heat required to change its own temperature:

$$C = Q/\Delta T (\text{J/K}) \qquad [5.1]$$

In particular, the heat capacity of the building envelope plays an essential role in determining energy efficiency and internal comfort of buildings, in both summer and winter seasons.

The energy performance of a building envelope is determined not only by its thermal resistance to the passage of heat (R-value) but also by its capacity to store and release heat during the day and night (thermal energy storage [TES]).

During summer, in daytime, good heat capacity in the building envelope allows slow accumulation of external heat (particularly direct solar radiation), followed by attenuated and delayed release of the heat inside the building during the night, when the air temperature is lower and cooling the rooms via natural ventilation is possible (thermal inertia—Fig. 5.1) (Table 5.1).

At the same time, good thermal capacity of the building envelope elements and internal partitions allows them to absorb heat present in interiors, reducing daily peak temperatures and maintaining temperature values closer to those of comfort

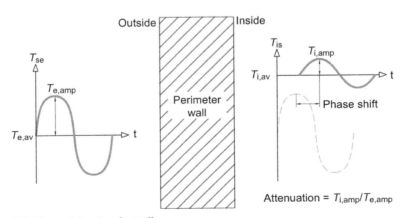

Figure 5.1 Thermal inertia of a wall.

Table 5.1 **Phase shift and attenuation performance of building envelope**[1]

Phase shift P (h)	Attenuation f	Performance
$P > 12$	$f < 0.15$	Best
$10 < P < 12$	$0.15 < f < 0.30$	Good
$8 < P < 10$	$0.30 < f < 0.40$	Sufficient
$6 < P < 8$	$0.40 < f < 0.60$	Mediocre
$P < 6$	$f > 0.60$	Poor

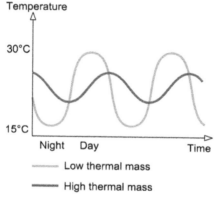

Figure 5.2 Internal temperature fluctuation and thermal mass.

(Fig. 5.2). Furthermore, since thermal comfort also depends on radiant temperature, an adequate heat capacity plays a positive role by keeping the surface temperature of the walls lower than that of air during daily peak times and reducing radiant asymmetry phenomena.

During winter, in daytime, the thermal mass inside the building may contribute to storing heat gain from incoming solar radiation, releasing it in the evening and at night when it is most needed (passive solar systems—Fig. 5.3). Furthermore, high thermal mass can prevent the rapid decrease of temperature in habitation units with nighttime heating setback, or heat loss in areas subject to frequent air changes and without heat recovery devices.

From an energy efficiency point of view, benefits achieved by high thermal mass result in the reduction of heating and cooling energy requirements and air-conditioning peak loads, and in their shift to off-peak electricity periods.

Traditionally, thermal energy storage in the building envelope is achieved by virtue of the change in the internal energy of its materials through sensible heat (sensible thermal energy storage). For this reason, due to the relatively low value of specific heat capacity (J/kg K) of most building materials, a high heat capacity of the building

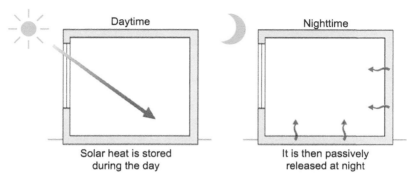

Figure 5.3 Passive solar gain system.

Table 5.2 Specific heat capacity and volume mass of main building materials[1]

Materials	Specific heat capacity (J/kg K)	Volume mass (kg/m³)
Still water (at 293 K)	4190	1000
Wood	1600	850/450
Cellular plastic materials	1450	80/10
Compact plastic materials	2200/900	2000/1050
Still air	1008	1
Natural stone	1000	3000/1500
Plasters and mortars	1000	2000/600
Concrete	1000	2400/1800
Waterproofing materials	1000	2100/1100
Porcelain	840	2300
Bricks	840	2000/600
Glass	750	2500
Metals	880/380	11,300/2700

envelope is normally obtained by resorting to elements of high surface mass (at least 200 kg/m²), using high specific weight materials such as stone, brick, or concrete with great thickness (heavyweight construction) (Table 5.2).

A body thermal capacity C results from its own mass m (kg), and the specific heat c_p of the material it is made of (J/kg K).

$$C = mc_p \qquad [5.2]$$

Therefore with equal specific heat of the material (c_p) and temperature difference (ΔT), a higher mass (m) results in a higher thermal capacity (C) and a higher heat energy stored (Q):

$$Q(J) = C\Delta T = mc_p\Delta T = mc_p(T_f - T_i) \qquad [5.3]$$

where ($T_f - T_i$) refers to the temperature swing.

Today, the positive effects of thermal heat storage in the envelope can be obtained by resorting to advanced building products using emerging smart materials such as phase-change materials (PCMs), able to accumulate and release the external heat through the phenomenon of phase transition while maintaining a constant internal temperature (latent heat storage).

Using PCMs, the ability to attenuate and shift the timing of air temperature fluctuations in a room does not depend on the traditional thermal capacity to accumulate external sensible heat without warming rapidly (sensible heat storage), but on the materials' ability to accumulate the external heat in the form of latent heat to undergo a phase transition (Fig. 5.4). During phase transition, PCMs are able to accumulate thermal energy of 100–250 kJ/kg in the form of latent heat; by comparison, the specific heat capacity of building materials varies from a maximum of 1.5–2 kJ/kg (wood and compact plastics) to a minimum of 0.8–0.3 kJ/kg (bricks and metals).

Overall heat storage capacity of a PCM is thus the sum of both the latent heat enthalpy in the phase transition temperature range and the sensible heat otherwise stored for lower or higher temperatures, expressed as:

$$Q = m\big[(c_p\Delta T)_{\text{sensible}} + (c_p\Delta T)_{\text{latent}}\big] \qquad [5.4]$$

It is thus possible to split heat capacity from body mass to obtain the same beneficial effects with lighter materials and significantly reduced thickness. Ultimately, through the use of PCMs, the building envelope tends to assume some of the characteristics of

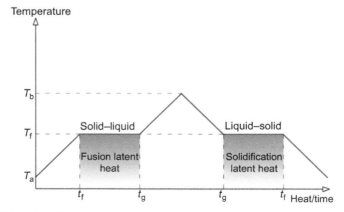

Figure 5.4 Phase transitions diagram.

human skin, whose main thermoregulation mechanism to maintain a constant body temperature is subtraction of heat as latent heat through evapotranspiration and sweating. PCMs exploit the phenomenon of phase transition to absorb or release latent heat without increasing their internal temperature. The energy supplied or subtracted from the material during the phase transition is used to break or form chemical bonds while keeping the kinetic energy of the body constant and hence its temperature stable.

During daytime, when air temperature exceeds a certain value (the so-called operating temperature at which the PCM begins the phase change, typically 23—26°C), the excess heat is used by the PCMs to undergo the phase transition (eg, from solid to liquid state) and thus subtracted from the environment. At night, when the temperature drops below the set point, the PCM returns the accumulated heat to the environment while undergoing the reverse phase transition (liquid—solid) (Fig. 5.5). This way it is possible to damp the temperature oscillations of an indoor environment and maintain a state of comfort. Obviously, for PCMs to perform their function, daily temperature must exceed the melting point to allow the phase transition, then drop below the melting point to start a new cycle the following day. If this is not allowed by environmental temperature conditions, systems containing PCMs whose cooling is induced by forced ventilation or using panel-integrated cooling coils are employed.

PCMs with different operating temperatures can be applied in any building element, enclosure, or partition as additional components or directly integrated in the construction materials, enhancing the energy storage capacity of new or existing building envelopes, especially in lightweight construction, and giving high performance in extremely reduced thickness and mass.

PCMs are also used in cooling systems as cold accumulators (with operating temperatures of 5—18°C), in heating systems as hot accumulators (temperatures around 55—60°C), or as high-temperature accumulators for solar cooling applications (over 80°C). They are also integrated in renewable energy systems, for instance to remove excess heat from photovoltaic (PV) panels or store solar thermal heat for subsequent release.

Thermal energy storage systems are becoming more and more attractive, and have been recognized by several national and international agencies as an effective means to

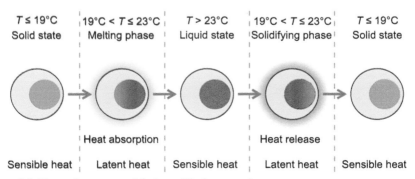

Figure 5.5 Phase-change material charge/discharge cycle.

reduce energy consumption and climate-altering gas emissions. Furthermore, TESs have a high potential in energy sectors, as they allow the shift of peak load demand to off-peak periods, thus leveling the gap between energy supply and demand and easing the load on energy grids. Thermal energy can be stored by means of a change in the internal energy of materials, in the phenomena of thermochemical heat, sensible heat, latent heat, or any combination thereof. Among these, latent heat TES (LHTES) systems utilizing PCMs can contribute to increasing energy performance and fostering higher energy efficiency in new and existing buildings.

5.2 Classification and technical specifications

The phase transition of a material may be through solid—solid, solid—gas, solid—liquid, or liquid—gas transformations.

PCMs used in building construction generally exploit solid—liquid transition, even if it is characterized by a lower exchange of latent heat than that achievable in liquid—gas phase change, as it involves a smaller volume variation resulting in more simplicity in the accumulation management system and greater affordability.

Depending on the type of material used for latent heat accumulation, PCMs are divided into two main categories (Fig. 5.6): organic PCMs, including paraffin waxes (C_nH_{2n+2}), nonparaffin waxes such as fatty acids ($CH_3(CH_2)_{2n}COOH$), and other organic nonparaffinic PCMs like sugar alcohols; and inorganic PCMs, including hydrate salts (M_nH_2O) and metallic materials with low melting temperatures. Eutectic PCMs can belong to both categories, as they consist of mixtures of different substances (organic and/or inorganic) whose overall melting point is lower than that of the individual substances. Both sugar alcohols and metallic materials find little application in buildings due to their unsuitable phase transition range (in the order of hundreds of degrees Celsius).

Each category has its own advantages and disadvantages for use in building construction from the thermal, physical, chemical, environmental, and economic points of view.

Regarding the thermal aspects, in general, for effective use in buildings, PCMs should have a melting point within a particular temperature range, high fusion latent

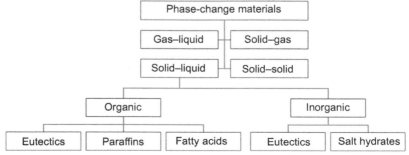

Figure 5.6 Phase-change material classification.[2]

Table 5.3 Phase-change materials: latent heat and fusion points[2]

Material	Fusion temperature (°C)	Latent heat (kJ/kg)
Paraffin	6–76	170–269
Nonparaffin	8–127	86–259
Fatty acids	17–102	146–242
Salt hydrates	14–117	68–296
Eutectics	15–82	95–218
Water	0	333

heat per unit mass (Table 5.3), high specific heat capacity to provide additional sensible heat storage, good thermal conductivity, fusion congruence, and limited change in volumetric capacity and vapor pressure of the material with regard to the operating temperature.

From a physical point of view, PCMs should be characterized by high density, reduced volume changes during phase transitions, sufficiently high crystallization rates, and limited, if any, supercooling phenomenon.

From a chemical point of view, PCMs should be characterized by high long-term chemical and thermal stability at the operating temperatures, and be compatible with building and containment materials. Moreover, they should be nontoxic, nonflammable, and nonexplosive.

From an environmental point of view, PCMs should have low impact throughout the entire life cycle, from production to decommissioning.

From an economic point of view, they should be characterized by wide availability on the market, easy procurement, and low cost.

There is not, today, a material with all the above characteristics. The most used materials are, as mentioned, paraffin (organic materials), hydrated salts (inorganic materials), and fatty acids (organic materials). For applications in cold accumulators, ice is also widely used.

5.2.1 Inorganic phase-change materials

Salt hydrates are a category of inorganic salts which contain one or multiple water molecules such that the resulting crystalline solid has a chemical formula of $AB \cdot nH_2O$. They are among the cheapest PCMs, with average material cost ranging from $0.13 to $0.46/kg for calcium chloride $CaCl_2 \cdot 6H_2O$.[3] Furthermore, they show good latent heat values and high thermal conductivity due to their higher density, combine a wide melting temperature range of 5–130°C with a fixed value for phase transition temperature, are not flammable, and are biodegradable and recyclable.

However, hydrated salts need careful preparation and additives for long-term stability, undergo considerable volume change during phase transition, can be corrosive to certain metals, and are prone to incongruent melting, phase segregation, and subcooling phenomena. Most of these issues have been addressed by research in the last decades, so today they are much more under control.

Incongruent melting occurs when water released in the crystallization process is insufficient to dissolve all the solid-state material present. Because of their higher density, less hydrated salts settle in the lower part of the accumulation, making the phase transition increasingly irreversible over time. This can be avoided by adding water to material to dissolve the entire anhydrous salt during melting (extra water principle), at the cost of some heat storage capacity loss.[4] Another strategy is thickening the material to gel form.

Subcooling is caused by the rate of nucleation of hydrated salt crystals, which is generally very low. This means that the latent heat, instead of being released at the melting temperature, is released at a lower temperature. Subcooling can be reduced by adding nucleating seed agents, with the added benefit of increasing the usually slow crystallization rate which affects most salt hydrates.

Phase segregation may be minimized by adding thickening materials, and the increased density also prevents nucleating agents from settling and losing effectiveness. Thickening agents include superabsorbent polymers, at 3–5% weight, or 2–4% weight of carboxyl methyl cellulose, depending on different hydrated salts. Subcooling in thickened hydrated salts can be further reduced by the addition of copper, carbon, or titanium oxide powder.[5]

Sodium sulfate decahydrate $Na_2SO_4 10H_2O$, commonly called Glauber's salt after its discoverer, Johann Rudolf Glauber, is the most widely used inorganic PCM and was the first to be employed in building applications; it has a melting point of 32.4°C with a high latent heat of 254 kJ/kg. Its use has been traditionally marred by extensive subcooling, phase segregation, and incongruent melting phenomena, but each problem has been successfully solved. Borax addition brought subcooling from 15 to 3–4°C, while extra water prevented incongruent melting.[3] Bentonite clay proved an effective and cheap thickening agent to avoid phase segregation.

The main manufacturers of hydrated salts include Rubitherm GmbH, Merck KGaA, Climator Sweden AB, Cristopia Energy Systems, PCM Energy, Mitsubishi Chemical, and EPS Ltd.

5.2.2 Organic phase-change materials

Organic PCMs are typically more expensive than inorganic ones: most paraffin waxes are byproducts from oil refineries and therefore available in abundant supply but at a relatively high price ($1.88–2.00/kg); fatty acid PCMs include stearic acid ($1.43–1.56/kg), palmitic acid ($1.61–1.72/kg), and oleic acid ($1.67–1.76/kg).[3]

Organic PCMs have a lower latent heat value and often a larger fusion range, can cause reactions with the concrete, and may be combustible (the latter problem can be solved through the use of appropriate containers).

Thermal conductivity is also inferior to hydrated salts, limiting heat exchange, and charging and discharging rates: heat transfer enhancement techniques include dispersion of high-conductivity particles (copper, aluminum, graphite, etc.) in the PCM, its impregnation in high-conductivity porous materials (graphite matrix or metallic foams), or the insertion of fibrous materials such as carbon fibers or nanofibers.[6]

On the other hand, they have the advantages of being simple to use, chemically stable, noncorrosive, having a more limited volume change during the phase transition, being free from undercooling phenomena and nucleation agents, and being recyclable.

Moreover, some organic PCMs such as fatty acids are obtainable from renewable or reused raw materials like animal fat and vegetation such as beef tallow, lard, palm oil, coconut, and soybean (biobased PCMs). Advantages of fatty acids include congruent melting and cooling, high latent heat of fusion, low cost, fire resistance, nontoxicity, very little subcooling and volume change, good chemical and thermal stability after a large number of thermal cycles, and rapid degradation when disposed of, reducing end-of-life environmental impact. Furthermore, melting temperatures can be freely adjusted by selecting the right combination of eutectic binary mixtures of fatty acids. Finally, fat- and oil-based PCM chemicals can be only half as expensive as paraffin PCM products currently produced from crude oil, and their manufacturing process releases far fewer CO_2 emissions.

However, their major drawback resides in their limited thermal conductivity, which reduces heat storage and release rates during phase transitions. One possible solution to this issue is the integration in the compound of nanomaterials such as carbon nanotubes of exfoliated graphite nanoplatelets, though they may reduce overall latent heat and lead to excessive supercooling.[7] Due to their acidic nature, fatty acids are also more aggressive than paraffin toward the surrounding environment.

Eutectic PCMs have a higher density than organics and give a compound with a single melting temperature. However, data on their thermal and physical properties is still limited due to their recent introduction.

Finally, both organic and inorganic materials are already taking advantage of the application of nanotechnology in the PCM sector, with more promising developments under way. Incorporation of nanomaterials can greatly improve several thermal storage characteristics, such as thermal conductivity (especially for inherently worse performing organic PCMs), freezing/melting rates, and thermal stability, even with limited nanomaterial additions. Thus more and more PCMs are becoming suitable for effective LHTES applications.

Organic PCMs benefit from the addition of nanomaterials such as copper, silver, or graphene nanoparticles or graphite nanofibers to enhance thermal conductivity and phase transition rates, at the cost of some loss of latent enthalpy (for instance, a 2% addition of graphene nanoparticles increases conductivity by 63% but reduces phase-change enthalpy by 8.7%).[8]

Regarding hydrated salts, nanoparticle addition is mainly researched as a method for addressing subcooling phenomena, but may have positive benefits also on other thermal storage performance factors: in the case of barium chloride, a 1.13% addition of titanium dioxide (TiO_2) nanoparticles results in an increase of thermal conductivity of 15.65%, with improved freezing and melting rates, nucleation effects, and overall heat transfer.[8]

Overall, nanomaterial application to PCMs can be beneficial in terms of enhancing thermal storage characteristics, achieving the production of materials with desired thermal, physical, and chemical properties, and reducing environmental impact. However, as with all technology breakthroughs, several challenges arise, such as the correct and safe disposal or discharging of synthetized nanomaterials and related byproducts, the hazardous nature of some chemical constituents needed for nanomaterial preparation, issues of nanoparticle agglomeration and aggregation within PCM suspensions, safety concerns regarding human contact with nanoparticles, and the lack of standardization in characterizing performance improvement of nano-embedded PCMs.[8]

5.3 Packaging and encapsulation methods

PCMs have been tested as a thermal mass component in buildings for at least six decades. The first experimental applications provided for PCM integration into building elements by direct incorporation or immersion.

In direct incorporation, the simplest and most economical method, liquid or powdered PCM was directly mixed with construction material blends such as concrete, plaster, or gypsum in the production phase, requiring no additional equipment.[8] Convenience was marred by PCM leakage issues, however, and undesired chemical interaction with the containment material, leading to its corrosion and deterioration over time.

In the immersion method, porous construction elements, such as gypsum boards, bricks or concrete blocks, were soaked in the melted PCM, absorbing it by capillary action. Again, the same problems of long-term leakage and incompatibility with building materials occurred.

Nowadays the most widely used and effective method is PCM encapsulation. This consists of inserting a PCM inside specific hermetic packages to contain the material in its liquid and solid states, prevent changes in its chemical composition, avoid interactions with the environment, increase compatibility with surrounding materials, and improve handling and reduce possible variations of the external volume. Encapsulation methods greatly influence final PCM product cost.

There are two types of encapsulation: macroencapsulation and microencapsulation.

5.3.1 Macroencapsulation

Macroencapsulation is carried out by enclosing PCMs in containers such as tubes, spheres, panels, and other shells which can act directly as heat exchangers or in turn be incorporated into other elements (Fig. 5.7). It is the most used form of encapsulation and involves containers larger than 1 cm.

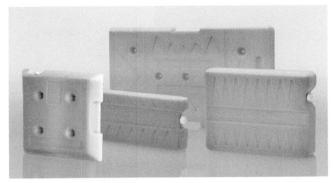

Figure 5.7 Macroencapsulated phase-change material.
Courtesy of va-Q-tec AG.

The material, shape, and size of the encapsulation package greatly influence PCM performance, affecting both thermal conductivity and the time lapse of the melting/solidification process.

Container shapes suitable for maximizing surface area per unit volume of storage material include flat-plate packages and cylindrical pipes, and radiating fins can be added to enhance thermal exchange with surrounding air or water. Encapsulation also helps to overcome safety issues such as paraffin flammability, which is can be addressed by fire-retardant pouches or aluminum foil. Larger containers may lead to nonhomogeneous phase transition, however, such as PCM freezing around the edges and slowing solidification of the inner regions, impeding stored heat release. Solidification time also increases proportionally to the square of the capsule radius.[9]

In addition, the more PCM is stored, the bigger becomes the issue of accommodating the expansion of the PCM upon phase transition, in particular with hydrate salts characterized by a greater volume change in the freezing/melting process. High melting temperatures and volume expansions also induce pressure buildup of air contained within the capsule. These issues can be addressed by depositing a sacrificial polymer layer between the envelope and the core PCM to provide a void to account for subsequent volume expansion. Other techniques involve flexible and selectively permeable polymer shells able to cope with the volume expansion and allow air diffusion, with mechanical resistance provided by a second metal shell added in the expanded state.[10]

Ordinary macroencapsulation packaging contributes around 20–35% of the PCM total cost, with some cheaper methods, such as sachet packaging akin to that found in food processing, accounting for 9–18%.[3]

5.3.2 Microencapsulation

The microencapsulation process provides for individual particles or droplets of solid or liquid PCM, constituting the core, to be surrounded or coated with a continuous film of polymeric material with high molecular weight (the shell), producing microcapsules in

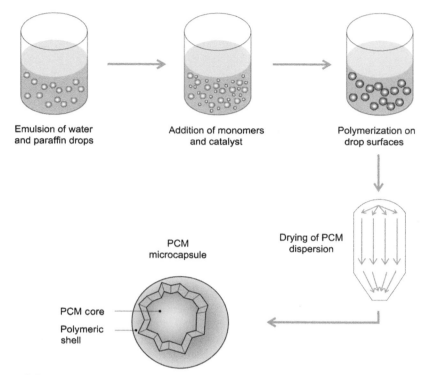

Figure 5.8 Phase-change material (PCM) microencapsulation process.

the micrometer to millimeter range (Fig. 5.8). These may be spherically shaped, with a continuous wall surrounding the core, or have more asymmetrical and variable shapes, with smaller droplets of core material embedded throughout the microcapsule. Solid, liquid, and gas states of matter may be microencapsulated, allowing liquid and gaseous materials to be easily handled as solids, as well as affording some measure of protection in handling hazardous materials.

An important factor that has to be controlled in this technique is the thickness of the polymeric capsule containing the PCM, which must be thick enough to sustain the PCM inside the capsule even after many thermal cycles, yet thin enough not to impede thermal exchange. Excessive thickness of the shell also reduces PCM mass ratio in the final product, which ranges from 85—90 to 50% according to particle size and manufacturing process. Furthermore, the coating material must not exhibit chemical interaction with the matrix material in which microcapsules are to be incorporated (Fig. 5.9).

Microcapsules, due to their size, are virtually indestructible, extremely safe from the risk of releasing harmful substances, and feature a remarkable heat exchange surface (1 g of encapsulated product has a specific surface of about 30 m^2).

The most widely used microencapsulation methods are reported in Table 5.4.

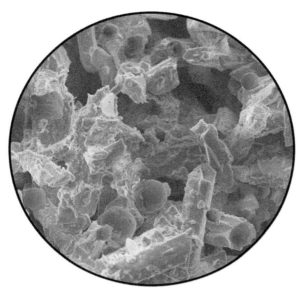

Figure 5.9 Microencapsulated phase-change material embedded in gypsum matrix. Courtesy of Datum Phase Change Ltd.

Table 5.4 Main phase-change material microencapsulation methods[11]

Physical methods	Chemical methods
Pan coating	Interfacial polymerization
Air-suspension coating	In situ polymerization
Centrifugal extrusion	Matrix polymerization
Vibrational nozzle	
Spray drying	

Microencapsulation appears to have considerable development scope, although at the moment it can only be applied to hydrophobic PCMs or materials that do not absorb or retain water on the surface, such as paraffin.[12]

Microencapsulation of salt hydrates is challenging due to their hydrophilic nature and inherent water fraction. A compatible shell material should act as a water/vapor barrier and be made of hydrophobic materials to prevent the salt hydrate from evaporating. Furthermore, organic polymers typically used as shell material in the encapsulation process are chemically incompatible with salt hydrates. Recent successful salt hydrate microencapsulation includes the method devised by Fraunhofer ISC, which injects a melted salt hydrate drop into a ceramic–organic hybrid polymer based on chemical nanotechnology known as ORMOCER, currently used in dental reconstruction.[5] The polymer forms a hard encapsulating layer around the drop and is then dried at ambient condition, while the salt hydrate drop freezes as it cools down. These microcapsules range from a few micrometers to a few millimeters in diameter. Currently the prototype is addressing water permeability and stability issues. Other techniques encapsulate Na_2CO_3 hydrate particles with an inorganic SiO_x layer via a sol–gel process in a tetramethyl orthosilicate solution.[5]

Microencapsulation also helps to improve the stability of the fusion–solidification cycle, since phase separation is confined to microscopic distances. It is, however, a more expensive process than macroencapsulation, accounting for around 45–65% of the final PCM cost.[3]

Microcapsules come in powder and slurry form and can be used as such, as thermal vector fluids, or be included in the mix of building products such as plaster, screed, concrete, gypsum, wood products such as medium density fibreboards (MDF) and oriented strand boards (OSB), or even in acrylic paints. Heat thermal storage is greatly improved, but for load-bearing materials some loss of mechanical strength is to be expected.

Among microencapsulated PCMs on the market, BASF Micronal (Figs. 5.10 and 5.11) is particularly interesting; it is made out of globules of highly refined formaldehyde-free paraffin wax ($\emptyset = 2-20$ μm) coated with an extremely rigid polymer layer (polymethyl methacrylate), and can be integrated in many construction products. The special coating protects the PCM inside, keeping it in a pure state and guaranteeing unchanged performance for a period of over 30 years (300 cycles/year for a total of 10,000 cycles). The very small size of the capsules (diameter of approximately 1/500 mm) also protects the material from any risk of breakage by impact or crushing.

According to the need, BASF Micronal is supplied as a liquid dispersion or powder with three different operating temperatures and latent heat values: 26°C and 110 kJ/kg for overheating protection during summer; 23°C and 100 kJ/kg for internal

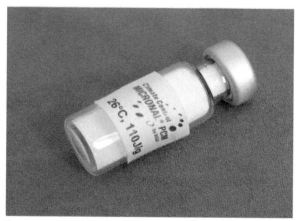

Figure 5.10 BASF Micronal microencapsulated phase-change material. Photo by the author.

Figure 5.11 BASF Micronal microencapsulated phase-change material. Photo by the author.

temperature stabilization in the comfort zone; and 21°C and 90 kJ/kg for use in surface cooling systems.

Microtek PCM microcapsules instead employ paraffins and eutectic mixtures to provide a switching temperature range from −30 to 43°C for most uses, with latent heat values ranging from 100 to 200 kJ/kg. Particle size is in the 17−20 μm range.

Yet another method of encapsulation resorts to the so-called shape-stabilized PCMs, in which a supporting material, such as high-density polyethylene (HDPE), is compounded with a PCM like paraffin wax. These two materials are finely mixed together while both in the liquid form, then cooled until solidification. The resulting mixture can keep its shape and mechanical resistance as long as the temperature is below the supporting polymer melting point, regardless of whether the embedded PCM is in liquid or solid phase, with good long-term performance expected over multiple thermal cycles. Support materials can also compound several substances with different roles, such as HDPE to give rigidity to the mix, styrene-butadiene-styrene to absorb paraffin wax in the liquid state, and other additives to enhance the otherwise low thermal conductivity—up to 53% using exfoliated graphite. The proportion of PCM mass can be up to 80%, thus total energy storage is comparable to that of traditional bulk PCMs. Applications are being tested in building walls, ceilings, and floors, and research is being done into solid−solid phase-change compounds.[13]

5.4 Functional model and building design

Inside the buildings, PCM-based LHTESs have different applications depending on building type and energy and microclimate objectives, and can be used in both new constructions and retrofit interventions. PCMs can both increase energy performance of the building envelope and improve the efficiency of systems and equipment.[5,8,14,15]

Concerning the building envelope, PCMs can be used in both passive and active systems.

Passive applications do not need additional mechanical devices or systems. Heat or cold are absorbed or released when the temperature reaches the PCM set point, without any user control. Passive systems may exploit solar gain as well as other internal thermal gains.

In active PCM applications, heat charging and discharging is achieved with the help of mechanical equipment such as forced air or water circulation via fans or opening fins; and active applications can be further classified according to the thermal energy exchange medium, such as water, air, or both. Some support strategies that take advantage of the environment conditions and employ low consumption devices are summer night forced air ventilation and the use of night radiative cooling. While active systems may indeed improve PCM performance, their application should be carefully evaluated as it is only justified if the benefits generated by the use of PCMs are greater than equipment costs.

As regards systems and equipment, PCMs can be used primarily as heat exchangers or be directly integrated in renewable energy systems. Examples of PCM integration in heating, ventilation, and air-conditioning (HVAC) systems include PCM heat exchangers used for free cooling the building, or combining solar heat pump systems with encapsulated PCMs.

Direct integration of PCMs in solar thermal and PV systems is also promising: solar air and water heaters and water storage tanks may be integrated with PCMs to improve their thermal stability, while PCM slurry may be used as a thermal exchange fluid to extract heat from building-integrated photovoltaics/thermal systems and use it to supply space heating, hot water, and ventilation systems.[8,16]

PCMs can be used in buildings as components, or be integrated into construction materials or storage units.

When the PCM element is inserted as an additional layer or part of a construction section, it is classified as a "component." When the PCM instead gives additional thermal storage properties to an established construction element, making it effectively multifunctional, it is classified as "integrated." The latter can be achieved by incorporating PCM into the construction material (eg, by adding PCM to cement or gypsum blends). These two categories may be used both in passive and active applications.

PCMs in storage units can only be applied to active systems; they are mostly integrated with HVAC or domestic hot-water systems, and are generally thermally insulated from air-conditioned spaces. Their functioning is not strictly dependent on meteorological changes as their thermal energy is transferred through forced fluid circulation, thus they have greater location flexibility compared to passive applications.

While the principle of PCM use is relatively simple, evaluating the effective contribution of additional latent heat capacity to improving the energy performance of the whole building is challenging; an efficient PCM application requires the identification of adequate support strategies, along with detailed dynamic simulations of the thermal behavior of the building in the expected conditions of use.

PCM behavior depends on multiple parameters, including climatic conditions such as solar radiation and ambient temperature, indoor thermal loads, building use and characteristics, design choices such as the presence of shading elements, and the type and amount of PCM chosen and its configuration, orientation, and location within the building. In particular, the main parameters to identify to select the appropriate PCM are phase transition temperature range, latent heat storage, and thermal conductivity.

The approach to PCM application should also be different depending on whether the main objective is to reduce the summer cooling loads or the winter heating loads.

For heating periods, the most important support strategy is taking advantage of solar direct gain: in this case, correct selection of switch temperature, location, and finishing material characteristics is paramount. For example, in floor applications, PCMs are most effective when directly exposed to solar radiation or the heated space, otherwise there will be a reduction in the quantity of heat transferred and a delay in the heat transfer process. The color and thermal conductivity of the finishing material also play an important role, as they affect the quantity of thermal energy received and the rate at which it is exchanged.

For cooling periods, adequate solar protection and nighttime ventilation are the most basic strategies. In climates characterized by significant difference between daytime and nighttime temperatures, night ventilation is a useful way to ensure full PCM efficiency, whether by opening windows, using ceiling fans, or other methods.[13,15] This strategy requires adequate thermal storage capacity, a correct switching temperature, and the guarantee that heat absorbed and stored during the day is released at night through ventilation, which can be achieved by design choices such as using an atrium, or resorting to active systems like mechanical ventilation, night radiative cooling, or evaporative cooling systems.

In both summer and winter use, switch temperature is an important determinant factor in the performance of PCM applications. An inappropriate phase-change temperature may hinder the effectiveness of the application or even make it completely useless. As stated, the choice of the switching temperature mainly depends on whether the main goal of the application is to prevent overheating or to save heating energy.[15]

PCMs that melt below 15°C are usually restricted to storing coolness in air-conditioning applications, while materials that melt above 90°C are used for absorption refrigeration. All PCMs with switching temperatures between 15 and 90°C can be applied in solar heating and for heat load leveling applications,[15] although most PCM applications have switch temperatures in the interior comfort range of 21.7–27.0°C. Moreover, different switching temperatures work best in different months of the year, so global effectiveness should be considered taking into account even midseason PCM behavior.

In fact, a PCM that performs well in heating periods may have a marginal effect in cooling periods, if any, and vice versa. To improve PCM performance in variable temperature conditions, authors have proposed the use of multiple PCMs with different switching temperatures in the same storage unit, such as a rooftop assembly with two PCM layers of different melting temperatures for good performance at any time of day and in various weather conditions.[17,18]

In addition to identifying the right switching temperature for the application, PCM quantity and thermal conductivity must be carefully evaluated, requiring an optimization process aimed at increasing the storage/release capacity using as little PCM as possible.

As noted, most energy stored in a complete daily cycle comes from latent heat absorbed by the PCM volume while the PCM is melting; after that, only sensible heat is absorbed, with a much smaller thermal capacity. Likewise, released energy is mostly proportional to the solidified volume of the PCM during the discharging period. While increasing PCM quantity may seem the easiest way to improve its effectiveness, it dynamic behavior may actually be negatively affected by excessive mass. Most PCMs have a low thermal conductivity, giving the inherent disadvantage of slowing heat transfer during the charging and discharging processes the more mass is added.

For example, considering a PCM building envelope application intended to take advantage of solar thermal energy for winter heating, the volume of melted PCM depends on the extent of the sunshine period and amount of solar radiation. In this case, overestimating the PCM mass may actually increase the time needed for heat to penetrate the PCM, making it longer than the sunshine period and preventing the melting process from being completed in the day cycle. Likewise, time needed for stored heat to be released indoors could become longer than the available discharge period, leading to an incomplete solidification process. Similar considerations apply to the use of PCMs to avoid overheating during summer.[5]

As suboptimal thermal storage is achieved when the PCM neither totally solidifies nor melts, its volume should be selected such that all PCM mass can be melted and frozen completely during a phase-change daily cycle.

In this picture, it is apparent that to understand the buildings' temperature response characteristics better and the economic feasibility of using PCMs, appropriate dynamic numerical simulation models and software to assess and optimize PCMs included in building walls, floors, roofs, or HVAC systems are necessary.

Nowadays there are several established models to analyze single PCM products, PCM-enhanced building envelopes, and whole buildings with PCM systems. The main building energy simulation tools for assessing dynamic energy performance include BLAST, BSim, CLim2000, DeST, DOE, ECOTECT, Ener-Win, Energy Express, Energy-10, EnergyPlus, eQUEST, ESP-r, IDA-ICE, IES, HAP, HVACsim, HEED, PowerDomus, RADCOOL, SUNREL, Tas, TRACE, TRNSYS, and WUFI. Only a few of those are able to simulate and analyze phase transition processes fully, usually by adding the phase-change effect to the energy balance equation as a latent heat generation effect modeled using temperature-dependent specific heat.[5,14]

In particular, EnergyPlus, ESP-r, and TRNSYS are recommended for their versatility and reliability.[14]

EnergyPlus is a widespread, free, open-source energy analysis and simulation program funded by the US Department of Energy and managed by the National Renewable Energy Laboratory. In addition to heating, cooling, lighting, ventilation, and other buildings energy flow models, it includes variable time steps and integration with heat and mass balance-based zone simulation, multizone airflow, thermal comfort, and natural ventilation. The EnergyPlus PCM module uses an implicit conduction finite-difference solution algorithm, which includes both phase-change enthalpy and temperature-dependent thermal conductivity.

ESP-r is another open-source advanced building energy simulation tool allowing detailed thermal and optical descriptions of buildings. The problem domain is discretized in control volume schemes, each with the relative conservation equation for mass, momentum, energy, etc. The effect of several factors, including weather, external shading, occupancy gains, HVAC systems, and LHTES, can be integrated. ESP-r results were reliable in a variety of PCM-related assessments and simulations.

The TRNSYS software specializes in dynamic simulation of transient systems based on transfer functions technique. Its modular nature includes multiple models and subroutines for various systems, including buildings. Energy simulation of buildings based on the storage and release effects of PCMs can be simulated using the active layer tool or by adding new modules representing the PCM application, such as PCM wallboards, external walls, or active ceiling panels.

However, the complex functioning of PCM application in buildings and the multiplicity of factors affecting their performance makes further research, and in particular long-term monitoring, irreplaceable to quantify the specific benefits of each application, notwithstanding positive results in some studies.

From a regulatory point of view, thermal regulation codes worldwide are still lacking any integration with PCM construction solutions. This issue must be addressed by developing standardized methodologies to take into account the latent heat loads from PCM phase-change processes in building projects, in particular for the many cases in which dynamic energy simulation throughout the year is not mandatory.

5.5 Building applications and products

Over the last decade many experimental, innovative, and new PCM systems and products have been used in both new and existing residential and service buildings. The possibilities provided by PCM microencapsulation and composite materials have facilitated the integration of latent thermal storage in buildings. PCM products can be used in different ways, integrated in construction materials or as additional building components in walls, ceilings, floors, glazing, and HVAC systems.

In particular, PCMs are integrated in construction materials including concrete and ready plaster blends, drywalls (gypsum boards, plaster panels) and composite panels, ceiling and floor tiles (microencapsulated or composite boards), insulation materials (microencapsulated, shape-stabilized, macroencapsulated), concrete bricks, structural concrete, and windows and glass bricks. Building components containing PCMs include pouched mats, panels, rusticated mats, bags, and blocks.

The use of PCMs in buildings may be desirable in new construction and retrofit interventions to improve the energy performance and internal comfort.

In new interventions, PCMs can be used:

- in lightweight constructions to provide adequate thermal mass of perimeter walls and internal partitions;
- in both lightweight and heavyweight constructions as a thermal storage medium in passive heating and cooling systems;

- in HVAC and renewable energy systems to improve energy efficiency by the integration of thermal energy storage.

In retrofit interventions, PCM application is viable in the following cases:

- in lightweight constructions to increase thermal mass of perimeter walls and internal partitions;
- in heavyweight constructions when interior insulation is added;
- in HVAC and renewable energy systems to improve energy efficiency by the integration of thermal energy storage.

5.5.1 Inner lining of walls

Many authors suggest that the best location for passive PCM applications is within the living spaces in walls and partitions.[5,8,14,15,19] PCMs can be applied as integrated finishing materials such as plaster, drywalls, or gypsum panels, insulation materials, or as separate components to add to the other wall layers (Fig. 5.12).

Among integrated products, drywall panels are one of the most common interior finishing materials, and microencapsulated PCMs have the advantages of easy application, proper heat transfer, and no need for protection against encapsulation destruction (Fig. 5.13). On the market are microencapsulated PCMs with switch temperatures of 21, 23, or 26°C. The maximum weight ratio of microencapsulated PCMs that can be incorporated in drywall panels is 30%.[5] Their efficiency depends on several factors: PCM incorporation method, wall orientation, climatic conditions, direct solar gains, internal gains, surface color, ventilation rate, the chosen PCM, temperature range over which phase change occurs, latent heat capacity per unit area of wall, etc.

Most drywall products available on the market (Table 5.5) employ BASF Micronal microencapsulated paraffin PCM. Such is the case of Knauf Comfortboard, which comes in a 125 × 200 cm size with 12.5 mm thickness for an 11 kg/m² surface mass. Thanks to PCM addition to the gypsum core, heat storage capacity is increased

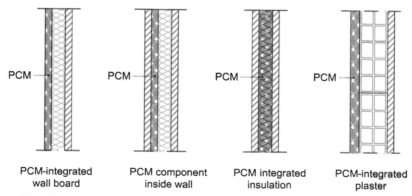

Figure 5.12 Phase-change material (PCM) integration in inner walls.

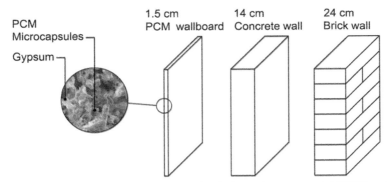

Figure 5.13 Gypsum board with microencapsulated phase-change material (PCM) and equivalent thermal capacity.

from 13 to 200 kJ/m² K at the phase-change temperature of 23°C, totaling 460 kJ/m² over the whole 10–30°C reference temperature range. National Gypsum ThermalCore also employs BASF Micronal with a 23°C switching temperature, achieving around 250 kJ/m² latent heat enthalpy. Its PCM-enhanced, mold-resistant gypsum core is wrapped in a fiberglass mat to improve fire resistance.

Better performance is provided by thicker and heavier wall slabs, like Saint-Gobain Rigips Alba Balance or Datum Phase Change ThermaCool Panels. The ThermaCool Panel is 100 × 60 cm, 25 mm thick, with 25 kg/m² surface mass, and achieves a total heat capacity of 760 kJ/m². Phase-change temperature range is 19–23°C, and fire rating is B-s2,d0, or nonflammable material.

Alba Balance boards come in two models with 23 and 25°C phase-change temperature, both 100 × 50 cm, 25 mm thick, and with 23 kg/m² surface mass. Specific heat is 26.7 kJ/m² K, which rises to 82 and 85 kJ/m² K for 23 and 25°C melting temperatures, totaling a respective heat capacity of 825 and 840 kJ/m² in the 10–30°C range. Both are fireproof with an A2-s1,d0 fire rating.

PCM-enhanced clay boards are available from EBB with 20% BASF Micronal content (3 kg/m² over a total surface mass of 15 kg/m²), achieving a 110 W h/m² heat capacity over the phase-change temperature range (set point 23 or 25°C according to the model specification).

All of these can be installed exactly as a normal wallboard, and can be cut, sewn, or perforated without any counterindication, although chiseling is to be avoided.

Another application concerns the use of premixed plaster with added microencapsulated PCM. Several products are available as ready-to-use blends, with thermal performance understandably dependent on application thickness, thus allowing a degree of freedom in tuning required energy storage capacity. Care must be taken to ensure all PCM can be engaged in the charging/discharging cycle. Therefore, since plasters have suboptimal thermal conductivity (0.3/0.5 W/m K range), excessively thick applications may prove ineffective. A premixed plaster with 30% microencapsulated PCM has a capacity to absorb, at its operating temperature (melting/solidification range),

Table 5.5 Main phase-change material (PCM) products for inner lining of walls

Product	Application	Composition	Total heat capacity (10–30°C)	Phase-change temperature (°C)	Fire reaction
Datum Phase Change ThermaCool Panel	Wall slab	Gypsum with BASF micronal	760 kJ/m^2	19–23	B-s2,d0
KnaufComfortboard	Wallboard	Gypsum with BASF micronal	460 kJ/m^2	23	n.d.[a]
National Gypsum ThermalCore	Wallboard	Gypsum with BASF micronal	250 kJ/m^2 latent heat only	23	ASTM Class B
Rigips Alba Balance 23	Wall slab	Gypsum with BASF micronal	825 kJ/m^2	23	A2-s1,d0
Rigips Alba Balance 25	Wall slab	Gypsum with BASF micronal	840 kJ/m^2	25	A2-s1,d0
EBB PCM Clay Board 23	Wallboard	Clay with 20% BASF micronal	396 kJ/m^2 over phase change	23	n.d.[a]
EBB PCM Clay Board 25	Wallboard	Clay with 20% BASF micronal	396 kJ/m^2 over phase change	25	n.d.[a]

[a]Non disclosed.

a quantity of heat 4.5 times higher than that of a traditional plaster (3 cm equal to 8 cm concrete or 14 cm gypsum).

Weber.Mur Clima is a PCM-enhanced plaster manufactured by Saint-Gobain. It comes in Clima 23 and Clima 26 specifications, respectively, with a 21−23 and 23−26°C phase-change temperature range. It is composed of gypsum, BASF Micronal PCM, mineral lightweight aggregates, and additives. The PCM component has an enthalpy of fusion of 100 kJ/kg, but the whole product is much lower at 18 kJ/kg, enough to achieve a thermal capacity of over 160 kJ/cm of application.

Another PCM plaster is Klima 544 TYNK, manufactured by Polish CSV. It is a pre-mixed blend of cement, lime, and mineral aggregates with 20% content of microencapsulated paraffin wax (BASF Micronal). Phase-change temperature is 26°C, and performance is 100 kJ/kg fusion enthalpy for the PCM and 18 kJ/kg for the plaster.

Other products include Scherff Clima-Akustikputz PCM and Maxit Clima 26.

In contrast, products such as ENERCIEL coating manufactured by Winco Technologies are engineered for thin applications of few millimeters. The coating contains a 50% proportion of microencapsulated, biobased PCM, manufactured by MCI Technology, with 184 kJ/kg enthalpy. A 3 mm layer is sufficient to achieve a thermal capacity of 184 kJ/m^2, further improvable by increasing thickness—for example, a 6 mm application has a 368 kJ/m^2 thermal capacity. Heat absorption range is 23−26°C, releasing heat at 18−22°C upon solidification.

PCM components usually provide higher PCM content and thus higher thermal performance, but require a finishing and protection layer which partly impedes thermal exchange with interior spaces. Available products come in panels, pouches, or rusticated mats, and employ organic as well as inorganic PCMs in macroencapsulated form.

Energain PCM panels, manufactured by DuPont, come as 5 mm thick slabs 1.2 × 1.0 m, with 4.5 kg/m^2 area mass (Fig. 5.14). They are composed of a 40%

Figure 5.14 DuPont Energain phase-change material panel.
Photo by the author.

ethylene polymer, 60% paraffin wax blend sandwiched between two 100 μm aluminum foils to provide protection against fire and wax migration and stiffening for easier installation. The paraffin melting phase is 18–22°C, peaking at 21.7°C. Overall thermal capacity is 630 kJ/m^2, further increasable by adding more panels on top of each other at the cost of a slower heat accumulation/release rate. Panels can be nailed on the wall studs over the insulation layer and cut to size or accommodate openings, as long as the seams are sealed with aluminum tape.

EBB PCM-Natural Fibre-Elements are biobased, 100% recyclable composite panels composed of 2 mm thick top and bottom layers made of wood fiber enclosing a core paper honeycomb mesh containing a mix of loam and PCM. The maximum board size is 110 × 300 cm with 1.6 cm thickness.

Products from US-based Phase Change Solutions employ a proprietary organic PCM called BioPCM, manufactured from nonfood palm oil byproducts and other plant materials, including coconut and soy. BioPCM has a thermal storage capacity of approximately 190 kJ/kg and is available in a variety of melting points: 22.8, 24.4, 26.1, or 29°C. It is available in multicouched flexible mats (BioPCmat) or in rigid shells made of highly conductive aluminum (ThermaStix). The BioPCmat product line comes in 1.22 and 2.44 m long, 42 cm wide mats. Installation is easy, as the mats are simply stapled to the studs over the standard insulation before installing the finishing layer, and can be cut to size as needed. Different thermal mass weights are available: 306, 578, or 1032 kJ/m^2 heat storage capacity.

Finally, the Dörken DELTA-COOL PCM product range employs hydrated salts and offers various forms of encapsulation: flexible pouches, translucent and opaque panels, spheres, boards, and dimple sheets. Most suitable for interior application is DELTA-COOL 24, which has a melting temperature range of 22–28°C and crystallization temperature of 22°C, for a melting energy of 158 kJ/kg and a cooling performance of 25–40 W/m^2 for a 4.0 kg board. Also available are DELTA-COOL 21 and DELTA-COOL 28, with different melting temperatures.

5.5.2 Heavyweight construction buildings with added interior insulation

PCMs can prove particularly useful in cases of energy retrofit of existing walls by using interior insulation systems.[2]

It is possible, for example, to intervene in existing buildings by applying internal insulation with a superimposed PCM lining, thus reducing dispersion and at the same time maintaining good thermal inertia to support the air-conditioning system. Application of thermal insulation panels inside a building to reduce energy loss during winter, especially in energy upgrading of historic buildings where working with an external insulation system is not possible, often excludes the preexisting thermal mass. This results in a quicker drop of the internal temperature during winter after turning the heating system off, and possible overheating of the interiors during summer because of the inability to dispose of excess heat through the mass of the walls. The application of a PCM lining above the insulation would overcome these drawbacks.

Furthermore, by combining PCM board with aerogel insulation (Figs. 5.15 and 5.16), it is possible to obtain a insulated double wall with minimum thickness (2–3 cm), capable of reducing energy loss, supporting heating systems, minimizing air temperature fluctuations, and reducing overheating during summer. A solution with comparable performance using traditional materials would demand an increase in wall thickness of up to 30 cm.

Integration of PCMs and insulation materials is a fertile field of research, and there are future prospects for PCM-enhanced plastic foams, which should combine foam's lightness and insulation properties with PCMs' thermal mass, and PCM blends with nanotechnology-engineered insulation materials such as aerogel for achieving higher thermal performance. The development of new high-performing building envelopes with latent-heat-based thermal mass and lightweight, compact characteristics has

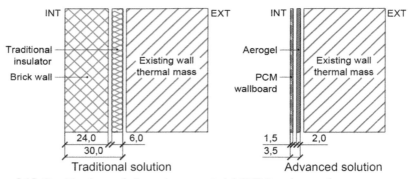

Figure 5.15 Combined use of phase-change material (PCM) and aerogel.

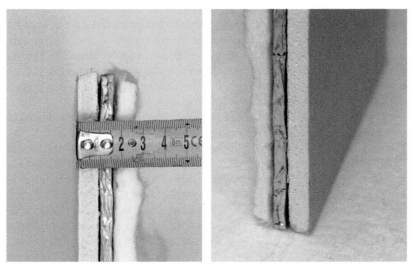

Figure 5.16 Combined use of phase-change material and aerogel.
Photo by the author.

been the focus of national and international institutions' programs. The European Union NanoPCM project, funded within the FP7-NMP research program, culminated in the development and patenting of encapsulated PCMs using nanomaterials as additives and polyurethane foam containing PCMs and nanomaterials, with payback periods as little as 2 years, thanks to high performance and optimized cost.

Similarly, PCM integration is being investigated in various insulators, such as nanoporous materials and reflective and radiant barriers: a partnership between the US Department of Defense and Aspen Aerogels developed a promising PCM-enhanced aerogel blanket.[5] Boeing has also patented a thermal insulation coating obtained by depositing aerogel particles, microencapsulated PCM, and a carrier binder solution on any suitable substrate.

5.5.3 Ceilings

Ceilings are one of the preferred surfaces for PCM integration, as they offer large areas available for heat transfer, less risk of damaging macroencapsulated PCMs by drilling or puncturing, and a heat exchange surface not affected by furniture or frames, as is the case of PCMs integrated in walls, or carpets in case of floor applications. These advantages are counterbalanced by a suboptimal heat exchange rate in the PCM solidifying phase, since ceiling horizontal surfaces do not favor convective motions when releasing heat and tend to accumulate warm air in the upper regions of interior spaces. Winter operation is thus less effective in aiding the heating system, and may require mechanical ventilation for better efficiency, whereas during hot months night heat release may be insufficient to recharge the PCM fully.[8]

Most PCM components and integrated products suitable for inner lining of walls, such as wallboards and plasters or panels/mats to be placed behind finishing, are also suitable for ceiling application, and are described in Section 5.5.1.

There are interesting PCM-integrated products specific for ceiling use, such as PCM ceiling tiles (Figs. 5.17 and 5.18). These have the advantage of ease of replacement of traditional modules, allowing retention of the original support structure, and are thus recommended for retrofit projects; integration with capillary heating/cooling systems is also viable. PCM cassettes may be easily moved within rooms to deal with higher thermal loads, or transferred to a new building.

Several products are available from different manufacturers, with dedicated surface finishing and acoustic performance for various applications (Table 5.6).

UK-based Datum Phase Change manufactures its ThermaCool ceiling tile in three product lines: light textured, acoustic (perforated for 54.2 dBA and Class C sound absorption), and metal composite, which also claims better thermal performance. Available size is a standard 600 × 600 mm and weight is 11 kg/m^2. The addition of BASF Micronal microencapsulated paraffin wax allows a total heat capacity over the 10–30°C range of 274 kJ/m^2, with a peak melting temperature of 23°C (Figs. 5.19 and 5.20).

Similarly, Armstrong CoolZone metal ceiling tile infill contains 25% Micronal by weight, and its heavy construction (25 kg/m^2) allows it to provide a total heat storage capacity of 490 kJ/m^2.

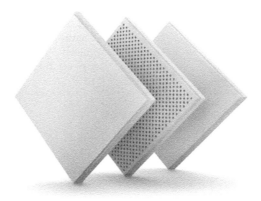

Figure 5.17 ThermaCool ceiling tiles with integrated phase-change material.
Courtesy of Datum Phase Change Ltd.

Figure 5.18 ThermaCool ceiling tile installation.
Courtesy of Datum Phase Change Ltd.

PCM-integrated ceilings are also well suited for integration into active systems. Datum Phase Change integrates its ThermaCool ceiling tiles and panels with Geoblue, a capillary-tube radiant heating and cooling system whose operation is enhanced by PCM and expanded graphite contained in the panels to ensure thermal conductivity and energy storage over a longer time lapse.

Another example is the Cooldeck system, in which macroencapsulated bags of ClimSel C24 hydrated salts PCM placed above the false ceiling are exploited to cool air passively without the need of air conditioning (Fig. 5.21).[20] When interior

Table 5.6 **Specifications of main ceiling tiles on the market with integrated phase-change material**

Product	Composition	Total heat capacity (10–30°C)	Phase-change temperature (°C)	Weight (kg/m^2)	Fire reaction
Datum Phase Change ThermaCool Tile	Mineral/metal with BASF micronal	274 kJ/m^2	19–23	11	B-s2,d0
Armstrong CoolZone	Mineral/metal with 25% BASF micronal	490 kJ/m^2	23	25	B-s1,d0

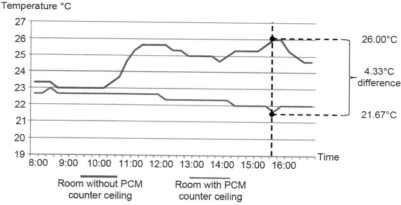

Figure 5.19 Daily fluctuation of the internal temperature of a room with or without a phase-change material (PCM) false ceiling.
Courtesy of Datum Phase Change Ltd.

temperature rises, ceiling fans ventilate indoor air in the cavity between the ceiling slabs and the PCM, creating a turbulent flow that increases the heat exchange rate and helps cool the air by releasing cold stored during the night in PCMs and the concrete building structure. At night, windows open and fresh air is pumped into the ceiling cavity to recharge the storage systems. An application in the town hall of Stevenage (UK) reported a 3–4°C decrease in maximum air temperature in the summer, with a lower cost than a conventional air-conditioning system.

A similar system was employed in the German prototype house in the 2009 Solar Decathlon competition (Fig. 5.22): in addition to inner lining of interior surfaces with PCM-enhanced Knauf Smartboard drywalls to dampen temperature fluctuations, an

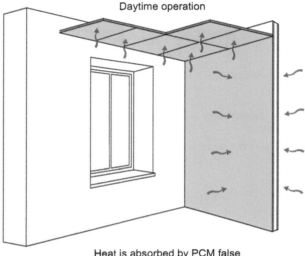
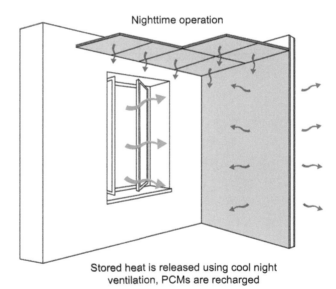

Figure 5.20 Phase-change material (PCM) interior operation.

active thermal storage system was installed in the ceiling to provide full daytime cooling.[14] The storage unit is composed of four 25 × 35 cm channels 11 m long, containing salt hydrate PCM macroencapsulated in polycarboante profiles (Dörken DELTA-COOL 28 with a 26°C melting temperature). PCM packages are isolated from the room by a 1 cm thick vacuum insulating panel lining the channels to preserve thermal storage when not needed. Channels are equipped with variable speed

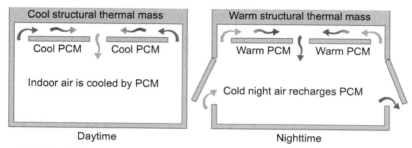

Figure 5.21 Cooldeck system operation. PCM, phase-change material.

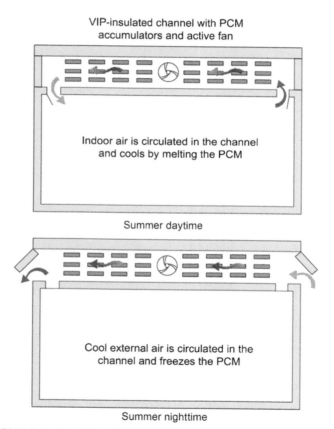

Figure 5.22 2009 Solar Decathlon German prototype operation. PCM, phase-change material; VIP, vacuum insulating panel.

ventilation fans, operable flaps, grills, and temperature sensors. In summer mode, during the day, the interior air circulates through the ceiling and decreases its temperature: the system capacity is equivalent to the energy demand for 1 day of cooling and its coefficient of performance varies from 9 to 15 depending on the ambience conditions. At night the cool exterior air blows through the ceiling and discharges the PCM, readying it for the next day's use. Winter operation similarly provides for PCM charging during the day using warm interior air by forced circulation. Stored heat is then released at night to increase indoor air temperature, discharging the PCM.

5.5.4 Floors

Floor application of PCMs is best recommended for heating periods, as PCM-integrated floor tiles or underfloor PCM components can take advantage of solar radiation entering through windows to provide passive free heating in the following hours.[5,14] Furthermore, their low position ensures the best warm air distribution during the heat-release phase, as upward convective motions are triggered and air in the whole space is affected. Care must, however, be taken to ensure the floor is free from carpets and other hindrances, and that paving materials are conductive, ie, stone and ceramic are preferred over wood tiles. Dark, matte colors are also best suited to harness solar gains.

In summer months, the use of PCMs in floor layers reduces overheating of floor areas directly exposed to solar radiation or important internal heat loads. PCMs absorb and store the excess energy, stabilizing the surface temperature around their own melting set point, then release the heat at night when the temperature drops.

Regarding PCM-integrated floor products, Ecocore raised floor panels, manufactured by US-based Tate Flooring, have a full steel construction with a cementitious infill mixed with vegetable-based biodegradable microencapsulated PCM. The PCM activates at 24°C, absorbing as much as 155 kJ during phase transition. The recommended position of PCM floor panels in large office spaces is the perimeter zone on the south, east, and west sides, to absorb solar peak loads and maintain a cooler interior. Winter operation should be carefully evaluated for office spaces, as stored heat is released at night when offices are empty and space heating in the morning may even take longer due to the additional thermal mass.

Floor application of PCMs has good potential for the implementation of active systems in addition to passive thermal storage. For example, the 2005 Solar Decathlon UPM Madrid house (Fig. 5.23) featured a PCM-integrated raised floor with dark gray ceramic tiles coupled with a metal container base filled with a eutectic mixture of paraffin with a melting point of 23°C, used mainly for winter heating periods.[14] This thermal storage floor takes advantage of the large glazed south facade, absorbing direct sun radiation and securing heat gain for subsequent nighttime heating: PCM tiles have shown surface temperature values 1.5−2.0°C above nonPCM reference tiles. Furthermore, the underfloor plenum hosts an active PCM thermal storage system for both winter and summer operation. Blow-molded HDPE containers containing a eutectic hydrocarbon paraffin mix gel with a switching temperature of 23°C efficiently exchange energy with circulating air blown by variable-speed fans controlled by the house automation system, together with opening of house windows for ventilation

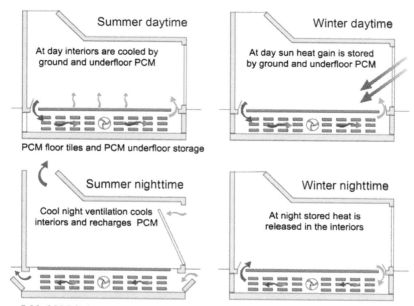

Figure 5.23 2005 Solar Decathlon UPM prototype operation. PCM, phase-change material.

and activation of a Peltier dehumidification system. In winter days, interior air warmed by the south-facing glazed façade is circulated in the floor plenum and PCMs are melted. The heat is later released by solidifying PCMs, warming circulating air and maintaining interior comfort when the sun is absent. In summer mode, exterior cool night air is blown on the storage units to freeze the PCMs, so that during the day they act as a heat sink to absorb excess heat and prevent overheating.

The system was further developed in the 2007 prototype: underfloor PCM was contained in metal multitubes with radiating fins, in which water from thermal solar and PV panels can be circulated for better energy exchange. In winter heating mode, surplus hot water produced by solar panels adds to direct and indirect solar gains to melt PCMs during the day, accumulating heat for subsequent use. For summer cooling, a copper pipe array behind the PV panels exploits them as nighttime radiative coolers, feeding a cold-water tank and ensuring PCMs can freeze and be recharged.

The underfloor thermal storage system of the Canada house prototype in the same competition was based on ordinary ceramic bricks soaked in a PCM mix (48% butyl stearate, 50% propyl palmitate, 2% fat acid). By soaking bricks in PCMs, their thermal capacity can be raised up to three times compared to standard bricks. The underfloor thermal storage system accumulates heat produced by solar and internal gains, PV/thermal panels, and the indirect gains provided by a water wall, and releases it when needed. For summer cooling, PCMs are charged at night by forced external air ventilation, and then provide cool air during the day.

PCMs can be easily integrated in underfloor electric radiant heating systems to act as heat storage, thus warming interiors even after the heating systems have been

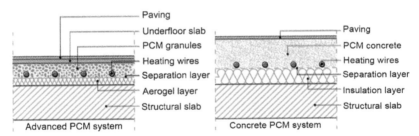

Figure 5.24 Underfloor phase-change material (PCM) application.

switched off (Fig. 5.24). In this application, the PCM is placed over the insulation and the heating cables, either in macroencapsulated containers such as granules, beads, or mats, which are more effective for maximizing thermal mass with low weight and thickness, or in microcapsules embedded in the concrete screed, which are more durable and suitable for direct floor installation on top. If enough thermal inertia is achieved, electric energy consumption can be entirely shifted from peak to off-peak demand, easing the energy load on the grid and allowing users to take advantage of cheaper energy. This system can reduce energy costs by about 30% compared with systems that use electricity at regular rates. Electricity costs can be further reduced by 15% through the use of a system equipped with a microcomputer that controls the surface temperature oscillations.[21]

While thermal mass is usually provided by thick, heavy concrete screeds, PCM systems are compact, lightweight, and can be dry-built, and are thus particularly suitable for integration in renovation projects. These benefits are further enhanced by employing aerogel for the insulation layer: thanks to its high thermal insulation properties in a reduced thickness, a complete advanced underfloor heating system can be achieved in just few centimeters.

5.5.5 Exterior walls/roof linings

PCMs can be positioned on the outer side of perimeter walls or roofs, where they are able to reduce the building envelope heat load during summer due to both solar radiation and the high outside air temperature. Using PCMs attenuates and shifts the incoming heat over time and reduces the radiating temperature of walls, improving the interior microclimate and reducing the envelope cooling requirements during summer.

External positioning is therefore particularly suitable in hot areas and in all cases where the existing wall does not have sufficient thermal mass to cope with the problems of summer overheating. PCMs can be applied as lining of a thermal insulation system or inside ventilated walls, next to either the cladding or the insulation, increasing heat dissipation toward the outside.

Suitable products for exterior use include CelBloc Plus from H + H Deutschland. This aerated concrete block gives enhanced latent heat storage thanks to microencapsulated BASF Micronal paraffin wax, in addition to good heat, fire, and sound insulation characteristics.

PCM bricks can also be obtained by soaking porous blocks in melted PCM or embedding macroencapsulated PCM containers in clay or cement bricks. A common building brick with cylindrical holes containing PCMs was devised for hot climate application, and experimental observation conducted to evaluate its potential to reduce daytime incoming heat flow by absorbing the heat gain before it reached interiors.[22] At night, the stored heat would be released to indoor and outdoor spaces. Results showed that three PCM cylinders placed at the centerline of the bricks could reduce incoming heat flux by up to 17.55%.

Other external PCM applications concern sloped and flat roofs: interior temperature fluctuations during the day may be reduced and postponed by placing a PCM panel with the switching temperature set 4–5°C above air temperature between the rooftop layer and the structural concrete slab.[8]

5.5.6 Use in glazing

Another interesting use of PCMs is inside glazing systems.[5] Swiss company GlassX has developed a triple-glazed glass (Fig. 5.25) with a selective polycarbonate layer that allows the reflection of solar radiation in the summer when the sun is high and its storage during winter in translucent slabs of hydrate salts (Dörken DELTA-COOL 28), embedded within the glazing, that allow the diffusion of light and give the effect of a translucent wall (Table 5.7).

It is thus possible to obtain a thermally insulated translucent wall that allows light radiation to pass through and increase the luminance levels of internal rooms and can

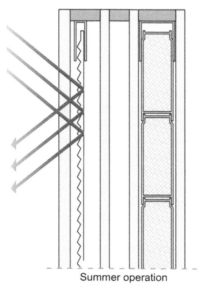
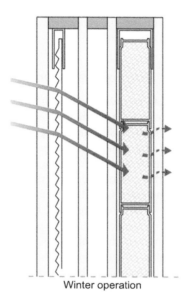

Summer operation   Winter operation

Figure 5.25 GlassX phase-change material glazing.

Table 5.7 GlassX cristal phase-change material (PCM) glass wall

Thermal and luminous parameters	Values
U (max)	0.48 W/m² K
Visible light transmission (VLT) (PCM in crystalline state)	8–28% (±3%)
VLT (PCM in liquid state)	12–44% (±4%)
Diffused VLT	29%
Solar heat-gain coefficient (SHGC), winter months, crystalline PCM	33%
SHGC winter months, liquid PCM	35%
SHGC summer months, crystalline PCM	6%
SHGC summer months, liquid PCM	9%
Accumulation capacity	1185 W h/m²
Accumulation temperature	26–28°C
Maximum size	3000 × 2000 mm

also, thanks to the presence of PCMs, act as a proper solar wall to accumulate the heat of incident solar radiation in the winter and release it during the night, maintaining high environmental comfort (Figs. 5.26 and 5.27).

Translucent PCMs are also well suited for integration in glass block walls for both perimeters and interior partitions. Hydrated salts are best recommended, as they have better translucency characteristics. Inclusion in perimeter walls in combination with translucent insulation materials allows selective optical transmittance of solar

Figure 5.26 Retirement residence community, Schwarz Architekten, Domat Ems CH (left and center), and GlassX phase-change material glazing section (right).
Courtesy of Gaston Wicky Photography© and GlassX AG.

Figure 5.27 Silence, Schwarz Architekten, St. Erhard CH.
Courtesy of Jürg Zimmermann Photography©.

radiation: visible light is transmitted and invisible infrared radiation is mainly absorbed and stored as heat.[14]

In interior vertical partitions, PCM glass blocks provide additional thermal mass to support heating systems in the hours when they are switched off.

Lastly, PCM-enhanced window shutters are gaining interest for their flexibility and optimized operation due to their movable configuration. During the day, shutters are open, capturing solar heat while not blocking heat gain passing through windows; at night, the shutters close and effectively transform the windows into heat radiators, releasing stored heat back inside and at the same time limiting heat loss through the glazing.[23]

5.5.7 Structural concrete

Another experimental application of PCMs in buildings is integration in structural concrete. Due to its inherent porosity, concrete is being tested as a viable carrier for PCMs in building walls and structures, with promising results: PCM integration of 1%, 3%, or 5% in mass increased heat capacity by 1.7, 3.0, and 3.5 times, respectively. PCM-enhanced concrete, also called thermo-concrete, is composed of open-cell cement and PCM, and may be fabricated in three main ways:

- by concrete immersion in a liquid PCM bath until saturation, more suitable for autoclaved concrete blocks but prone to melted PCM leakage;
- by impregnation of porous aggregates such as expanded clay or shale with PCMs in vacuum condition before their addition to the concrete mix;
- by directly adding microencapsulated PCMs in the concrete mixing and casting process.[24]

PCM concrete obtained through the immersion technique shows compressive strength comparable to that of a normal concrete, or even superior when the embedded paraffin wax solidifies. The addition of PCM microcapsules to the concrete mix instead significantly decreases the compressive strength, up to a 13% drop for every additional percent of microencapsulated PCM by total concrete weight, due to the disparity in strength between the microcapsules and the other concrete constituents.[25]

5.5.8 Use in mechanical systems

High latent heat values and a wide range of phase transition temperatures make PCMs a suitable choice for heat storage systems integrated with heating and ventilation systems, in particular if compared to water-based storage devices which are cumbersome and difficult to integrate with smaller systems.

PCMs as storage equipment are particularly promising for small power heating/cooling appliances, especially those installed in well-insulated habitation units where the majority of heat is required to compensate for ventilation heat losses rather than heat loss through walls and windows. Used in combination with solar energy, other renewables, or heat pumps on low-tariff night electricity, PCMs are able to balance energy flows and help achieve an economic and energy-efficient heating method. Macroencapsulated PCMs in plates, pellets, or slabs have a large heat exchange area, are suitable for most containers, and have three to five times higher thermal capacity compared to traditional thermal storage materials such as gravel, stone, or sand, allowing for lighter and smaller devices with the same performance. Another promising integration is within domestic hot water and heating systems, again balancing heat supply throughout the day to reduce energy consumption and cost.

5.6 Conclusions and future trends

The use of PCMs in buildings is an effective solution both in new constructions and in retrofit interventions to improve energy performance and internal comfort. PCMs can be used in both lightweight and heavyweight constructions, to increase thermal mass where needed and as a thermal storage medium in passive heating and cooling systems and air-conditioning equipment. The combined use in inner insulation methods of nanoporous insulating materials and PCMs can improve thermal insulation without losing the benefits of high thermal mass, and in a reduced thickness.

There are great opportunities for the development of new construction solutions with PCMs in emerging economies in high need of housing, and in the ever-necessary energy upgrading of existing buildings in developed countries. Research should focus on new types of PCMs, new methods of incorporating them into building materials, and enhanced heat transfer techniques and how to integrated these in the design of passive LHTES systems, in particular those related to harnessing solar thermal energy for heating during winter and those optimized to reduce the overheating problem during summer. Lastly, great opportunity lies in the research field related to the development of hybrid and adaptable systems to tackle both winter and summer issues at the same time.

Further research, and in particular long-term monitoring, is required to quantify the specific benefits of PCM applications, in view of the multiplicity of factors affecting their performance.

The incorporation of various nanoproducts is already bringing important advantages to both organic and inorganic PCMs, greatly improving key performance such as thermal conductivity, freezing/melting rates, and thermal stability. This will allow more and more PCMs in the future to become suitable for effective LHTES applications.

Organic PCMs have been the most widespread solution until now due to their long-term stability, ease of application, and microencapsulation possibilities. PCM integration into buildings is still niche, however, and for applications to grow, issues such as high cost, high flammability, and lower enthalpy per unit volume must be addressed. Using these criteria, hydrated salts fare better than paraffin, and recent developments may finally make them the PCM of choice. In fact, traditional hydrated salt challenges such as sub-cooling and incongruent melting have been successfully addressed, and microencapsulation is on track to become a viable solution. Costs have also been falling steadily.

Technological advances in hydrated salts and the diffusion of cheaper, renewable, and eco-friendly alternatives to paraffin, such as biobased PCMs, will lead to the development of low-cost, easy, and reliable applications to ensure PCM use becomes commonplace in the near future.

Heat energy accumulators, such as those described in this chapter, tend to release stored heat gradually over time, which can be recommended for interior climate-control applications but may be limiting for energy generation or waste energy recovery systems. For these last applications, a heat storage material which can store energy for a prolonged period and release it when effectively needed may be preferable. Researchers at the University of Tokyo have developed a ceramic material able to absorb different types of energy, such as light, electric current, and heat, store them up to 230 kJ/L (approximately 70% of the latent heat energy of water at melting point), and subsequently release them as heat when a pressure of 600 bar is applied.[26] This behavior is caused by the chemical composition of the ceramic, namely, stripe-type lambda-trititanium pentoxide (λ-Ti_3O_5), which is able to undergo a reversible, heat-releasing solid—solid phase transition and transform into beta-trititanium pentoxide (β-Ti_3O_5) upon pressure application. Beta-trititanium pentoxide is then able to revert back to its lambda precursor by gradually absorbing electricity, light, and heat. This heat storage ceramic is made of easy-to-supply and environmentally friendly elements, and may find use in solar heat power generation systems, domestic energy storage, and industrial heat waste, as well as electronic components and sensors.

References

1. Casini M. *Costruire l'ambiente. Gli Strumenti e i metodi della progettazione ambientale.* Milano: Edizioni Ambiente; 2009.
2. Casini M. Smart Materials and nanotechnology for energy retrofit of historic buildings. *Int J Civ Struct Eng* 2014;**1**(3):88—97.

3. U.S. Department of Energy. *Cost analysis of simple phase change material-enhanced building envelopes in southern U.S. Climates.* Golden, CO: U.S. Department of Energy; 2013.
4. Cabeza LF. *Advances in Thermal energy storage systems methods and applications.* London: Woodhead Publishing; 2014.
5. Kosny J. *PCM-enhanced building components.* Boston: Springer; 2015.
6. Fukai J, Hamada Y, Morozumi Y, Miyatake O. Improvement of thermal characteristics of latent heat thermal energy storage units using carbon-fiber brushes: experiments and modeling. *Int J Heat Mass Transf* 2003;**46**(23):4513−25.
7. Yu S, Jeong S, Chung O, Kim S. Bio-based PCM/carbon nanomaterials composites with enhanced thermal conductivity. *Sol Energy Mater Sol Cells* 2014;**120**(B):549−54.
8. Kalaiselvam S, Parameshwaran R. *Thermal energy storage technologies for sustainability.* London: Academic Press; 2014.
9. Barba A, Spiga M. Discharge mode for encapsulated PCMs in storage tanks. *Sol Energy* 2003;**74**(2):141−8.
10. Alam TE, Dhau JS, Goswami DY, Stefanakos E. Macroencapsulation and characterization of phase change materials for latent heat thermal energy storage systems. *Appl Energy* 2015;**154**:92−101.
11. Tyagi VV, Kaushik SC, Tyagi SK, Akiyama T. Development of phase change materials based microencapsulated technology for buildings: a review. *Renew Sustain Energy Rev* 2011;**15**(2):1373−91.
12. Schossig P, Henning H-M, Gschwander S, Haussmann T. Micro-encapsulated phase change materials integrated into construction materials. *Sol Energy Mater Sol Cells* 2005;**89**:297−306.
13. Zhang YP, Lin KP, Yang R, Jiang HFD. Preparation, thermal performance and application of shape-stabilized PCM in energy efficient buildings. *Energy Build* 2006;**38**(10):1262−9.
14. Soares N, Costa JJ, Gaspar AR, Santon P. Review of passive PCM latent heat thermal energy storage systems towards buildings' energy efficiency. *Energy Build* 2013;**59**:82−103.
15. Rodriguez-Ubinas E, Ruiz-Valero L, Vega S, Neila J. Applications of phase change material in highly energy-efficient houses. *Energy Build* 2012;**50**:49−62.
16. Ma T, Yang H, Zhang Y, Lu L, Wang X. Using phase change materials in photovoltaic systems for thermal regulation and electrical efficiency improvement: a review and outlook. *Renew Sustain Energy Rev* 2015;**43**:1273−84.
17. Farid MM, Kim Y. Thermal performance of a heat storage module using PCM's with different melting temperature: experimental. *J Sol Energy* 1990;**112**(2):125−31.
18. Pasupathy A, Velray R. Effect of double layer phase change material in building roof for year round thermal management. *Energy Build* 2008;**40**(3):193−203.
19. Kuznik F, David D, Johannes K, Roux JJ. A review on phase change materials integrated in building walls. *Renew Sustain Energy Rev* 2011;**15**(1):379−91.
20. Mehling H, Cabeza LF. *Heat and cold storage with PCM: an up to date introduction into basics and applications.* Berlin: Springer-Verlag; 2008.
21. Konstantinidis CV. *Integration of thermal energy storage in buildings* (Master Thesis School of Architecture). Austin: University of Texas; 2010.
22. Silva T, Vicente R, Soares N, Ferreira V. Experimental testing and numerical modelling of masonry wall solution with PCM incorporation: a passive construction solution. *Energy Build* 2012;**49**:235−45.
23. Weinlaeder H, Koerner W, Heidenflder M. Monitoring results of an interior sun protection system with integrated latent heat storage. *Energy Build* 2011;**43**(9):2468−75.

24. Pons O, Aguado A, Fernandez AI, Cabeza LF, Chimenos JM. Review of the use of phase change materials (PCMs) in buildings with reinforced concrete structures. *Mater Constr* 2014;**64**(315).
25. Ling TC, Poon CS. Use of phase change materials for thermal energy storage in concrete: an overview. *Constr Build Mater* 2013;**46.** P:55–62.
26. Tokoro H, Yoshikiyo M, Imoto K, Namai A, Nasu T, Nakagawa K, et al. External stimulation-controllable heat-storage ceramics. *Nat Commun* 2015;**6**.

Advanced building skin 6

6.1 Cool roofs

To limit the roof absorption of solar radiation, which simultaneously warms inhabited rooms underneath and heats the surrounding air, vegetated roofs (green roofs) can be used to counter the overheating by providing additional thermal mass, vegetation shade, and evapotranspiration. It is also possible today to resort to innovative roof coating materials able to maximize solar reflectance (SR) values, reducing both the urban heat island (UHI) effect in densely populated areas and the summer overheating of building envelopes.

In fact the focus of most countries' building regulations on winter energy performance only, along with nonoptimal behavior of traditional materials (bitumen or asphalt roofing, shingles, or clay tiles), has accentuated the UHI effect—the process which increases the average urban temperature compared to that found in rural surroundings. The particular morphology of urban areas—about 20—25% of urban areas are covered by roofs and 30—45% by pavements[1]—has increased energy storage capacity and reduced heat exchange with the surroundings; consequently, a greater amount of energy is trapped within the city and the urban environment grows hotter. In the city center, the large concentration of built-up areas and road paving, together with the high thermal conductivity of certain materials such as reinforced concrete, determine absorption of 10% more solar energy than a corresponding area covered by vegetation. The temperature difference between town and country peaks a few hours after sunset and is minimal in the early afternoon. At night the situation does not improve: the heat accumulated during the day is released as infrared radiation and intercepted by facing buildings, rather than dispersed into space. A first and immediate consequence is increased electricity consumption for summer cooling.[2] This, combined with the production of dust and the emission of pollutants, further contributes to the temperature rise, turning the city into a real heat island.

Key actions for mitigating the UHI effect are increasing pavement reflectivity by replacing or topping dark asphalt with clear concrete or aggregates where possible,[3] or making it more permeable to exploit cooling by evaporating stored moisture, planting trees to provide shade, and finally replacing conventional roofs with vegetated or cool roofs.

At the same time, solar-reflective coatings can sensibly improve the thermal insulation of buildings.

When the building envelope is exposed to solar radiation, outgoing or incoming thermal flux is different from that which would occur as a result of the normal temperature difference between indoor and outdoor air. In such cases, heat flow is calculated using the concept of "fictitious temperature" (or temperature in the sun), defined as the uniform temperature that the external environment should have in the absence of direct solar radiation, so the heat flow transmitted is the same as what is actually found in the real world.

The value of the fictitious temperature t_{fs} is obtained as the sum of the air temperature and a second term which accounts for solar radiant power absorbed from the building envelope.[4] In particular:

$$t_{fs} = t_e + \frac{a_s}{h_e} \cdot I \qquad [6.1]$$

where t_e is the outside temperature (°C); a_s is the solar radiation absorption coefficient of the outside surface (dimensionless); h_e is the adduction factor of the outside surface (W/m²K); and I is the intensity of the solar radiation incident on the surface (W/m²).

Total thermal energy transmitted through an opaque surface exposed to the sun is thus obtainable from the equation:

$$\Phi = U \cdot A \cdot (t_{fs} - t_i)[W] \qquad [6.2]$$

where U is the thermal transmittance of the surface (W/m²K); A is the surface area (m²); and t_i is the internal air temperature (°C). It is evident that the higher the absorption coefficient of the surface exposed to the sun, the higher is the fictitious temperature and thus incoming heat flow. In the summer, in places characterized by high values of solar radiation, this can lead to significant phenomena of interiors overheating (Fig. 6.1).

At the same time, the adoption of cool roofs has to be carefully evaluated in light of the climatic characteristics of the area of intervention. In warmer climates, the benefits obtained in terms of reducing energy consumption for summer cooling are significant (up to 25% for low-insulated roofs)[5] and, given reasonable intervention costs, make these investments cost-effective even in the short term (less than a year in many cases). In colder climates, on the other hand, part of the solar gain during winter is lost, and poor absorption of solar heat by the roof can prevent it from drying quickly, undermining indoor health in very rainy climates.

This does not stop cool roofs from being profitable in most temperate regions, because prolonged snow cover makes the underlying roof color irrelevant and heating loads and expenditures are typically more pronounced in evenings (especially in residential buildings), while the benefit of a darker roof in winter is mostly realized during daylight hours.

The building's surrounding context should also be taken into consideration: in densely built areas, sunlight reflected by cool roofs may bounce into the higher windows of neighboring taller buildings, causing unwanted heat and glare. In these cases, cool roof reflectance should be carefully evaluated to provide energy and comfort improvement without significantly affecting surrounding buildings.

Energy-saving benefits of cool coatings are also reduced (savings potential up to 13%)[5] for roofs already well insulated against the cold, notwithstanding other positive effects such as reduced UHI effect or extended service life of roof components, since lower average temperature slows down heat-related degradation like color fading.

Advanced building skin

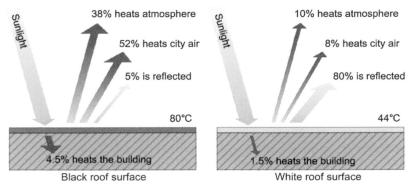

Figure 6.1 Response to sun radiation of black and white roofs (air temperature 37°C).

6.1.1 Standards and regulations

Cool roof performance is normally measured using SR or the solar reflex index (SRI).

SR, or albedo, measures the ability of a material to reflect incident solar radiation without absorbing it. To be classified as a cool roof, a material must have a minimum SR value of 0.65.

The SRI instead measures a surface's ability to stay cool in the sun by reflecting solar radiation and emitting thermal radiation. The SRI is determined by comparing the surface temperature in referenced conditions to standard black (SR = 0.05, emissivity = 0.90: SRI = 0) and white (SR = 0.80, emissivity = 0.90: SRI = 100) surfaces:

$$\text{SRI} = \frac{T_{\text{black}} - T_{\text{surface}}}{T_{\text{black}} - T_{\text{white}}} \qquad [6.3]$$

The temperature T_{surface} is obtained analytically from the material's SR and thermal emissivity values: a high SR value decreases the energy absorbed from the sun, while high thermal emissivity allows the body to reach the point of thermal equilibrium (equivalence between energy absorbance and energy released into the environment) at lower temperatures. The Green Building Council takes into account SRI values for granting Leadership in Energy and Environmental Design (LEED) credits related to the reduction of UHI effect for roofing (minimum SRI values of 78 for flat roofs and 29 for pitched roofs, for at least the 75% of the roof area) and paved areas.

The main initiative for the certification of cool roof materials is the Cool Roofs Rating Council (CRRC) Product Rating Program, established in 1998, which introduced uniform and stringent procedures for evaluating and labeling the SR and thermal emittance of roofing products, publishing them in the CRRC online Rated Products Directory. Manufacturers are encouraged to test their products using the CRRC-1 method and list them in the ever-growing directory. CRRC rating also takes into account performance deterioration (aged values) as a result of ultraviolet (UV) light degradation, thermal cycling, biological growth, and particulate accumulation.

Performance loss in terms of SR over time can be drastic: a white roof's albedo value may drop from 80 in its clean state to 60 in a soiled state, thus annual washing or recovery treatments just 2/3 years after installation are often required to maintain minimum cool roof performance values.[1]

In the United States several programs are used to promote cool roofs, especially within redevelopment of existing buildings. The ENERGY STAR Roof Product Program, developed in 1998 by the Department of Energy and the Environment Protection Agency, requires minimum albedo values of 0.65 (0.50 after 3 years) for flat or slightly sloped roofs and 0.25 (0.15 after 3 years) for inclined roofs to be certified.

The United States has also made solar-reflective roofs mandatory in more and more building codes, such as in California, where they have been mandatory in its Energy Code since October 2008 (Title 24), and New York, which compels all roof interventions to meet the ENERGY STAR requirements (Local Law 21 of 2011) and has implemented the NYC °CoolRoofs information and promotion initiative within the PlaNYC program. The city of Chicago has deployed an incentive campaign for the adoption of cool roofs, the Chicago Energy Conservation Code. Finally, local building codes such as Florida, Washington DC, and Hawaii prescribe or foster cool roofs, and the National US model codes ASHRAE 90.1 and 90.2 allow reduced roof insulation if a rated cool roof is used. Cool roofs are now also increasingly attracting the attention of Japan, China, India, Brazil, and Europe, where the European Cool Roof Council was established in 2011. In Italy, minimum reflectivity requirements (0.65 for flat roofs, 0.30 for sloped roofs) were introduced in 2015 for new constructions as well as renovations of existing buildings.

6.1.2 Products and specifications

Cool roofs usually merely translate into white roofs, since a simple white painting of the roofing materials is enough to ensure good values of reflectance (>0.8) and thermal emissivity ($0.5-0.9$ depending on the support).

The market for white cool roof products consists today mostly of organic (especially acrylic) paints based on titanium dioxide (TiO_2), applied as a simple coating, or sheet waterproof roofing products with bright white (Derbigum Brite) or metallic (Soprema Sopralast 50 Alu) top finishes. More advanced products employ nanotechnology to reach high performance: Ama Composites Thermogel acrylic paint is modified with aerogel and achieves 0.84 SR, 0.90 emissivity, and an SRI value of 106 according to American Society for Testing and Materials standards. The best cool roof coatings combine SR values around 90% with thermal emittance around 95%, and their diffuse appearance reduces unwanted glare phenomena.

Where esthetic needs or requirements dependent on the architectural context do not allow white roofs, it is possible to use specific coatings based on polyvinylidene fluoride, silicate polyester, or acrylic resins with silicon oxide, such as ARC Cooltile IR or BASF Ultracolor, or to add normal paints with specific inorganic pigments based on spinel (blue-black dye) and rutile (yellow-orange dye), produced by companies such as Heubach, Altiris, and Asahi Kasei. These can ensure high levels of reflectance and thermal emittance in a wide range of tones, even darker ones (cool colors).

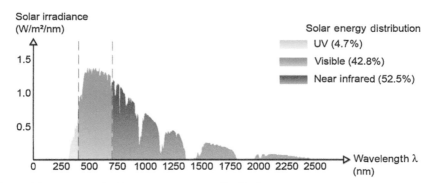

Figure 6.2 Normalized solar spectrum energy distribution.

The key point in the operation of cool-colored coatings is the energy distribution across the whole light spectrum. Global energy can be described as the sum of the different UV, visible light, and near infrared (NIR) light wavelength components (see Fig. 6.2), as in

$$E_{SOL} = 0.047 E_{UV} + 0.428 E_{VIS} + 0.525 E_{NIR} \qquad [6.4]$$

It is apparent that more than 52% of solar radiation energy lies in the NIR range, and as such high SR can only be obtained if NIR reflectance is very high. A theoretical ideal cool roof would have, neglecting the UV, SR in the visible component of 43% and an NIR reflectance of 52%, achieving a global SR value of 95%, while a traditional black paint has an SR of only 6%, being mostly opaque in both visible and NIR ranges. At the same time, a black surface engineered to be more reflective in the NIR range while maintaining its low visible reflectance—thus its black appearance—may well achieve an SR of $5\%_{VIS} + 50_{NIR} = 55\%$, a nearly tenfold increase.[6]

Cool-colored coatings can in fact achieve the same spectral response as traditional materials, such as terracotta tiles, in the visible range (wavelength 400–700 nm) which characterizes the color response of a surface. At the same time they are able to reflect much more energy in the NIR spectral band (about 700–2500 nm), which as noted includes more than 52% of solar radiation energy.

Most products on the market achieve this effect by exploiting a top layer capable of reflecting the visible component of the desired color but transparent to the other wavelengths, which can thus reach a previously deposited substrate characterized by a high infrared reflectance, such as TiO_2, and then be reflected into the atmosphere (see Fig. 6.3). This NIR-transparent colored layer contains pigments normally in a nanoparticle form.[6]

Another solution is based on oxide-coated aluminum flakes suspended in a transparent binder, constituting a high reflective, color-neutral coating. Each flake is coated with a nano-thin oxide layer with thickness between 20 and 200 nm, corresponding to different optical interference phenomena which impart the desired color appearance. As these oxides are largely transparent to NIR wavelengths, reflection in that range is carried out by the aluminum flakes. Multiple oxide layers can even impart a variable

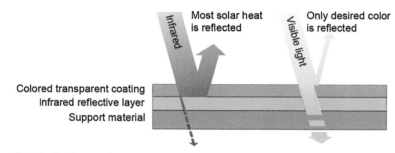

Figure 6.3 Cool color coatings.

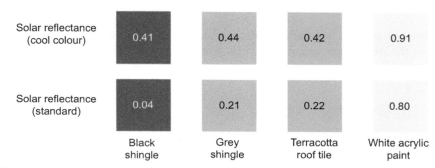

Figure 6.4 SR values of cool and regular colors.

color appearance according to the view direction (dichroic effect), for instance, by sol—gel coating the metal flakes with SiO_2 before applying a second iron oxide layer.[6]

Thanks to cool-color technology, building materials in traditional colors such as gray or brick red can reach albedo values of 0.4—0.5, whereas common products do not exceed values of 0.1—0.2 (Figs. 6.4 and 6.5). For uncoated metal roofing, inherently characterized by high reflectivity, it is instead convenient to act on emissivity by matting shiny surfaces to release excess heat at lower temperatures and prevent overheating.

For granular coatings such as roof shingles, products are available such as 3M Cool Roofing Granules, characterized by a reflectance up to four times higher than the traditional granules, thanks to their coating with infrared reflective pigments.

Finally, for roofing brick tiles, companies such as MCA Clay Roof Tile, Entegra, and Eagle Roofing Products manufacture glazed tiles with albedo values from 0.32 up to 0.82, which meet the CRRC requirements (Table 6.1).

Today research on cool roofs is focusing on prolonging useful life and dirt resistance, and on recyclable or otherwise more environmentally friendly products based on vegetable oils and resins and waste byproducts of other industrial processes.

The theme of dirt and aging resistance is sensitive for most manufacturers, as cool roof rating protocols require aged performance to be tested by exposing samples to environmental conditions for up to 3 years, discouragingly lengthening the rating

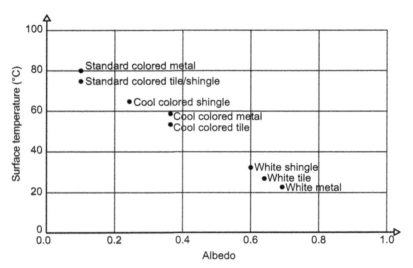

Figure 6.5 Surface temperature of different steep sloped roof surfaces (air temperature 37°C). Redrawn from Global Cool Cities Alliance. *A practical guide to cool roofs and cool pavements*. New York: GCCA; 2012.

process for new products. Accelerated simulation protocols are being researched to predict aged performance accurately in a much shorter timeframe and give a quicker rating process, in particular for materials that resist dirt pickup and/or chemically alter and remove deposited dirt to help improve the performance of aged roofs. Dirt pickup can be reduced by using materials that are smooth and reducing the use of plasticizers that can leach to the roof surface. Dirt can be chemically altered and removed by incorporating photocatalytic compounds such as TiO_2. Another potential benefit of using photocatalytic materials is the reduction of ground-level ozone precursors. Standards organizations, governments, and research centers are also at work to identify and establish cool roof building code criteria to promote adoption in appropriate conditions, such as hot climates, urban areas, or emerging economies where they can prevent the need for air conditioners.

6.1.3 Emerging technologies

Research on cool roofs is today focusing on achieving the highest reflectivity levels and taking advantage of other strategies such as radiative cooling to enhance cool roof effectiveness.

6.1.3.1 Radiative sky cooling

The same principles of reflectance and emissivity can be further applied to enhance cool roof performance by making them not only impervious to solar heating but even colder than the surrounding air, by removing heat from the roof and pumping it into the sky.

Table 6.1 **Performance of most common roofing materials on the market**

Material surface	SR	Thermal emissivity	SRI
Black acrylic paint	0.05	0.90	0
New asphalt	0.05	0.90	0
Aged asphalt	0.10	0.90	6
Asphalt shingle (weathered wood)	0.08	0.91	4
Asphalt shingle (white)	0.21	0.91	21
Cool asphalt shingle (weathered wood)	0.25	0.90	28
Cool asphalt shingle (white)	0.41	0.92	48
Clay tile	0.22	0.86	20
Cool clay tile	0.42	0.90	48
Concrete	0.35−0.45	0.90	38−52
White concrete	0.70−0.80	0.90	86−100
Aluminum	0.61	0.25	56
White coating on metal roof	0.65	0.85	82
White bitumen sheet roofing	0.76	0.81	93
White acrylic paint, 1 coat	0.80	0.90	100
Aerogel acrylic paint	0.84	0.90	106
White silicone coating	0.87	0.90	110
White polyvinyl chloride roll	0.91	0.85	115
White acrylic paint, 2 coats	0.90	0.95	115

The concept of sky cooling exploits the fact that all bodies exposed to the sky radiate heat toward it, only for it to be absorbed by the atmosphere and clouds: short wavelengths are mostly absorbed by water vapor, while longer ones are absorbed by carbon dioxide. However, radiation emitted in a specific range of wavelengths, 8−13 μm, is not absorbed by the atmosphere and is free to reach outer space, provided air humidity is not too high. This interval is called the "atmospheric window" or "sky window," and is the main channel through which the Earth ultimately loses the energy gained from the sun (Fig. 6.6).[6]

The sky window gradually narrows as the atmosphere thickens, so rays passing close to the horizon have more difficulty in bypassing the atmosphere than those going straight up, making horizontal surfaces such as roofs more suitable for radiative cooling applications. Furthermore, the sky window is diminished by higher concentrations

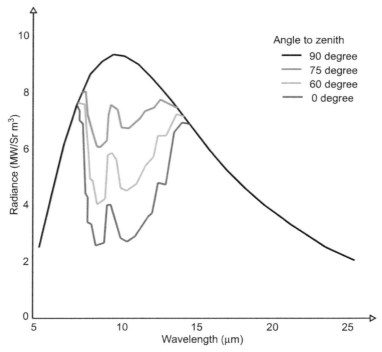

Figure 6.6 Solar irradiance according to zenith angle and wavelength.
Redrawn from Smith GB, Granqvist CG. *Green nanotechnology: solutions for sustainability and energy in the built environment*. Boca Raton, FL: CRC Press; 2011.

of water vapor and carbon dioxide, such as in cloudy weather or much-polluted areas. In fact, a narrower sky window is one of the factors contributing in global warming.[6]

As such, coatings able to limit heat gain, thanks to their high reflectance, and at the same time able to emit radiation in the sky window, can effectively lose heat, dispersing it in deep space where it is lost forever. Night-time cooling potential is even more pronounced as it is not impeded by incoming radiation, and is already used in passive water condensation systems to harvest dew from air vapor. Research is currently under way to exploit this phenomenon, and some products on the market claim additional performance, thanks to this.

Sydney University of Technology[7] has tested a coating based on a sandwich structure of modified polyesters on a substrate of silver which can absorb only 3% of the incident solar radiation and irradiate the infrared in wavelengths of 7.9−14 μm—within the sky window—dispersing it directly into space. The temperature of the surface exposed to the sun stays 11°C below the air temperature, leading to the formation of condensation on the surface which then helps to wash away the accumulation of dirt and preserve the efficiency of the cold roof.

The same concept is exploited by Australia-based Skycool's coating for metal roofs, which combines a high SR value (90%) with a 94% emittance in the sky window, passively radiating incoming heat away without transmitting it inside the building. Provided the sky is clear, the paint also works at night, enhancing nighttime cooling of interiors. Application testing carried out by Sydney and Queensland Universities showed a 50% air-conditioning cost reduction during summer compared to an almost identical noncoated reference building, with the understandable drawback of some increase in heating consumption during the winter due to reduced solar gain and increased night-time heat loss.[8]

Researchers at Stanford University have developed an innovative prototype of a photonic radiant coating, consisting of seven layers of silicon and hafnium oxides on a thin support of silver. The internal structure is similarly designed to radiate infrared rays at a frequency able to project them directly beyond the atmosphere, without heating surrounding air. This way it is possible to reflect 97% of incident solar energy and lower the temperature of the air near the surface by more than 12°C. Development is now focusing on increasing manufacturing size and on systems to direct part of the heat inside the building to the radiating surface, thus creating a complete passive cooling system.[9]

Radiative cooling was also demonstrated in Paris as part of the initiatives for the UN COP 21 Climate Change Summit, with three interactive stations combining different types of membranes and computer simulations to show visitors potential climate responses according to wind speed, humidity, and air temperature and compare them using the Universal Thermal Climate Index.

6.1.3.2 Thermochromic cool roofs

There are interesting applications in cool roofs for thermochromics technology, producing paints and coatings able to vary their color and reflectance performance according to temperature, thus combining heat load reduction in summer with heat gain exploitation during winter. Thermochromic paints and coatings available on the market are usually based on organic leuco dye mixtures composed of three elements: a cyclic ester which determines the final color of the product in its tinted state; the color developer, a weak acid that imparts the reversible color change to the thermochromic material and is responsible for the color intensity; and finally the third component, usually an alcohol or an ester with a fixed melting point, which controls the color transition temperature.[10] When the temperature rises the thermochromic layer becomes translucent, revealing a clear, highly reflective TiO_2 layer beneath. The thermochromic mixture is microencapsulated and isolated from the paint base and binders to preserve its properties over time.

Researchers at Athens University compared summer performance of traditional, "cool roof," and thermochromic paints in several colors; in the case of a black coating, the thermochromic paint produced surface temperatures up to 15.3°C lower than traditional paint and 8.3°C lower than "cool-colored" paint.[10]

The thermochromic roofing tile Thermeleon, patented by MIT graduates, is based on a phase-change polymer gel layer over a dark, heat-absorbing substrate. At lower temperatures the solution is clear and colorless, allowing the black backing to absorb most solar radiation. When temperatures rise above a fixed set point, in a range of

0−100°C, the polymer phase separates from the gel, turning the solution into a mixture of polymer and solvent with different refractive indexes, and the tile becomes white and light scattering. Tests showed that the tile could absorb 70% of solar radiation in winter mode and reflect 80% in summer mode.

6.1.3.3 PCM color coatings

Cool roof coatings can take advantage of the added thermal storage potential provided by phase-change materials (PCMs), with beneficial effects in maintaining thermal comfort and temperature stability in the indoor environment.[11] Microencapsulated PCM pigments with particle size in the 17−20 μm range, containing paraffin wax with 170−180 kJ/kg latent enthalpy, can be mixed to any color paint material. Application of these PCM color coatings proved effective in reducing surface temperature by up to 7−8°C compared to conventional color coatings of the same tint, a performance slightly higher than correspondent cool color paints. As such, PCM paints can be considered a viable method to incorporate latent heat storage systems in the roof surface.

6.1.3.4 Lenticular cool roof

Finally, another solution to the white roof/dark roof esthetic issue may come from "lenticular printing" technology, as devised by Case Western Reserve University researchers, which utilizes a layer of micro semicylindrical lenses to orient light toward one or another background image according to the observer's point of view, akin to a three-dimensional (3D) effect picture or postcard. It is thus possible to achieve perspective-dependent color-changing roofs which appear white if viewed from above, and thus react as a white surface reflective to solar radiation when the sun is high in the sky, while showing another color (eg, dark brown, black, or brick red) when viewed from street level to preserve the building appearance.

6.2 Green walls

An interesting form of interaction between the building envelope and the surrounding environment is the use of vegetable species as a cladding system for opaque walls. The adoption of a vegetable skin gives a dynamic image and a naturalized perception of the envelope, according to the varying environmental conditions, through the natural modifications of the vegetable essences.[4]

Green coverage of vertical walls has usually been achieved by planting a few vigorous climbing species on the ground or in perimeter flowerpots, then growing them along the sides of the building on purpose-built trellises or wired meshes, or directly on the building surface, causing damage to the facade materials and attracting animals.[12] Instead, green wall systems available today allow for semiintensive growing of a much greater range of plants, pose no risk to the facade integrity, and are designed to minimize plant maintenance.

During summer, wall vegetation reduces the heat flows entering through the building envelope thanks to shading and the absorption of radiant energy needed for photosynthesis and thermal energy used in evaporation—transpiration processes. The phenomenon of phototropism (self-orientation of the leaves toward sunlight) helps make this system even more effective through self-regulation depending on the angle of the sun. As for ventilated facades, green walls are more effective for elevations exposed to the east and west which are subject to more summertime direct sunlight.

Significant parameters that influence energy performance are average leaf size, leaf area index (LAI), defined as the one-sided green leaf area per unit of ground surface area, and growing substrate water retention to maximize evapotranspiration. LAI values higher than 3 or 4, related to larger plants with bigger leaves, usually yield maximum cooling potential while tiny-leafed plants such as mycrophylls have a lower ability to shed heat.[13]

Experimental tests have shown energy consumption reductions up to 50% compared to reference buildings without any shielding device.[14]

During winter, the surface of evergreen species contributes to the limitation of heat loss due to convective phenomena induced by cold air currents that skim the envelope.

These benefits add to those of UHI reduction and improvement of air quality, thanks to the purification of air and filtration of dust operated by the plants.

Regarding air quality, recent studies show that the presence of vertical gardens in large cities reduces air pollution (Fig. 6.7), especially along so-called "street canyons"—long street corridors lined with tall buildings. Because of atmospheric conditions characterized by little air movement, high concentrations of pollutants such as nitrogen oxides (NO_x) and particulates (PM_{10}), particularly dangerous to human health are often observed in street canyons. Vertical hanging gardens reduce the presence of NO_x by 40%, going as far as 60% for particulates.[15]

Green walls also show promise concerning acoustic comfort for building users and street occupants, providing benefits in both noise insulation and noise absorption. It has been observed that green wall behavior toward sound waves is differentiated

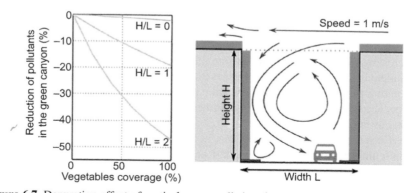

Figure 6.7 Depuration effect of vertical green walls in urban centers.
Redrawn from Pugh T, Mackenzie R, Whyatt JD, Hewitt CN. Effectiveness of green infrastructure for improvement of air quality in urban street canyons. *Environ Sci Technol* 2012; **46**(14):7692—9.

according to their components: foliage and branches dissipate sound energy by vibrating and converting it into heat, and are most effective for higher frequencies; while the soil substrate absorbs sound through its own bigger mass and by interference between incoming sound waves and those it partially reflects, and is particularly efficient in limiting traffic noise (0.10 kHz).[16]

Mass is known to be paramount in sound behavior, so heavier green wall systems such as metal or polymer boxes containing topsoil are expected to perform better than lighter ones based on geotextile felts without substrate.

Laboratory testing of a module-based green wall system made of recycled polyethylene, with a growing substrate of coconut fiber precultivated with 40 cm high *Helichrysum thianschanicum* plants, yielded interesting results, with a weighted sound absorption coefficient $\alpha = 40$, enough to reduce traffic noise reverberation sensibly on the streets or for acoustic correction of interiors, and a sound reduction index Rw of 15 dB. The latter value is low compared to glazing (30 dB) and wall (50 dB) insulating performance, but is to be considered as an addition to an existing load-bearing structure providing the bulk of the sound isolation. Sound reduction can be further enhanced by sealing the joints between the green wall panels to prevent sound leakage, achieving 18 dB sound insulation.[16]

Green walls may significantly reduce facade noise reflections in urban street canyons, and thus prevent traffic noise from diffracting over rooftops and reaching inner courtyards: simulations show how they are particularly effective if applied to rigid brick facades, as opposed to already mildly absorbent surfaces, and that vegetation on the upper half of the walls is better at trapping noise inside the street canyon.[17]

Still, the creation of a "green wall" requires considerable attention in design.

First it is essential to correctly identify the plant species to use according to the climate conditions and the facade's exposure.

In particular, issues relate to seasonal growing cycles (deciduous or evergreen), size, growth rate, weight per square meter of surface, shape and density of leaves, need for maintenance, recurring weather adversities, and the method of fixing plants to the wall (direct or independently supported). Normally, to reduce maintenance costs, about 80% of the vegetable species used in vertical walls are evergreen, leaving about 20% of seasonal plants to give the impression of a living work which changes color and appearance with the passing seasons.

In recent years highly innovative green wall systems have been developed, consisting of modular components used as actual surface cladding elements of the wall assembly. The appearance obtained is essentially that of a vertical garden, and gives the building skin a changing and strong character in esthetic, tactile, and olfactory terms.

Systems on the market are mainly of two types: the green wall and the living wall.

The green wall is typically constituted by elements between 200 and 1000 mm wide, 200 mm high, and 85 mm thick, with an organic mat made of *Sphagnum* (a moss characterized by excellent values of lightness and hygroscopicity), soil, and natural fibers or inorganic media to house precultivated plant species held together by a cage structure in galvanized steel (30 × 30 mm). The modules are attached to a galvanized steel support grid fixed to the wall at a distance of 20–50 mm to allow air circulation (Figs. 6.8 and 6.9).

Figure 6.8 Green wall module installation.
Courtesy of Peverelli s.r.l.

Figure 6.9 Growing of green wall modules.
Courtesy of Peverelli s.r.l.

Similar systems employ lightweight recycled plastic and come as framed boxes or boxes with precut holes which may be subdivided into smaller cells.

The modules are easily interchangeable, and the fixing system allows maintenance without having to intervene in the entire structure. The wall is completed by an integrated pressurized irrigation network consisting of perforated pipes with a water recovery and recycling system. Water is recovered at the base of the wall, conveyed to a collection tank, and then redistributed in the green wall.

UK-based Treebox Rain Garden's system further simplifies the vertical garden by forgoing pressurized irrigation in favor of a self-watering method that requires no power to operate. Rainwater is collected in storage tanks hidden behind the planting, then supplied by capillary action via a wicking rope. At times of high rainfall the system reduces water supply so as not to overwater plants and save capacity. A full water tank can sustain the plants for up to 6 weeks, so the system may still require artificial refilling in the most arid climates.

The living wall system (Fig 6.10), designed by botanist Patrick Blanc and now available in several commercial versions, instead employs a highly technological growing medium based on expanded polyurethane foam with a thickness of 20–140 mm, sandwiched between two layers of polyamide felt in which are formed the pockets for holding the plants, without the need of topsoil, and the housing for the watering system pipes (Fig. 6.11).[18]

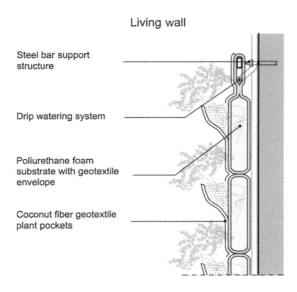

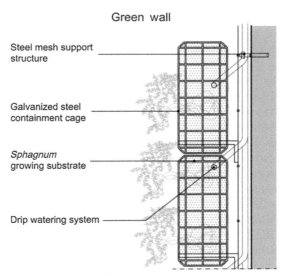

Figure 6.10 Living wall and green wall systems.

Figure 6.11 Living wall growing felt and irrigation system.
Courtesy of Poliflor s.a.s.

A hydroponic system distributes water mixed with nutrients, and monitors the development of the plants with thousands of sensors while controlling humidity levels, pH, and the underlying nourishment substrate; this solution saves water and energy, minimizing the environmental impact of the roof garden. The system, similar to the human circulatory apparatus, allows the living wall to grow both outdoors and indoors.

There are now many examples of green walls using one of the two systems described above, in both new constructions and redevelopments of existing buildings, all aimed at improving formal quality as well as energy upgrading the building envelope.

Among the most recent interventions of particular interest are those at Museé du Quai Branly in Paris by Jean Nouvel/Patrick Blanc (2006; Fig. 6.12), Caixa Forum in Madrid by Herzog & de Meuron/Patrick Blanc (2007; Fig. 6.13), the Athenaeum Hotel in London by Patrick Blanc (2009; Fig. 6.14), L'Oasis d'Aboukir in Paris by Patrick Blanc (2013; Fig. 6.15), the facades of the Palacio de Congresos Europa in Vitoria-Gasteiz in Spain (2014), with a hanging-garden-like living wall of approximately 1500 m^2 decorated with 33,000 plants, and Guildford Town Center in British Columbia by Green over Gray, the largest vertical garden in North America (2014). Green wall was also prominently showcased in the US Pavilion in EXPO Milan 2015 by James Biber (Fig. 6.16). Around the world more and more buildings are integrating vegetation in their architecture, achieving proper vertical gardens such as One Central Park in Sydney by Nouvel/Blanc (2015), and the Bosco Verticale in Milan by Stefano Boeri Architects (2014; Fig. 6.17), which won the 2014 International Highrise award and the CTBUH 2015 Best Tall Building Worldwide award.

Interior green walls can be integrated with heating, ventilation, and air-conditioning systems to act as "biofilters" or "active living walls" for air purification from volatile organic compounds and indoor pollutants: natural phytoremediation is exploited by

Advanced building skin 235

Figure 6.12 Musée du quai Branly, Jean Nouvel/Patrick Blanc, Paris. Photo by the author.

Figure 6.13 Caixa Forum, Herzog & de Meuron/Patrick Blanc, Madrid. Courtesy of Patrick Blanc.

drawing polluted exhaust air through the root system of the plants, where beneficial bacteria actively degrade the harmful substances. A continuous flow of water is required to trap suspended pollutants and drive them to the rhizosphere surrounding the plant roots. Testing demonstrated promising reduction of pollutants sprayed on the biofilter after a single pass through the living wall—up to 80% of formaldeyde, 50% of toluene, and 10% of trichloroethylene, some of the most common indoor pollutants.[19]

Figure 6.14 Athenaeum Hotel, Patrick Blanc, London.
Courtesy of Patrick Blanc.

Figure 6.15 Oasis d'Aboukir, Patrick Blanc, Paris.
Courtesy of Patrick Blanc.

Another interesting green wall system was developed by Arup to allow crops to grow inside a double glazing façade, combining sun shading and agricultural production from the plants with the thermal benefit of a state-of-the-art glazed double skin facade. Dubbed the "vertically integrated greenhouse" (Fig. 6.18),[20] the system is designed around two rows of trays suspended on a continuous cable system. The trays host the crops to be

Advanced building skin

Figure 6.16 Detail of US Pavilion at Expo 2015, Milan.
Photo by the author.

Figure 6.17 Bosco Verticale, Stefano Boeri, Milan.
Photo by the author.

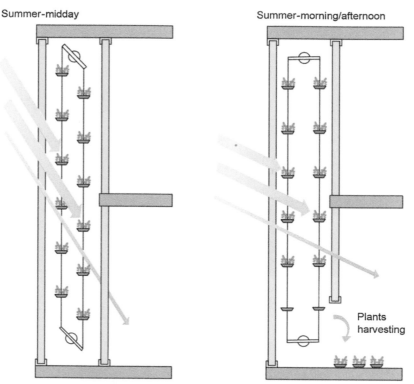

Figure 6.18 Vertical integrated greenhouse.

cultivated, which are hydroponically fed through a thin film of water enriched by nutrients that passes from one tray to the next before being recovered at the bottom and reused. A computerized motor scrolls the suspension cable lift, bringing crops to the bottom for easy harvesting or maintenance. Plant position toward the sun and spacing can be varied in a way similar to a Venetian blind, tracking solar elevation throughout the day and regulating irradiation of the interiors. This versatility allows seasonal modes of operation: during winter, trays will be aligned to let in as much solar radiation as possible, and the greenhouse will be closed to act as a heated buffer for better thermal insulation; at night, exhaust air from the building can be ducted to the plants to maintain temperature and be purified. In summer the tray position will shield occupants from the sun while gathering solar rays and cooling air via evapotranspiration, and the greenhouse facade air inlet and outlet will open to provide heat extraction via a stack effect.

Along with the development of technological systems for vertical gardening, research is testing alternative solutions such as organic concrete for facades and the futuristic biosynthetic leaves of the "Silkleaf" project by researcher Julian Melchiorri, which can produce oxygen in the presence of water and sunlight.

Thanks to the union of two different types of cement-based compounds, the biological concrete developed by a team of researchers at the Polytechnic University of

Catalonia is able to change color over time, absorb CO_2, and insulate from cold and heat. This innovative concrete is characterized by the presence of living organisms on its surface, capable of assuming different pigmentations over the years and capturing a part of the CO_2 present in the air. To develop the concrete, Catalan scientists started from two different cement-based materials and transformed their chemical properties, pH, and some physical characteristics. The first compound is made from traditional concrete based on already carbonated Portland cement, with a pH very close to 8. The second material is a magnesium phosphate cement composted of a hydraulic conglomerate with a pH already acid enough without needing further treatment, usually used in medicine and dentistry as a biological concrete with remarkable, all-natural regenerative abilities. By mixing these two compounds together, a completely new material is obtained, formed by a series of layers of different porosity suitable for the proliferation of selected biological organisms (microalgae, fungi, lichens, and mosses) and able to store the water necessary for their survival, while expelling superfluous water that could cause structural damage. In addition to generating a unique color variation on facades, the organic layer that covers this biological concrete will capture a large percentage of CO_2 and will act as a filter between the interior and exterior to enhance the thermal insulation of the building.

6.3 Environment-adaptive skin facades

The interest in smart insulation technologies also focuses on passive adaptive facades, ie, systems capable of changing their configuration autonomously without the need for external control and power supply. These systems usually employ smart materials able to vary their configuration according to external environmental stimuli such as air temperature, solar heat (SABER [self-activated building envelope regulation] facade system, thermobimetal ventilating skin, passive deployable insulation), or air humidity (water-reacting facade, Hydromembrane). Evaporative facade systems are made of materials able to accumulate great quantities of water and release it as vapor, subtracting heat from the envelope and thus cooling the interiors (Hydroceramic systems, Cool bricks, TiO_2 evaporative cooling).

6.3.1 SABER breathing facade

The SABER (self-activated building envelope regulation) breathing facade system was developed by the BIOMS team of researchers at the University of California, Berkeley. A bioengineering approach, inspired by the human skin, devised a thin multifunctional smart surface intended to wrap around buildings, capable of autonomously opening and closing its pores, and regulating air intake according to external climatic conditions of heat, humidity, and spectral selectivity. The membrane's dehumidification properties also allow it to control indoor air humidity.

The key element of the system lies in a geometrical network of poly (*N*-isopropylacrylamide)—a temperature-responsive phase-change hydrogel capable of swelling or shrinking at a given temperature, releasing or absorbing water vapor with a 90% volume variation. A customized 3D printing process with multiple prototype syringes produces asymmetrical pores (pore diameter/film thickness $= 0.05-0.70$) to induce a stress gradient. Matching the critical temperatures required for human physiological microenvironments, controlled displacement is accomplished to yield closed pores at $T = 20°C$ and a pore diameter of 1.5 mm at $T = 40°C$. The material mechanism takes 20 min to open and complete closure is reached in 10 min.

Further integrated with light-sensing lenses, this membrane of microscale valves can thus be fine tuned to open at the desired climatic conditions and allow natural ventilation. Its energy-free operation makes it particularly suitable for use in developing tropical countries and disaster relief operations.

6.3.2 Thermobimetal

Other promising materials for passive facade system application are thermobimetals, which consist of two metals with different thermal expansion coefficients laminated on top of each other. The Smart Thermobimetal Self-Ventilating Skin prototype devised by DO|SU Studio Architecture is able to open its pores to allow ventilation when exposed to heat and UV rays: thermobimetal panes react by expanding and contracting unevenly at different rates, creating tensions capable of bending the surface and eventually varying the configuration of the building envelope.

6.3.3 Passive deployable insulation

The deployable external insulation project at Bartett School of Architecture exploits the physical properties of the PCM paraffin wax. This adaptive facade system is composed of insulating blinds which open during the day to let air and natural light in and close at night to ensure maximum thermal insulation when the temperature drops. The insulating blinds are operated by self-actuated pistons containing paraffin wax. The wax melts or solidifies when a temperature set point is reached, dramatically changing its volume and thus operating the pistons to open or close the blinds. The temperature threshold for piston operation can be carefully tuned, acting on the wax composition to suit this adaptive facade to different climate zones.

6.3.4 Water-reacting facade

Air humidity is the environmental parameter considered by the water-reacting architectural surface, devised by product designer Chao Chen. It takes inspiration from the pinecone, which is able to open and close its shingles according to air humidity.

Veneer and nylon/polyester/styrene are laminated to create a bilayer material with different hygroscopic expanding ratios that may bend or stretch depending on moisture. This way it is possible to achieve facades capable of letting air and light through in good weather, while at the same time protecting the building from elements like rain. In rainy climates the operation would be reversed: when it rains, the water-reacting tiles open up to let light in, closing when they sense sunshine to protect inner walls from solar radiation and prevent overheating.

6.3.5 Hydromembrane

The Hydromembrane, developed at Institute for Advanced Architecture of Catalonia (IAAC) in Barcelona, is a humidity-sensitive composite system which can be used in different potential applications at macroscale or microscale, from building facades to clothes. The Hydromembrane reacts to moisture with aperture deformations, and a secondary cooling effect appears due to its components' property of water absorption and evaporation.

The membrane is organized in six layers consisting of three materials with significantly different physical and chemical features: triple-stacked hydromorph connectors/actuators, an elastic textile membrane with silicone coating, and an anchoring silicone grid structure.

Key to the system are its hydromorph actuators, created by imbuing a laser-cut double-sided textile tape 0.5 mm thick with a high-absorbing liquid such as vinegar or water with color pigments. The resulting devices show shape-memory properties along with high volume expansion, swelling up to 60% of their original size with enough force to induce membrane deformation, evaporative cooling, and water storage up to 12 h.

The laser-cut 0.1 mm thick textile membrane, coated with a 0.5 mm film of silicone, is connected to the silicone grid by the hydromorph actuators, which are also employed to fix the substructure to the building and as a final external layer.

With enough air humidity, the triple-stacked hydromorph connectors absorb moisture, swell, and bend each textile tile to allow natural ventilation. Their high water retention property allows slow release of absorbed water by evaporation in the passing air, controlling indoor heat and humidity (Fig. 6.19).

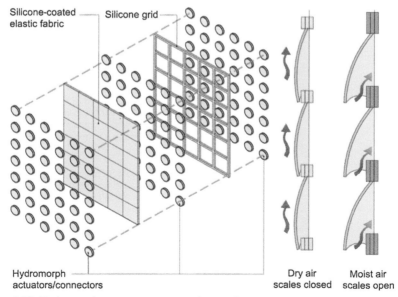

Figure 6.19 Hydromembrane components and operation.

6.3.6 Hydroceramic

Students at the Digital Matter Intelligent Constructions studio at Barcelona's IAAC have devised Hydroceramic, an architectural element that employs globes of hydrogel, a material able to absorb up to 500 times its weight in water, usually made of insoluble polymers of hydroxyethyl acrylate, acrylamide, or polyethylene oxide (Fig. 6.20). By sandwiching hydrogel pellets between two breathable ceramic layers and a water supply fabric, a large evaporating surface is achieved, in contact with both interior and exterior, which can be applied on walls and roofs. Testing has shown a reduction of internal temperature of more than 6°C after 20 min, with an increase in air humidity of 15.5%. This solution is already weather responsive, since the evaporation and thus the cooling effect are greatest for higher temperatures and decrease when the surrounding air is already cool. This system claims to be inexpensive in manufacture (estimated cost €28/m^2) and operation (requires 1 L/m^2 water to function, which can be supplied directly by rainwater or from a harvesting tank), saving up to 28% of power consumption.

6.3.7 Cool bricks

Cool bricks are interlocking modular blocks designed by studio Emerging Objects (Fig. 6.21). Their 3D-printed, porous ceramic structure is able to accumulate rainwater like a sponge while remaining air permeable and then to cool interiors by evaporating water into incoming airflow, following a principle proven millennia ago by evaporative windows in Muscat, Oman.

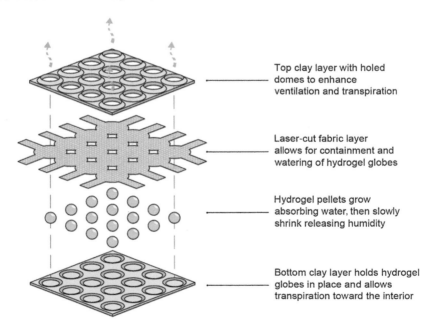

Figure 6.20 Hydroceramic.

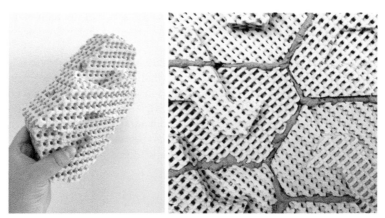

Figure 6.21 Cool bricks.
Courtesy of Emerging Objects.

6.3.8 TiO₂ photocatalyst evaporative shell

Evaporative cooling may also be exploited by spraying a continuous water layer on building surfaces to cool them by subtracting heat via its evaporation. A recent innovative method takes advantage of a TiO_2 photocatalyst, which when irradiated by the sun causes the surface to become highly hydrophilic, minimizing the amount of water consumption to form the water film. This way it is possible to cover the whole building with little water supply, as the water layer is just 0.1 mm thick. Testing showed a temperature drop of 15°C on window glass and 40−50°C on black roof-tile surfaces on a clear day in the middle of summer, a promising result that could significantly reduce electricity consumed for air-conditioning or avoid its need altogether.[21]

6.4 Conclusions and future trends

In the last two decades building skin has become an innovative research topic due to the need to reduce building energy consumption and environmental impact and improve internal comfort and architectural integration.

The need to reduce the effects of cities overheating in the summer, as well as mitigate the increase in energy consumption for air cooling, is greatly spreading the use of reflective coatings (cool roofs) as building skin cover and technological solutions to integrate vegetable species as cladding systems, especially in hot climate and Mediterranean countries.

The latter can also bring important advantages regarding both indoor and outdoor air quality.

Today, cool roof development is focused on prolonging useful life and dirt resistance of coatings, and on creating more recyclable and environmentally friendly products based on ecological or recovered materials. Research on innovative radiative cooling surfaces is also showing promising results. The scope of research on green and living walls is instead focused on growing substrate materials and hydroponic watering systems.

The greatest future potential is in realizing building skin facades able to adjust their physical properties and energetic performance dynamically according to changing demands from indoor and outdoor conditions, without the need for external control or power supply (environment-adaptive skin facades).

Related research is concentrated on the application of suitable smart materials (PCMs, memory alloys, hydrogel) capable of varying their configuration according to external environmental stimuli such as air temperature, solar heat, or air humidity.

References

1. Global Cool Cities Alliance. *A practical guide to cool roofs and cool pavements.* New York: GCCA; 2012.
2. Wang X, Kendrick C, Ogden R, Maxted J. Dynamic thermal simulation of a retail shed with solar reflective coatings. *Appl Therm Eng* 2008;**28**(8−9):1066−73.
3. Clatrans Division of Research and Innovation. *Cool pavements research and technology.* Berkeley, CA: Institute of Transportation Studies Library; 2011.
4. Casini M. *Costruire l'ambiente. Gli Strumenti e i metodi della progettazione ambientale.* Milano: Edizioni Ambiente; 2009.
5. IEA. *Technology roadmap: energy efficient building envelopes.* Paris: OECD/IEA; 2013.
6. Smith GB, Granqvist CG. *Green nanotechnology: solutions for sustainability and energy in the built environment.* Boca Raton, FL: CRC Press; 2011.
7. Gentle AR, Smith GB. A subambient open roof surface under the mid-summer sun. *Adv Sci* 2015;**2**(9).
8. Bell JM, Smith GB. Advanced roof coatings: materials and their applications. In: *Proceedings CD and summary book of the CIB 2003 international conference on smart and sustainable built environment. Brisbane, Queensland.* Brisbane. CIB; November 2003. W116.
9. Raman AP, Anoma MA, Zhu L, Rephaeli E, Fan S. Passive radiative cooling below ambient air temperature under direct sunlight. *Nature* 2014;**515**:540−4.
10. Karlessi T, Santamouris M, Synnefa A. Development and testing of thermochromic coatings for buildings and urban structures. *Sol Energy* 2009;**83**(4):538−51.
11. Kalaiselvam S, Parameshwaran R. *Thermal energy storage technologies for sustainability.* London: Academic Press; 2014.
12. Perez G, Rincon L, Vila A, Gonzalez J, Cabeza LF. Behaviour of green facades in Mediterranean continental climate. *Energy Convers Manag* 2011;**52**(4):1861−7.
13. Stav Y, Lawson GM. Vertical vegetation design decisions and their impact on energy consumption in subtropical cities. In: Pacetti M, Passerini G, Brebbia CA, Latini G, editors. *The sustainable city VII: urban regeneration and sustainability.* Ancona, Italy: WIT Press; 2012.
14. Perez G, Coma J, Martorell I, Cabeza LF. Vertical greenery systems (VGS) for energy saving in buildings: a review. *Renew Sustain Energy Rev* 2014;**39**:139−55.
15. Pugh T, Mackenzie R, Whyatt JD, Hewitt CN. Effectiveness of green infrastructure for improvement of air quality in urban street canyons. *Environ Sci Technol* 2012;**46**(14): 7692−9.

16. Azkorra Z, Perez G, Coma J, Cabeza LF, Bures S, Alvaro JE, et al. Evaluation of green walls as a passive acoustic insulation system for buildings. *Appl Acoust* 2015;**89**:46−56.
17. Van Renterghem T, Hornikx M, Forssen J, Botteldooren D. The potential of building envelope greening to achieve quietness. *Build Environ* 2013;**61**:34−44.
18. Blanc P. *The vertical garden: from nature to the city*. New York: Norton & Company; 2012.
19. Afrin S. *Green skyscraper: integration of plants into skyscrapers*. Master thesis. Stockholm: Kungliga Tekniska Högskolan: Department of Urban Planning and Environment; 2009.
20. Sheweka SM, Mohamed NM. Green facades as a new sustainable approach towards climate change. *Energy Procedia* 2012;**18**:507−20.
21. Hashimoto K, Irie H, Fujishima A. TiO_2 photocatalysis: a historical overview and future prospects. *Jpn J Appl Phys* 2005;**44**(12):8269−85.

Part Three

Smart windows

Advanced insulation glazing

7.1 Advanced low-emission glazing

The thermal insulation of transparent closures is a prerequisite in the pursuit of thermal comfort of interiors and the reduction of energy consumption in buildings. Currently, over 40% of all windows in the European Union are still single glazed, with thermal transmittance U_g values higher than 5.5 W/m²K, causing high dispersion of energy and local discomfort due to the low temperature of the glass panes in winter.[1] Another 40% are untreated double glazing with an air-filled cavity, with transmittance values lower than single glazing (3.3–2.7 W/m²K) but still higher than standards set by law today.[1] According to the Lawrence Berkeley National Laboratory, energy lost through windows accounts for 4–5% of the total annual consumption of energy in the United States, at a cost of about $50 billion a year.

To reduce heat loss through glazing, one can intervene in both conduction and irradiation heat transfer mechanisms.

For conduction, increasing the thickness of the glass pane does not affect thermal performance. Heat transfer is regulated primarily by convective motions and irradiation, and most of the thermal load is localized in correspondence to the adduction thermal resistances. Therefore, even doubling the thickness of the glass pane does not result in a significant reduction of its transmittance: a 5 mm thick float glass has an U_g transmittance equal to 5.88 W/m²K, while a 10 mm thick pane has a U_g equal to 5.71 W/m²K.

To reduce the passage of heat it is necessary to interpose an air gap between two glass panes (double glazing), to give greater resistance to heat flow. In fact, still air has a conductivity equal to 0.026 W/mK, compared to a glass value equal to 1 W/mK. Thus if a single 10 mm thick glass pane has a transmittance of 5.71 W/m²K, a double-glazing unit comprising two glass panes, each 4 mm thick, separated by a 6 mm gap (4–6–4) has a transmittance of 3.3 W/m²K.[2]

Cavity thickness should not be more than 15 mm, otherwise convective motions to increase heat transfer between the two panes are triggered, reducing energy performance. In particular, thermal resistance of the insulating glass unit increases with interspace thickness up to 15 mm, remains constant up to about 20 mm, and is reduced for greater thicknesses.

Dividing the air cavity into two separate hollow spaces (triple glazing) interrupts convection and further reduces the transport of energy: a double-glazing unit composed of two glass panes and a 12 mm cavity (4–12–4) has a transmittance of 2.9 W/m²K; with three 4 mm thick float glass panes and two cavities of 12 mm each (4–12–4–12–4), transmittance drops to 1.9 W/m²K (Fig. 7.1).

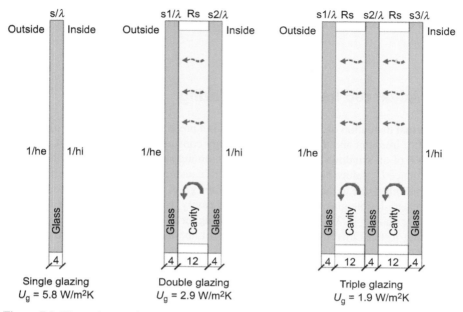

Figure 7.1 Thermal transmittance of a single-glazed, double-glazed, and triple-glazed window.

Austria, Germany, and Switzerland have the highest market share for triple glazing, usually with two low-emissivity (low-e) surfaces, at 54% of total window sales. New constructions and the residential sector have the highest market penetration. Overall, the majority of windows sold in the European Union are still double glazed.[3,4]

Superior performance is achieved by filling the cavity of the insulated glass unit (IGU) with gases with lower conductivity than air (argon, xenon, krypton, or mixtures thereof), which, being heavier than air itself, also reduce interior convective motions (Table 7.1).

In choosing the filling gas, economic considerations weigh heavier than thermal effects. Although xenon and krypton possess better thermal properties, the most used gas is argon, which can be easily produced from the environmental air where it is contained in a proportion of about 1%.[6]

To reduce further the flow of heat through the windows, however, acting only on conduction by using multiple-glazing units is insufficient; it is also necessary to intervene in the phenomenon of irradiation. Glass has a high absorption coefficient for wavelengths emitted by bodies at room temperature, 2–30 μm (thermal infrared (IR) radiation), and as a consequence has high emissivity for the same wavelengths (the normal emissivity for uncoated soda lime glass and borosilicate glass is equal to 0.837). Internal heat radiates outward through the glass in the form of IR radiation, and the heat flow is proportional to the emissivity (Table 7.2).

Surface coatings with a thickness of 0.01–1 mm can improve the physical properties of glass regarding radiation; depending on layer thickness and composition,

Advanced insulation glazing

Table 7.1 Main properties of IGU filling gases at 10°C temperature[5]

Property	Filling gas				
	Air	Argon	Krypton	Xenon	SF_6
Volume mass (kg/m^3)	1.232	1.699	3.560	5.689	6.360
Thermal conductivity (W/mK)	2.496×10^{-2}	1.684×10^{-2}	0.900×10^{-2}	0.529×10^{-2}	1.275×10^{-2}
Specific heat (J/kgK)	1.008×10^3	0.519×10^3	0.245×10^3	0.161×10^3	0.614×10^3
Dynamic viscosity (kg/ms)	1.761×10^{-5}	2.164×10^{-5}	2.400×10^{-5}	2.226×10^{-5}	1.459×10^{-5}

Table 7.2 Main energy parameters in glazing[7]

Glazing	Typical section	Typical values		
		U_g	Solar heat-gain coefficient (SHGC)	Visible light transmission (VLT) (%)
Single pane	Float glass 4–8 mm Laminated glass 6–10 mm	5.9	≈0.85	≈0.90
Double glazing	Glass panes 4–8 mm Cavity 12–16 mm Emissivity (ε_n) 0.837 Filling: air	≈2.7	≈0.78	≈0.82
Double glazing with low-e coating	Glass panes 4–8 mm Cavity 12–16 mm Emissivity (ε_n) 0.03–0.05 Filling: air, argon	≈1.1–1.3	≈0.62	≈0.80
Triple glazing with low-e coating	Glass panes 4–8 mm Cavity 12–16 mm Emissivity (ε_n) 0.03–0.05 Filling: air, argon	≈0.6–0.7	≈0.55	≈0.70

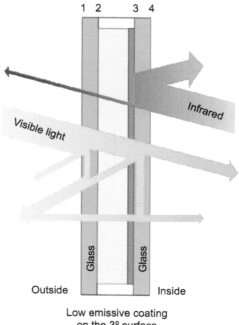

Figure 7.2 Double glazing with low-e coating.

energy transmission (ET) can be reflected or absorbed and emissivity reduced. By coating at least one of the four glass surfaces of a double-glazing unit with a layer of material highly reflective to IR radiation (low-e), it is possible to obtain insulating glazing with high thermal performance. Instead of radiating part of the absorbed heat outward and dissipating it, the glass pane reflects it toward the interior, reducing overall transmittance (Fig. 7.2).

The coating is applied on the glass surfaces facing the cavity to avoid damage and oxidation. For the purpose of thermal insulation, the position of this layer related to the air cavity (on the interior or exterior pane) is irrelevant, but it can affect the amount of solar energy that penetrates into the interior. Since these windows are used in cold climates, where the goal is to allow the greatest possible amount of solar energy to contribute to the heating of interiors, the coating is placed on the surface of the inner pane of the double glazing, on the cavity side (the so-called third position starting from the outside), thus optimizing solar gain. Improvements in thermal insulation can only be achieved with a reduction of the solar factor. The low-e coating absorbs the part of solar radiation that it is no longer possible to accumulate, even through thermal adduction. A reduction in the VLT factor must also be noted.

Thanks to these coatings, low-e glazing reflects toward interiors up to 90% of the radiation produced with a wavelength greater than 2.5 µm, preventing it from escaping, while the transmission of sunlight remains largely unchanged. The coefficient of visible

Table 7.3 **Examples of advanced insulating glass units available on the market**

Advanced insulating glass unit	Composition	U (W/m²K)	VLT (%)	ET (%)	SHGC
Untreated double glazing	4—6 air—4	3.5	81	72	0.77
Saint-Gobain Climaplus	4—16 argon—4	1.2	80	53	0.64
Pilkington Optitherm S1	4—16 argon—4	1.0	70	42	0.48
AGC Thermobel Top N+	4—15 argon—4	1.1	78	n.d.	0.60
Guardian ClimaGuard Premium	4—10 krypton—4	1.0	80	n.d.	0.63

Table 7.4 **Example of typical composition of a low-e coating**[8]

Layer	Material	Function in low-e coating
Protective coating	SiN_xO_y	Chemically and mechanically resistant top layer
Antireflective layer	SnO_2	Protective and antireflection coating
Interface layer	TiO_x	Suboxidic layer
Metal layer	Ag	High IR reflectance layer
Seed layer	ZnO	Optimizes growth of silver layer so the metal can be very thin
Antireflective layer	SnO_2 or TiO_2	Increases light transmittance of low-e coating
Glass substrate	—	—

transmission is equal to 70—80% and the SHGC to 50—70% (Table 7.3). High thermal performance insulating glass shows no difference with double glazing; the layers have neutral colors so they are practically invisible.

Low-e coatings usually employ multiple layers of metals and dielectric thin films, the latter consisting of doped oxide semiconductors. These are arranged in a dielectric/metal/dielectric layer scheme (see Table 7.4).[8] Metals such as silver, gold, and copper have high IR reflectance, with silver being frequently used because it combines high IR reflectivity and color neutrality.

Thin dielectric films are designed to protect the metal film, increase its light transmittance, and serve an antireflective purpose. In addition to protection, they act as seed and sacrificial layers: the seed layer undercoats the metal film and optimizes its crystalline growth, and the sacrificial layer gives a suboxidic interface with the metal film. Frequently used dielectric materials are TiO_2, SnO_2, SiO_2, ZrO_2, ZnS, ZnO, $SnBO_2$, In_2O_3, Si_3N_4, and Bi_2O_3.

These coatings may be further articulated in two variations according to the choice of the antireflective layers:[8]

- symmetrical coatings with two antireflective layers of the same refractive index (RI) (both SnO_2, with 2.0 RI);
- asymmetrical coatings with dielectric antireflective layers of different refractive index (TiO_2 with 2.5 RI and SnO_2 with 2.0 RI)—this composition creates a low-e coating with higher light transmittance than a similar symmetrical coating.

High-performance low-e window coatings are designed with double or triple metal layer films. In addition to seed, antireflective, and sacrificial layers, the metal film is further separated into two or three thin layers that alternate with dielectrics. These multilayer dielectric/metal/dielectric coatings serve as broadband IR reflective coatings, reducing thermal radiation losses through the glass and also providing solar control due to reduction of solar IR transmittance (Figs. 7.3 and 7.4).

Today, low-e coatings are found in the high-range IGUs of all major manufacturers (Saint-Gobain Climaplus, Pilkington Optitherm, AGC Thermobel Top N, Guardian ClimaGuard) in multiple configurations, and are also engineered for solar control, acoustic insulation, or safety. Top-of-the-line products may even take advantage of nanotechnology, employing coatings obtained by layered deposition of nanoparticles of metallic oxides and nitrides via magnetic-enhanced cathodic sputtering under vacuum conditions (Saint-Gobain SGG Nano).

Furthermore, while gold-based reflective coating technology is very expensive for window glazing and is not color neutral, new nanomaterials offer gold nanoparticle coatings with optimized reflectance and color.

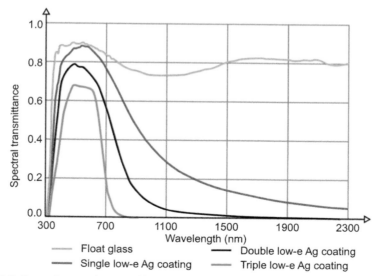

Figure 7.3 Spectral transmittance of low-e coated glass.
Redrawn from Casini M. *Costruire l'ambiente. Gli Strumenti e i metodi della progettazione ambientale*. Milano: Edizioni Ambiente; 2009.

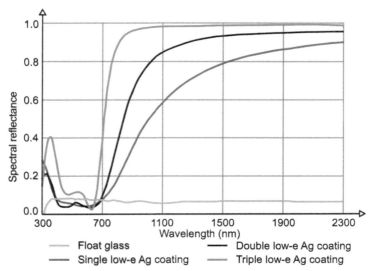

Figure 7.4 Spectral reflectance of low-e coated glass.
Redrawn from Casini M. *Costruire l'ambiente. Gli Strumenti e i metodi della progettazione ambientale.* Milano: Edizioni Ambiente; 2009.

Depending on the thickness of the cavity and the filling gas, low-e glazing has thermal transmittance values between 1.7 and 1.0 W/m²K for double glazing and values lower than 0.7 W/m²K for triple glazing. Superior performance can be reached by combining more coatings: for example, Pilkington K Glass OW provides a magnetron selective coating on the first wall and a low-e pyrolytic coating on the fourth, totaling a thermal transmittance of 0.8 W/m²K; and Guardian ClimaGuard 1.0 triple glazing reaches a transmittance U of 0.4 W/m²K thanks to the low-e coating on the second and fifth surfaces.

7.2 Suspended film glazing

Another approach to multilayer glass solutions employs polymeric films instead of glass panes as interior cavity partitions.[9] Called suspended films, these can be effective in allowing larger cavities for better thermal transmittance without adding excessive weight compared to triple- or quadruple-glazed windows, which often require reinforced frames and may for the same reason have size constraints (Fig. 7.5).

Suspended films may be enhanced with low-e coatings and are treated to resist ultraviolet (UV) degradation, as well as being heat shrunk to avoid wrinkling over time.

Performance is on par with multilayered glazing, with thermal transmittance U values around 0.80–0.60 W/m²K for double-glazed IGUs with a single film and special applications reaching U values up to 0.28 W/m²K (Hurd Ultra-R window with double glass panes, three coated suspended films, and four krypton-filled

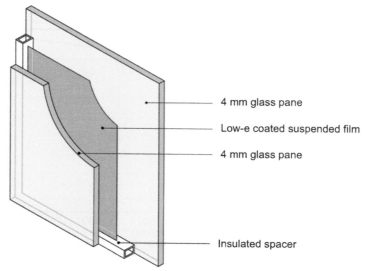

Figure 7.5 Suspended film glazing.

cavities). Noise insulation performance is also notable, with R_w values up to 50 dB (Visionwall glazed facade) compared to 40 dB for traditional double-glazed units.

The ability to reach high thermal insulating values without adding too much weight or increasing size makes suspended film insertion a viable solution for upgrading existing windows without replacing frames or glass panes. An example is the window retrofit intervention on the Empire State Building, which was carried out on site by disassembling the 6500 double-glazed window units, inserting the suspended coated film, and reassembling the glass panes with argon and krypton filling. The intervention managed to upgrade thermal transmittance U from 2.80 W/m²K to 0.95 W/m²K and halve heat gain, while reusing 96% of original windowpanes and frames.

Suspended film glazing is mainly available in the United States, where it is manufactured by Hurd, Alpenglass, and Visionwall.

7.3 Vacuum insulating glass

Vacuum insulating glass (VIG), already widespread in Asia, consists of two glass panes separated by a thin cavity (0.2 mm) at a pressure of 10^{-2} mbar (Fig. 7.6). Due to the vacuum inside the cavity, heat flows caused by conduction and convection are completely blocked. The high pressure exerted on the two plates makes it necessary to insert microspacers in the cavity. The thermal performance of available products (Pilkington Spacia) is comparable to conventional double glazing (1.1 W/m²K), but obtained with a total thickness of only 6.5 mm, compared to 15–20 mm usually required, allowing the energy upgrade of single-glazed windows without the need to change existing frames.

Advanced insulation glazing

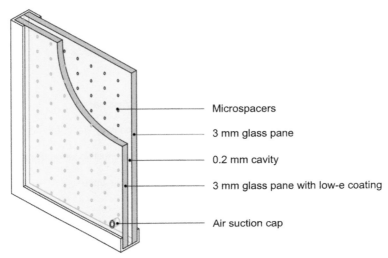

Figure 7.6 Vacuum insulating glass.

Research is now focusing on perimeter sealing and spacers with the aim of reaching an internal pressure in the order of 10^{-3} mbar, which would allow a thermal transmittance U of 0.4 W/m²K.[10] Nevertheless, more research and development is needed to achieve commercially viable high-performance vacuum glazing systems. In the United States several research efforts are under way to develop a vacuum glazing process compatible with the main manufacturing processes in window production and viable in all climates, but a successful product has not yet been developed. China similarly has been working on vacuum glazing systems, but there are limitations on product applicability and large-scale production is yet to be initiated.

7.4 Monolithic aerogel insulating glazing

A further development in the technology of VIG is represented by evacuated monolithic aerogel insulating glazing (Fig. 7.7). These windows consist of two glass panes separated by a cavity of 14—20 mm at a pressure of around 10 hPa filled with a monolithic aerogel slab ($\lambda = 0.010$ W/mK).

Monolithic aerogel is still difficult to supply in large format and has little market penetration. Airglass AB, one of the few suppliers, manufactures only on demand and size is limited to 60 cm wide × 60 cm long × 3 cm thick. This implies that is still not possible to manufacture large-size single panes, so monolithic aerogel glazing needs multiple frames to be realized.

Experimental results on energy performance are promising, showing excellent thermal and solar ET characteristics (evacuated aerogel between two layers of low-iron glass with 15 mm glass distance had a U value below 0.7 W/m²K and solar

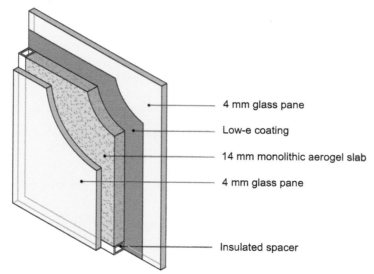

Figure 7.7 Monolithic aerogel insulating glazing.

transmittance of 76%[11]; an aerogel glazing with 20 mm glass distance reached a U value below 0.5 W/m²K combined with an SHGC above 0.75,[12] along with good noise insulation (+3 dB R_w compared to traditional double glazing),[13] even better than available granular aerogel fillings. Additional testing has shown thermal transmittance values for a 4−14−4 IGU of 0.6 W/m²K for evacuated monolithic aerogel filling at a pressure of 10 hPa, compared to U values of 1.1 W/m²K for granular filling.[13] Furthermore, a monolithic IGU has the main advantage of much better visible light transmittance (VLT = 60% compared to 27%), along with higher SHGC (0.70 compared to 0.32), and allowing vision through.[13] Vision is hazy, however, due to internal Rayleigh scattering, appearing bluish to the viewer and giving a red tint to transmitted light.

7.5 Advanced window frames

Frames of windows constitute in most cases the weak point of transparent closures from the energy point of view, in terms of both airtightness and thermal transmittance of the building components. A frame with poor insulating characteristics may in fact undermine the overall performance of the window even in the presence of advanced thermal insulation glass (see Table 7.5). To improve the thermal performance of the frame, one may resort to inherently less conductive materials such as wood or polyvinyl chloride (PVC), or, in the case of metal frames, interrupt the continuity of the material with polyurethane or polyamide membranes and insulating bars (thermal breaks).

Metal frames remain essential for most demanding structural requirements, and research on increasing insulation performance of windows has led to improvement

Table 7.5 Frame and global thermal transmittance of different window frames on the market

Frame material	Window frame thermal transmittance U_f (W/m^{2}K)	Window global thermal transmittance U_w (W/m^{2}K)[a]
Aluminum without thermal break	5.2–7.0	2.27–2.63
Steel	3.4–4.2	1.91–2.07
Aluminum with thermal break	2.0–3.8	1.63–1.99
Wood	1.8–2.4	1.59–1.71
PVC	1.2–2.2	1.47–1.67
Aluminum with aerogel	0.57	1.34

[a]Values for a 1.2 × 1.5 m window with 20% frame surface and $U_g = 1.3$ W/m^{2}K.

of frame insulation with state-of-the-art thermal breaks (40 mm polyamide strips), optimized triple-gasket chamber design, and foam filling, achieving thermal transmittance values of 2.0 W/m^{2}K and below. By comparison, the first nonthermal-break aluminum window frames introduced in the market in the 1930s had a transmittance of 7.8 W/m^{2}K, while frames with simple thermal breaks from the 1970s reached 4.8 W/m^{2}K (Fig. 7.8).[3,4,7]

A further step involves the integration of innovative materials such as aerogel to improve thermal break characteristics. Poland-based Aluprof succeeded in inserting aerogel bars inside its MB 86 aluminum frame, achieving a thermal transmittance value U_f of 0.57 W/m^{2}K, almost four times less than a traditional thermal break aluminum frame. When used in combination with an advanced insulating glass, window global thermal transmittance U_w can reach 1.10 W/m^{2}K with an U_g of 1.0 W/m^{2}K.

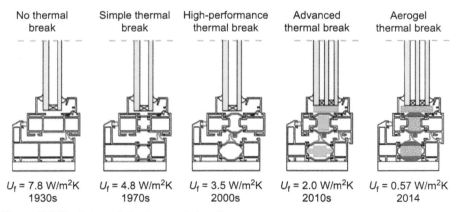

Figure 7.8 Evolution of aluminum window frames.

Nevertheless, inefficient window frames with no thermal break are widespread in the existing building stock around the world and are still being installed in many countries, especially those outside the Organisation for Economic Co-operation and Development.

7.6 Glazed double-skin facades

The widespread use of fully glazed facades for formal and interior daylight requirements, especially for buildings related to tertiary or service activities, has led to the development of technological solutions aimed at increasing thermal and acoustic performance to improve thermal comfort and reduce energy consumption for air conditioning (Figs. 7.9 and 7.10).

The most innovative solutions are represented by systems with a double glass envelope (double-skin facades) consisting of two distinct glazed walls separated by an air cavity with thickness varying from 10 to 120 cm, inside which solar shading systems are generally placed.[6,14–19]

The internal glazing is constituted by an openable hinged or sliding window frame with insulating glass, which allows natural ventilation and access to the cavity for inspection and maintenance. The outer wall generally consists of independent slabs or suspended glazing (with punctual or mullion and transom fixing), usually a single pane of low-iron extra-clear glass to allow maximum transparency, and is integrated with the ventilation openings. Shading systems can be positioned inside or outside the cavity using an active (louvers or blinds) or passive (thin sun screens) type.[20] Light redirection systems such as reflective blinds or anidolic mirrors may also fit inside the cavity to extend natural light reach inside.

The system is completed by devices for adjusting the air flow in relation to external and internal climatic conditions, placed horizontally to correspond with floor slabs,

Figure 7.9 Glazed double-skin facade, interior view, Chicago Art Institute, RPBW, Chicago. Photo by the author.

Advanced insulation glazing 261

Figure 7.10 Glazed double-skin façade, exterior view, Chicago Art Institute, RPBW, Chicago. Photo by the author.

vertically on the outer glazing, or on the cavity borders. Usually the valves operate automatically, controlled by local control systems based on temperature and/or radiation sensors. Integrated thermal dampers and heat recovery systems may extend fresh air ventilation in summer and winter months.

User control override may be provided for ventilation by either regulating valves or opening windows when the external climate is appropriate, and for operating integrated blinds to control daylight. Most advanced systems also give advice to occupants to avoid disrupting thermal and ventilation balance.[21]

The double-skin glazing system is designed as a selective envelope with the main function of dynamically manage heat flows to obtain improved interior comfort and a secondary function to protect the inner wall. The system also allows integration with air conditioning for better energy saving (Fig. 7.11).

During winter, solar radiation triggers a greenhouse effect inside the cavity that heats the air. Heat is distributed to the interiors directly by convection and indirectly through the inner glazing. Solar-heated air may be distributed laterally around the building by corner ventilators to equalize the thermal load on the different sides. Solar blinds may be folded to allow maximum solar gain or turned on their absorbing side to increase air temperature in the cavity.

Summer–natural ventilation
Heat is extracted from cavity by stack effect

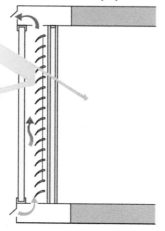

Summer–mechanical ventilation
Cavity is cooled with exhaust air

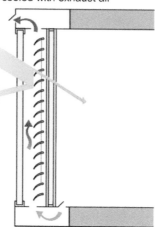

Winter–natural ventilation
Outside air is preheated before immission

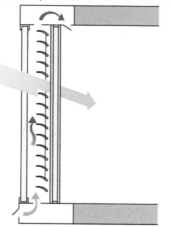

Winter–mechanical ventilation
Exhaust air is re-heated and recirculated

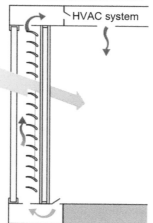

Figure 7.11 Operation modes of double-skin glazed facade.

The cavity is ventilated by return air from the interiors, which is extracted at the base of the air cavity, preheated, and sent to air-conditioning devices through recovery ducts in the upper part of the facade. Using the air cavity as a static air buffer, by sealing the heat in the cavity can reduce energy consumption by 20–30% according to the building's latitude.[22]

In summer the free circulation of air in the cavity produces a current flow (stack effect), which, if properly supported by shading systems located in the protected air gap, contributes to the reduction of interior overheating. The cavity between the two skins is ventilated by outdoor air that enters via the lower external inlets and is ejected from the top by stack effect or radial fans that are activated when the temperature in the ventilation chamber exceeds a predetermined threshold. The cavity can also be cooled with conditioned exhaust air, exploiting its temperature difference with outside air. External vents and inner facade windows may be left open overnight to allow night-time cooling of interiors.

In relation to the characteristics of the cavity, the system is articulated in two main types: a continuous (on the entire surface) or discontinuous (in channels, corridor or cells) air gap.

The first type is characterized by a single cavity space spanning the entire height of the building. The air between the interior and exterior glazing flows from the ground floor to the top floor without interruptions that may halt the circulation.

The second type involves a division of the interspace by vertical and/or horizontal baffles in correspondence with each mullion and/or floor slab that connects the two facades. The single cells are overlaying and completely independent in the management of energy flows, although this tends to decrease the efficiency of the stack effect, whose flow rate is directly proportional to the cavity height.

In both summer and winter the presence of the air cavity keeps the temperature of the inner glass pane closer to that of interior air,[23] with a consequent reduction of irradiation phenomena which may cause local discomfort, thus increasing the usability of interiors. The presence of the outer pane also protects the internal glazing from atmospheric elements and greatly reduces noise levels, allowing interior windows to be kept open for natural ventilation.

The main disadvantages of these systems are a possible increase in the summer thermal load if the system is not well designed, acoustic transmission through the cavity between the adjacent spaces (for continuous glazing), and the propagation of fires through the air gap.[24]

Recent developments in double-glazing technology include closed-cavity facades (CCFs), where the air gap between the internal and external panes is completely sealed with no air exchange between interior and exterior environments (Fig. 7.12). An example is Permasteelisa's MFREE-S facade system, which employs an external 6 mm extra-clear glass pane, a narrow air cavity (150–600 mm), and insulating, low-e coated double glazing on the inside: integral to the system operation are electrically controlled blinds located in the air gap, which are not in contact with any weather element or uncontrolled air, and a central air distribution system that supplies dry, clean, low-pressurized air (15 mbar) to each CCF unit to prevent condensation. Outside conditions are constantly monitored electronically, and the quantity of dry air is adjusted accordingly to keep energy consumption to a minimum. Sun protection systems are thus virtually maintenance free, do not require any cleaning, and do not pose soiling or contamination risks, making this system highly suitable for healthcare applications. University College Hospital's Cancer Centre in London features a 2500 m^2 CCF, preferred to a traditional double-skin façade for hygienic reasons to avoid pollution of blinds, that earned it a Building Research Establishment Environmental Assessment Methodology (BREEAM) Excellence Hospital label.

Furthermore, since the air cavity does not require frequent access for cleaning and maintenance, windows on the inside can be effectively eliminated, removing safety

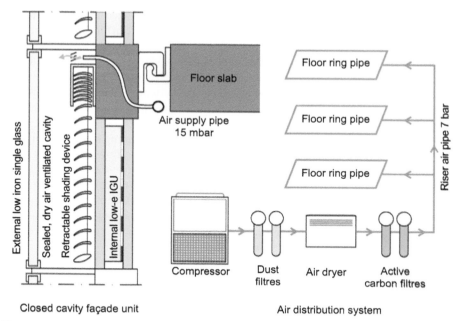

Figure 7.12 Closed cavity façade.

and security issues and freeing up leasable space. The controlled environment within the cavity gives optimal conditions for operating dynamic screening systems, including light guidance and retroreflection systems, achieving best results for radiation, daylight, and glare control even with highly transparent low-iron glazing. If required CCF windows can be opened to evacuate smoke in case of fire or for natural aeration. Finally, CCFs claim notable performance results: $U_w = 0.90$ W/m²K, SHGC $= 0.08$, VLT $= 0.04$ with integrated blinds down and $U_w = 1.20$ W/m²K, SHGC $= 0.48$, VLT $= 0.63$ with integrated blinds up, and excellent aerial noise control, with sound-absorbing values R_w up to 58 $(-2, -4)$ dB.

7.7 Heating glazing

Released to the market in Finland in the 1980s and now produced by Saint-Gobain Glass, with Quantum E-Glas, the British IQ Glass, and the American Vitrius Technologies and ThermiqueTech, self-heating glazing exploits a semiconductive aluminum oxide coating which heats up by Joule effect when an electric current is applied and irradiates heat in one direction of another according to the positioning inside the double-glazing unit, without compromising the characteristics of VLT (70%).

By placing the coating on the second surface it is possible to direct heat toward the outside of the building, preventing snow accumulation during the winter to ensure light access to windows and skylights and enable lighter support structures. Positioning the

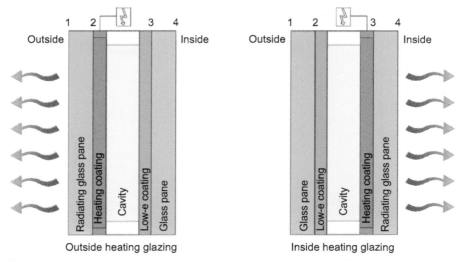

Figure 7.13 Double glazing with heating coating on the second and third surfaces.

heating coating on the third surface of the double-glazing unit instead directs the heat toward the interior, creating a radiant surface and preventing the condensation of water vapor on the glass surface even in wet rooms such as steam baths or pools (Fig. 7.13). This way the negative effect of asymmetrical radiation of cold glass panes can be averted.

Finally, the coating may be interposed within laminated glazing to be used as an interior vertical partition with a heating function.

Use for winter heating of interiors requires 100–300 W/m², providing occupants with a pane surface temperature of 40°C. To prevent surface condensation and cold drafts 50–100 W/m² is sufficient, even at temperatures of −35°C, while avoiding snow accumulation requires 350–600 W/m². According to the configuration, glass panels are available in sizes up to 2400 × 5500 mm. Compared to forced-air heating systems, radiant glazing claims to provide conditions of comfort to occupants at 3–5°C lower air temperatures.

Further developments of heating glazing technology include replacing the aluminum oxide or indium tin oxide coating with nanoparticle high-conductive coatings, such as Cima Nanotech SANTE, that promise faster heating rates, lower consumption, and higher color neutrality and resistance to moiré. Carbon nanotube coatings are also being considered for their excellent thermal and electric conductivity characteristics.

7.8 Fire-resistant glazing

Fire-resistant glazing, namely glass able to maintain its integrity in the presence of fire, is usually composed of two or more toughened glass panes assembled with an intumescent glue-based interlayer (Fig. 7.14). In the event of excessive heat, the first layer of

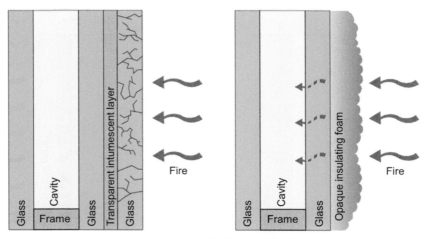

1. Glass pane facing the fire breaks due to heat stress
2. Exposed interlayer reacts to heat developing an opaque insulating foam

Figure 7.14 Fire-resistant glazing.

glass facing the fire cracks and exposes the intumescent interlayer beneath, which reacts by developing an opaque, heat-insulating shield to protect the other glass layers and prevent fire spread while limiting temperature rise on the other side of the partition. While giving unprecedented freedom in designing glazed architectural spaces, fire-resistant glazing suffers from high costs due to its energy-intensive manufacturing process and is prone to yellowing, hazing, or cracking over time due to UV degradation of the organic interlayer and syneresis phenomena.

In the last few years alternative foaming fire-resistant materials based on silica nanoparticles (SiO_2), such as fumed silica dispersions (Evonik Aerosil), have been introduced by several manufacturers.[25] Thanks to their nanoscale size (70 nm), they are very reactive and ensure high performance in extremely reduced interlayer thickness, allowing slender, lighter glazing which also boasts higher light transmission and UV resistance (Saint-Gobain SGG Contraflam).[26] Butt-joined all-glass partitions are also achievable for frameless applications.

Fire rating goes from 30 to 120 min (EI30–EI120) according to the number of glass panes and nanosilica gel interlayers. According to Swiss-based Interver, 3 mm of its proprietary intumescent layer can withstand continuous flames over 1000°C for more than two hours.[24]

7.9 ETFE transparent closures

An alternative to glass for transparent closures is a plastic material, ethylene tetrafluoroethylene (ETFE).[27,28] Devised in the 1940s by DuPont and developed in the 1970s for the aviation industry, by the end of the 1990s it had found wide application in construction because its interesting technical features give designers the ability to develop innovative technology solutions that integrate design and sustainability.

ETFE is a partially fluorinated copolymer—poly(ethene-co-tetrafluoroethene), according to the classification provided by the International Union of Pure and Applied Chemistry—of the same family as Teflon (polytetrafluoroethylene), which is characterized among fluorinated plastics for its combination of good mechanical properties, lower density, easy workability, and improved radiation resistance.

ETFE is sold under different names, such as Tefzel by DuPont, Fluon by Asahi Glass Company, Neoflon by Daikin, Dyneon by 3M, and Texlon by VectorFoiltec, and is used in architecture as sheets and in the form of inflated air cushions for the construction of roofs or transparent facades of residential, commercial, or industrial buildings, sports, education, and health facilities, museums, train stations, airports, zoos, and botanical gardens.

Some of the first creations made with this polymer were the Eden Project in St. Austell in Cornwall (2001) by Nicholas Grimshaw, the Allianz Arena in Munich, Bayern (2005) by Herzog & de Meuron/Arup (Figs. 7.15 and 7.16), and the National Aquatics Center (Water Cube) in Beijing (2008) by PTW Architects/Arup (Figs. 7.17 and 7.18). Other renowned examples of projects featuring ETFE surfaces are London's King's Cross Station (2012) by John McAslan + Partners, Canary Wharf Cross-Rail Station (2015) by Foster + Partners, Manchester Victoria Station (2015; Figs. 7.19–7.21), Anaheim Regional Transportation Intermodal Center, California (2014) by HOK (Fig. 7.22), the Olympic Stadium in Baku, Azerbaijan (2015) by Heerim Architect & Planners (Fig. 7.23) and the Tottenham Hale Bus Station (2015) by Landolt + Brown Architects (Fig. 7.24).

Characteristics which make this material a valid alternative to glass are, in particular, its high transparency to solar radiation (>90%), lightness (its surface mass is 1% of that of glass), mechanical resistance even in very high temperatures (up to 200°C), resistance to UV rays and air pollution (it does not stiffen, yellow, or

Figure 7.15 Allianz Arena, Herzon & de Meuron/Arup, Munich.
Courtesy of Ulrich Rossmann/Arup ©.

Figure 7.16 Allianz Arena, Herzon & de Meuron/Arup, Munich.
Courtesy of Ulrich Rossmann/Arup ©.

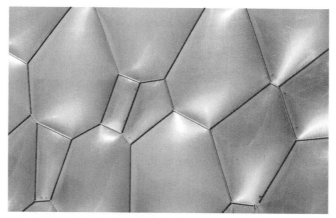

Figure 7.17 National Aquatics Center, PTW Architects/Arup, Beijing.
Courtesy of Ben McMillan/Arup ©.

deteriorate over time), high sound-absorption values (it does not generate echoes and reverberations in indoor environments or require acoustic correction interventions), lower installation costs (25–70% less), and self-cleaning characteristics when exposed to rain (thanks to the low friction coefficient and antiadhesive properties) that reduce maintenance costs. The material is certified as Class B1 fire resistant according to DIN 4102 and has a melting point of 265°C; in case of combustion the material shrinks on itself and does not propagate the flames through falling fragments, and allows the evacuation of smoke to the outside.

ETFE also gives a high degree of thermal insulation and solar control, is 100% recyclable, and has a low level of embodied energy (27 MJ/m^2 for an ETFE foil compared to

Advanced insulation glazing

Figure 7.18 National Aquatics Center, PTW Architects/Arup, Beijing.
Courtesy of Chris Dite/Arup ©.

Figure 7.19 Aisle of Victoria Station, Manchester.
Courtesy of Sam Peach/Vector Foiltec ©.

200 MJ/m² for a 6 mm float glass) and carbon (a 1 m² ETFE system has a global warming potential of 40.7 kg CO_2/kg, up to 80% of which is related to the aluminum structure)[29] compared to traditional transparent systems. Its resistance to atmospheric elements ensures a service life of 50–100 years. The raw polymer is spun into a thin, resistant film (thickness 80–300 μm), and packaged in rolls up to 3500 mm wide.

In typical use, two or more layers of ETFE are welded together in panes or inflated directly on site with low-pressure air (220 Pa), forming "air cushions" in the most various shapes to build, with aluminum frames, tensile structures and roofs elements (Figs. 7.25–7.27). Alternatively, thanks to its ability to increase in length up to three times without losing elasticity, it is employed as a component of the external vertical walls, wrapping the building as a transparent second skin.

Figure 7.20 Front view of Victoria Station, Manchester.
Courtesy of Sam Peach/Vector Foiltec ©.

Figure 7.21 Side view of Victoria Station, Manchester.
Courtesy of Sam Peach/Vector Foiltec ©.

Figure 7.22 Anaheim Regional Transportation Intermodal Center, HOK, Anaheim, CA.
Courtesy of John Linden/Vector Foiltec ©.

Advanced insulation glazing 271

Figure 7.23 Olympic Stadium, Heerim Architect & Planners, Baku.
Courtesy of Altkat Architectural Photography/Vector Foiltec ©.

Figure 7.24 Tottenham Hale bus station, London.
Courtesy of Sam Peach/Vector Foiltec ©.

The aluminum frame accounts for 65—80% of the weight of an ETFE shell structure, but even so the total mass required for a roof structure when using an ETFE solution as the transparent system component is up to seven times less the mass required for an equivalent glass structure.

ETFE building envelopes can reach large dimensions and cover surfaces greater than those obtainable with conventional technologies, and its high tear resistance (>300 N/mm) and ability to withstand a tensile elongation of more than 300—400% allow the material to cope with considerable flexing (deformation) of the support structure. ETFE envelopes also have high thermal insulation properties that allow the realization of large transparent shells without compromising the energy performance of the building. The low values of thermal conductivity of the material (0.24 W/mK),

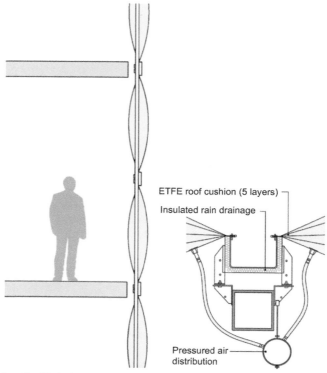

Figure 7.25 Detail of Ethylene tetrafluoroethylene (ETFE) facade and ETFE cushion support structure.

Figure 7.26 ETFE roof installation, King's Cross Station, London. Courtesy of John Sturrock/Vector Foiltec ©.

Figure 7.27 Detail of ETFE roof pressured air distribution, King's Cross Station, London. Courtesy of John Sturrock/Vector Foiltec ©.

together with the presence of air cavities between each layer of the membrane, provide U values lower than 1.2 W/m²K in the case of a cushion composed of five layers, and up to 0.3 W/m²K with the insertion of aerogel mats.

Total energy transmittance in accordance to ISO 15099 amounts to 70–80% for a three-layer ETFE cushion.

Moreover, the presence of pressurized cushions allows almost complete elimination of heat losses due to air infiltration. The multilayer composition of ETFE offers designers many opportunities to vary the transparency of the building envelope and control solar radiation, optimizing the levels of thermal comfort and daylight. Cushions can be designed and built to transmit or reflect selectively the different wavelengths of the solar spectrum by using UV or IR filters, or specific surface treatments, in response to varying external conditions. Through the impression of particular graphic patterns on the membranes of the different layers of material, and with sophisticated pneumatic systems to vary the overlapping of graphic motifs, it is also possible to modify solar radiation and brightness within interiors and the outside and inside visual appearance of the envelopes themselves (Fig. 7.28). The ETFE facade of the Media Tic building in Barcelona is able to vary the transparency of the air cushions by pumping smoke generated from liquid nitrogen.

The possibility of printing graphic motifs, varying the characteristics and colors of the surfaces, and laminating light-emitting diodes (LEDs) and light strips in each layer of the membrane allows interesting visual solutions, including multimedia ones. Surfaces can be designed to receive the projection of images, video, or coloring effects; each layer can be arranged to transmit, reflect, or diffuse images, using the entire membrane shell the same way as a visual screen. Low-voltage lighting strips and LEDs of different colors, intensity, and size may be laminated in the layers, transforming the membrane shell into a light source and a communication vehicle at the same time.

Particularly interesting, finally, is the integration of ETFE with photovoltaic systems for the production of energy. Solar cells are manufactured on sheet substrates by a continuous deposition process and directly laminated in the upper layers of the cushions. The result translates into solar cells with unique lightness and flexibility that, encapsulated in the membranes of ETFE, provide a technology that is reliable

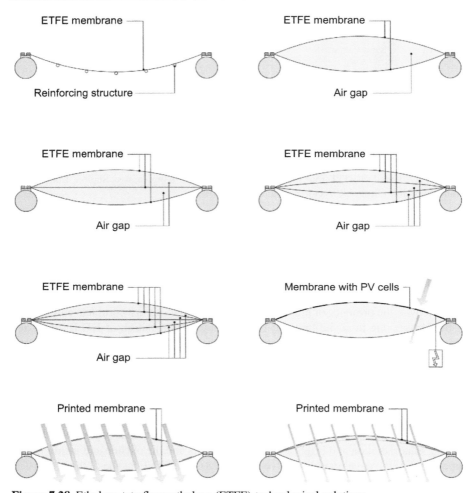

Figure 7.28 Ethylene tetrafluoroethylene (ETFE) technological solutions.

and durable over time. Pass diodes connect across each cell, allowing the modules to produce continuous power even if partially obscured. Each cell consists of three layers deposited in sequence. The cell on the bottom is capable of absorbing red light, the middle absorbs green light, and the top one absorbs blue light.

7.10 Conclusions and future development

Transparent surfaces of the building envelope are the most critical element in environmental comfort design and the energy balance, as they represent the weakest element of the interface between internal and external environments in terms of thermal resistance.

In fact, windows installed in most existing buildings, as well as many newly installed windows in hot climates and developing countries, have poor performance, with low thermal resistance and high sensitivity to solar radiation. Advanced insulating glazing, by contrast, can even outperform well-insulated walls and offer great potential for passive heating in moderate climates.

However, the cost-effectiveness of installing highly insulating windows depends significantly on heating requirements, thus the lowest U values are only viable where climates are colder and energy prices are higher. For most cold countries, this indicates a performance below 1.1 W/m^{2}K. If the goal is instead the achievement of zero-energy buildings, or "energy-plus" windows able to harvest more passive heating than energy losses annually, then an even lower transmittance (below 0.6 W/m^{2}K) is required, along with higher SHGC values to provide the best energy balance.[30]

For all regions of the world, a baseline window performance of 1.8 W/m^{2}K is recommended, with lower levels in colder countries.[3,4] For new buildings in cold or moderate climates, to improve thermal comfort and reduce energy consumption for air conditioning, the most effective solutions are represented by systems with a double glass envelope (double-skin facades) or ETFE closures as an alternative to glass.

Still, to make high-performance windows with U values below 0.6 W/m^{2}K market viable, significant additional research and development will be required. While several independent efforts are globally under way, this work can be accelerated by better international collaboration. Furthermore, the potential market for these high-performance windows is huge in cold climates such as North America, Northern Europe, Russia, China, Japan, and Korea, with additional smaller markets in cold southern hemisphere climates. Some mixed or moderate climates can also benefit from advanced insulating glazing.

Actual cost-effectiveness of such advanced window designs will be achieved once they reach market maturity, with expected price premiums of $50–120/m^2. This goal will, however, take time and significant market conditioning, requiring better information to make builders and policymakers look beyond the specific energy-saving benefits of windows and consider whole-system efficiencies. For instance, much higher-performing windows significantly reduce the costs of heating, ventilation, and air-conditioning or thermal distribution equipment, yielding an overall economic saving. This systems perspective is often already considered by advanced builders and researchers, but has yet to be implemented on a wider scale or incorporated in most building codes.

Lastly, research is focusing on the development of more affordable options for retrofitting existing windows, such as cheaper low-e window films, low-e storm or interior panels, highly insulating window-frame caps, and lower-cost insulated shades.

References

1. TNO Built Environment and Geosciences. *Glazing type distribution in the EU building stock*. TNO Report TNO-60-DTM-2011–00338. The Hague (NL): TNO; 2011.
2. ISO 10077-1:20013 Thermal performance of windows, doors and shutters – calculation of thermal transmittance – Part 1: General.

3. IEA. *Technology roadmap: energy efficient building envelopes*. Paris: OECD/IEA; 2013a.
4. IEA. *Transition to sustainable buildings: strategies and opportunities to 2050*. Paris: OECD/IEA; 2013b.
5. EN 673. *Glass in building. Determination of thermal transmittance (U value). Calculation method*. 2011.
6. Casini M. *Costruire l'ambiente. Gli Strumenti e i metodi della progettazione ambientale*. Milano: Edizioni Ambiente; 2009.
7. VITO. (Flemish Institute for Technological Research NV). *LOT/32-Ecodesign of window products: Task 4-Technology*. Boeretang: VITO; 2014.
8. Makhlouf ASH, Tiginyanu I, editors. *Nanocoatings and ultra-thin films*. Cambridge: Woodhead Publishing Limited; 2011.
9. Jelle BP, Hynd A, Gustavsen A, Arasteh D, Goudey H, Har R. Fenestration of today and tomorrow: a state-of-the-art review and future research opportunities. *Sol Energy Mater Sol Cells* 2012;**96**:1−28.
10. Energy Conservation in Buildings and Community Systems. *Vacuum insulation panel properties & building applications*. ECBCS Annex 39. St Albans (UK): AECOM Ltd.; 2010.
11. Schultz JM, Jensen KI, Kristiansen FH. Super insulating aerogel glazing. *Sol Energy Mater Sol Cells* 2005;**89**:275−85.
12. Schultz JM, Jensen KI. Evacuated aerogel glazings. *Vacuum* 2008;**82**:723−9.
13. Buratti C, Moretti E. Transparent insulating materials for buildings energy saving: experimental results and performance evaluation. In: *Third international conference on applied energy. Perugia, 16−18 May 2011*; 2011.
14. Wigginton M. *Glass in architecture*. London: Phaidon Press; 1996.
15. Hertzsch E. *Double skin façades*. Munich: Peschke; 1998.
16. A report of IEA SHC Task 34 ECBCS Annex 43. In: Poirazis H, editor. *Double skin façades a literature review*. Lund: Lund University; 2006.
17. Kurt R, Tyson L, James B. Double-skin façades. *ASHRAE J* 2007;**49**(10):70−3.
18. Schittich C, Staib G, Balkow D, Schuler M, Sobek W. *Glasbau atlas*. Munich: Detail; 2006.
19. Hegger M, Fuchs M, Stark T, Zeumer M. *Energy manual*. Monaco: Birkhäuser; 2008 [Edition Detail].
20. Gratia E, De Herde A. The most efficient position of shading devices in a double-skin facade. *Energy Build* 2007;**39**(3):364−73.
21. Wigginton M, Harris J. *Intelligent skins*. Oxford: Elsevier Architectural Press; 2006.
22. Xu L, Ojima T. Field experiments on natural energy utilization in a residential house with a double skin façade system. *Build Environ* 2007;**42**(5):2014−23.
23. Hong T, Kim J, Lee J, Koo C, Hyo Seon Park HS. Assessment of seasonal energy efficiency strategies of a double skin façade in a monsoon climate region. *Energies* 2013;**6**(9): 4352−76.
24. Shameri MA, Alghoulb MA, Sopianb K, Zain MFM, Elayebb O. Perspectives of double skin façade systems in buildings and energy saving. *Renewable Sustainable Energy Rev* 2011;**15**(3):1468−75.
25. Atwa MH, Al-Kattan A, Elwan A. Towards nano architecture:nanomaterial in architecture − a review of functions and applications. *Int J Recent Sci Res* 2015;**6**(4): 3551−64.
26. Leydecker S. *Nano materials: in architecture, Interior architecture and design*. Basel (CZ): Springer Science + Business Media; 2008.
27. LeCuyer A. *ETFE. Technology and design*. Boston: Birkhauser; 2008.

28. Monticelli C, Campioli A, Zanelli A. Environmental load of EFTE cushions and future ways for their self-sufficient performances. In: *Proceedings of the international association for shell and spatial structures (IASS) symposium*. Spain: Universidad Politecnica de Valencia; 2009.
29. Vector Foiltec GmbH. *Environmental product declaration texlon system*. Berlin: Institut Bauen und Umwelt (IBU); 2014.
30. Arasteh D, Selkowitz S, Apte J, LaFrance M. Zero energy windows. In: *Proceedings of the 2006 ACEEE summer study on energy efficiency in buildings. Pacific Grove CA*; 2006.

Light and solar control glazing and systems

8.1 Antireflective glazing

The need to maximize glass transmittance, be it in the visible range to harvest as much light as possible, especially in the case of multiglazed windows in facades, or total energy transmittance to ensure the best yield of window-integrated photovoltaics (PVs), has led to the development of extra-clear glass with exceptionally low iron content. Products in this field include AGC Sunmax, Pilkington Optiwhite, Guardian Ultrawhite, and SGG Planiclear, with light transmission values up to 92% and energy transmission values up to 91%. Nevertheless, even these highly transparent glass panes still reflect part of the incoming radiation, in particular solar rays coming at an angle and diffused light, preventing it from reaching the interiors. One of the most appreciated glass properties is therefore antireflection, which reduces unwanted reflections and glare and increases both light transmission through a surface and contrast.

Today, nanotechnology-enhanced antireflective coatings claim unmatched levels of energy transparency, effectively eliminating unwanted reflection.

The application of nanoscience and nanomaterials allows direct intervention in the same order of magnitude of light waves and controls their behavior to an unprecedented degree. In fact, visible light wavelength ranges from 400 to 700 nm, while most nanoparticles used in coating may be as small as a tenth of that size. For instance, the iridescent effect of surfaces such as foam bubbles or oil drops in liquid is determined by the constructive or destructive interference between reflected rays, which can reinforce one another leading to brighter colors or annihilate themselves to give a darker appearance. Among others effects, particles in the nanoscale range cause the surface appearance to change depending on viewing angles or deformation strain (dichroic effects).

Thin films, sheets, or coatings to exploit the light-controlling properties of nanomaterials are obtained by chemical or physical vapor deposition on a number of substrates, mainly glass or polymeric films.

Antireflective coatings normally employ several transparent thin-film layers with varying refractive indices, obtained by varying concentrations of metal oxide nanoparticles in a polymer matrix, to produce destructive interference for reflected light waves and constructive interference for transmitted ones (Fig. 8.1). Such results are obtainable for specific ranges of light spectrum—ultraviolet (UV), visible, or infrared (IR) light—and incident light angles, and broader bands are possible, although extremely expensive. Of course, by carefully controlling the thickness of these layers the opposite effect can be achieved, obtaining high-reflective glazing with low light transmittance or even different behavior according to incident light frequency and angle.[1]

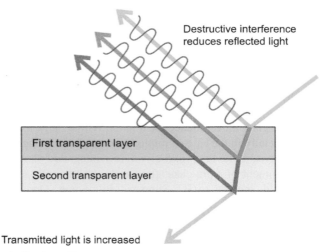

Figure 8.1 Antireflective effect.

For instance, Germany-based Centrosol's HiT antireflective glass reaches visible light transmittance (VLT) of 99% and energy transmittance levels over 97%, thanks to a two-side nanotechnological coating made of porous SiO_2 which also has hydrophilic self-cleaning properties.

Similarly, dichroic effects, or angular dependence of reflected and transmitted light waves, are obtainable by accurately selecting nanoscale thin films. Dichroic behavior of several architectural glasses is produced by the cumulative build-up of layered thin-film depositions, each with different light transmission, reflection, and absorption qualities.

8.2 Self-cleaning glazing

Widely available on the market is the so-called self-cleaning glazing, with products which claim excellent self-cleaning characteristics.

Self-cleaning properties of a surface derive from its capacity to bond with water (surface wettability), and can be obtained through two opposites processes: one based on the surface's hydrophobicity (water-repellent action), and the other on its hydrophilicity (water-attractive action). The wetting of a solid with water depends on the relation between the interfacial tensions (water/air, water/solid, and solid/air). The ratio between these tensions determines the contact angle θ between a water droplet and a given surface. A contact angle of 0 degree means complete wetting, and a contact angle of 180 degrees corresponds to complete nonwetting (Fig. 8.2). Therefore, a solid surface whose water contact angle is smaller than 90 degrees is considered hydrophilic, whereas a contact angle higher than 90 degrees makes it hydrophobic. For contact

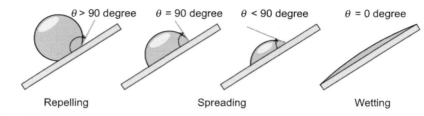

Figure 8.2 Surface wettability and contact angle θ.

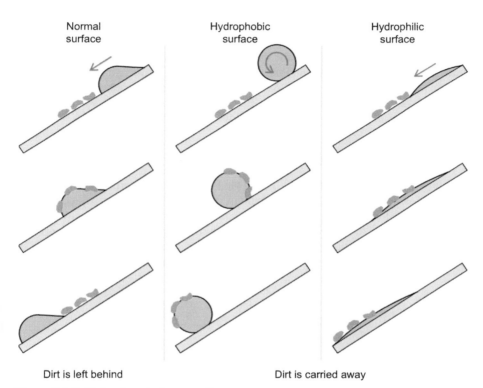

Figure 8.3 Self-cleaning glazing operation.

angles above 120 degrees the surface's wettability state is considered superhydrophobic, and in contrast when the contact angle is lower than 60 degrees the surface is considered superhydrophilic.[2]

The two main approaches to make surfaces self-cleaning either prevent dirt and water from attaching to the glass surface (water-repellent action) or degrade organic and inorganic dirt and let it wash away by rainwater (water-attractive action). According to these two processes, self-cleaning glazing products on the market can be divided in two main technologies, exploiting coatings with superhydrophobic (easy to clean) or superhydrophilic characteristics (Fig. 8.3).

8.2.1 Superhydrophobic nanotechnological glazing

While glass is inherently hydrophobic by itself, its surface is still rough and microscopically porous, and thus subject to dirt and bacterial deposit.[3] Superhydrophobic coatings employ nanoscale polymers or silica nanoparticles (SiO_2) to fill discontinuities and provide a perfectly smooth surface.[4] This way water droplets have an exceptionally high contact angle (θ), up to 180 degrees, and are instantaneously flushed away, carrying dirt particles with them without leaving any mark (Fig. 8.4). Superhydrophobic coatings are difficult to apply during the glass-slab manufacturing process, and alternative physical processing methods such as ion etching and compression of polymer beads or chemical methods such as plasma-chemical roughening still result in hazy and fragile products. On-site application is simple, however (some of these coatings come in do-it-yourself kits for self-curing of desired surfaces), with the main drawback of the limited duration of the hydrophobic effect, thus coatings might require reapplication after less than a year. Several nanotech-based hydrophobic coatings are currently available (Nanotech Coatings, Balconano, Diamon Fusion), and can also be used for curing mirrors, tiles, kitchen worktops, and the like.

8.2.2 Superhydrophilic photocatalytic glazing

First introduced to the market in 2001 with the Pilkington Activ product range, hydrophilic self-cleaning glass employs a titanium dioxide (TiO_2) coating which exploits two characteristics of this material: its photocatalytic properties and its actual hydrophilicity.[5] Exposed to the sun's UV rays, the TiO_2 thin film reacts by oxidizing organic material deposited on the glass surface, such as dirt, bacteria, algae, pollutants like nitrogen oxides (NO_x), and particulates (PM_{10}).[6] The photocatalyst effect also makes

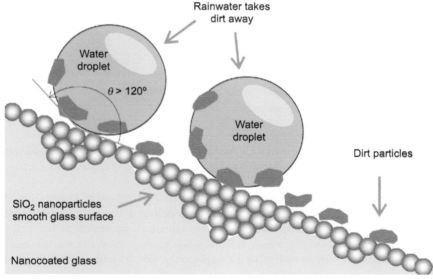

Figure 8.4 Hydrophobic glass operation.

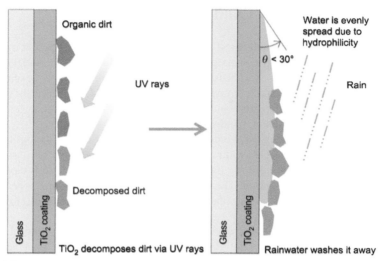

Figure 8.5 Hydrophilic glass operation.

the glass surface highly hydrophilic,[7] allowing rainwater to spread in an even layer across the glass pane and wash away decomposed dirt, leaving no mark (Fig. 8.5).[8] TiO_2-based surfaces can maintain their hydrophilic properties indefinitely as long as they are irradiated by UV light. After illumination stops, superhydrophilic behavior remains for about two days.

Current titanium-based self-cleaning glazing available on the market includes Pilkington Activ and Saint-Gobain Glass Bioclean, which employ a 15–30 nm layer of crystalline anatase TiO_2 vapor deposited on to soda lime silicate float glass, and PPG Sunclean, where TiO_2 coating is applied to the hot glass pane via a pyrolytic process. Self-cleaning glass usually has around 80–90% UV transmission, with the rest used for photocatalysis. Factory-applied TiO_2 coatings have no duration issues and maintain operation throughout the glass pane's lifetime. However, since their operation relies on both UV rays and rainwater, they are best suited for difficult-to-reach windows exposed to sun and weather, such as roof lights, full-glazed facades, or conservatories. Furthermore, photocatalysis only degrades organic, carbon-based dirt, so it is ineffective against salt deposits, sand, or paint splashes. The price premium is 15–20% against an untreated glass pane.

Another noteworthy property of photocatalytic glazing is its inherent resistance to misting up in cases of superficial condensation, since water droplets are prevented from forming because of its superhydrophilicity (antifogging effect). Misting of glazing is caused by diminutive water droplets which form as vapor cooling down and condensing on colder surfaces. These droplets block or distort vision through and reflection, making the glass appear hazy or translucent. This undesired effect is prevented on hydrophilic surfaces, as condensed water immediately spreads in a thin continuous layer without impeding vision. Moreover, the superhydrophilic layer allows glass to dry without leaving any droplet marks behind.[9]

8.3 Light-redirection and optical systems

Light comfort control is one of the most central issues in designing transparent enclosures, as it directly influences energy consumption for interior artificial lighting and users' level of satisfaction at home and work. The issue of light distribution and glare prevention concerns in particular educational, commercial, and tertiary buildings, since these usually have larger floor spaces and bigger windows, as well as special requirements for work activity (video terminals, drawing, handwriting, projection rooms, etc.).[10,11]

In wider rooms, enlarging the windows is not sufficient to ensure daylight distribution in the most recessed areas, while presenting issues of excessive heat loss or heat gain over the seasons and requiring light control to prevent workstations closer to the windows from excessive lighting. The practical depth of the daylighting zone of a window is usually limited to 1.5 times the window head height, with a negligible contribution thereafter.

An optimal strategy to exploit daylight best is to separate fenestration glazing according to its main purpose: view glazing allowing vision through is necessarily at eye level and requires careful control of visible light and glare, while daylighting glazing is placed on the upper part of windows and features extra-clear glass to ensure deepest reach of light.

Another way to harvest sunlight from outside and diffuse it in interiors in an inoffensive way is to employ exterior devices to bounce light from the sky toward the room ceiling (Fig. 8.6). These are placed above eye level to establish a glare-free environment. Exterior sunscreens can also act as light reflectors, or specific anidolic mirrors with a custom-curved shape can be employed. Prism solar protection panels can shield against annoying direct light while allowing indirect lighting. Exterior light reflectors are very effective, since they can see a bigger portion of the sky. However, they can be cumbersome, alter the building shape, and require frequent cleaning, as well as possibly being costly to install.

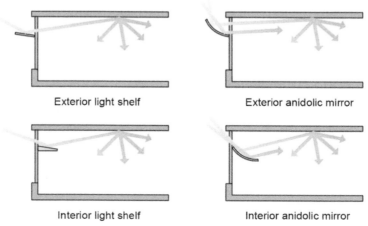

Figure 8.6 Reflective light-redirection systems.

Devices to reflect and direct light can also be installed on the inside of glazed walls, using architectural light shelves or, again, anidolic mirrors. This way they are more accessible for cleaning and maintenance and do not need to withstand weather elements, but gather less light and their bulk makes them hard to integrate in the interior architecture.

8.3.1 Light-redirection louver systems

Interior louver systems are the least intrusive and easiest to integrate, even in existing buildings.[12] They can be placed inside the double-glazing cavity itself, thus requiring no maintenance to keep reflective surfaces glossy; but limited space availability offers many design challenges. Louvers can either be metallic, redirecting light by reflecting it, or acrylic, exploiting material refraction to bounce light upwards (Fig. 8.7). They come as strips, prismatic panels, or laser-cut hollow panels. Folded or curved metallic louvers can be custom shaped to reflect higher sunrays back outside and bounce them in when the sun is low on the horizon, avoiding some of the heat impact in summer while allowing more light in during winter. The best placement for light-redirection louvers is in the upper section of the window, allowing vision through at eye level while efficiently redirecting light in near the ceiling (see Fig. 8.8).

Windows equipped with light-redirecting systems can dramatically improve light reach in interiors, achieving 400 lux at distances up to 6 m into the room space, enough for artificial lighting to be switched off (Fig. 8.9).

The main light-redirecting products on the market include SGG Lumitop, an argon-filled, double-glazing pane with 3.5×12 mm polymethylmethacrylate (PMMA)

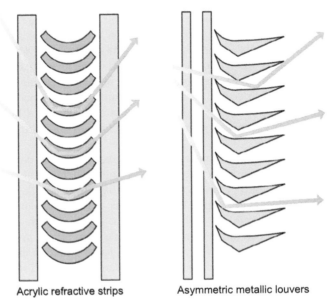

    Acrylic refractive strips   Asymmetric metallic louvers

Figure 8.7 Light-redirection louver systems.

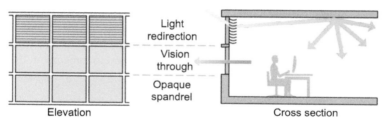

Figure 8.8 Light-redirecting facade structure.

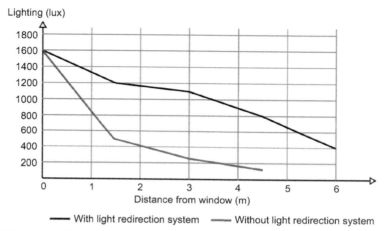

Figure 8.9 Light-redirection system efficacy (SGG Lumitop).

sinusoidal elements integrated into the cavity and able to redirect direct sunrays of solar altitude between 15 and 65 degrees, and Lightlouver, an interior double-curvature metallic light reflector suitable for upgrading existing buildings. Okalux has several light-redirection models tailored to specific uses and orientation (Okasolar, Okaflex).

Light-redirecting systems are gaining attention in the market since their relative simplicity and reasonable cost allows a quick return on investment, and they are considered a viable way to reduce energy consumption for artificial lighting (the US Department of Energy estimates a potential 345 TBtu saving if combined with lighting controls).[12]

Research is at present focusing on active light-redirection systems controlled by intelligent sensors, such as electrically operated butterfly blinds or micro electro-mechanical systems (MEMS) micromirror arrays contained within the insulated glass unit air cavity.

The Smartlight device, developed by researchers at the University of Cincinnati, employs an array of electrofluidic optical cells as the light-redirecting medium. Powered by embedded PVs, these can change their shape from prisms to lenses and vary light distribution according to user control. The device is hosted in the top section of the window, and can be replicated on internal transom windows to steer light deeper inside the building, or redirect solar rays toward the ceiling to provide room ambient lighting.

Light-redirecting films that are light, inexpensive to manufacture, and suitable for retrofit of existing windows are entering the market. 3M Daylight redirecting film employs a prism microstructure to steer 80% of daylight upward, extending the natural lighting zone by 2 m for every 25 cm of treated window and claiming 52% saving in lighting energy consumption.

8.3.2 Tubular solar conveyors

Tubular optical systems are able to transport daylight inside windowless or oversize interiors without resorting to artificial lighting.[13] Such systems are called "anidolic" because their function is to transmit light but not images.

The main products on the market include Italian Solarspot and Power Lux, Danish Velux Sun Tunnels, and Solatube (Fig. 8.10) and Monodraught from the United States.

Optical systems can be divided into passive and active systems. Both are generally formed by three components: a collector, which captures the light, a reflective device that effectively transports it, and another that diffuses it.

The collector is composed of a transparent dome in unbreakable material (shockproof methacrylate), protected from deterioration induced by UV rays (Fig. 8.11). In passive

Figure 8.10 Solatube light redirection system.
Photo by the author.

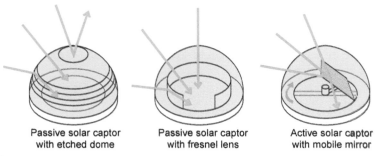

Figure 8.11 Light-redirection system captor domes.

systems the dome can be etched with refractive light-redirecting prisms to optimize light harvesting during the whole sun path, redirecting low-angle sunlight toward the reflective tube for maximum light capture in winter months and rejecting overpowering summer midday sunlight to prevent overheating and glare, and may also feature an internal reflector (Solatube). Other models combine a smooth, clear dome with an internal optical refractive and reflective device based on the principle of the Fresnel lens, able to intercept and deflect light radiation diffused from the whole horizon arc (360 degrees): morning and afternoon low-level light is refracted into the conveyor tube while higher midday light has unobstructed entry (Solarspot). Compared to prismatic domes, these are more effective in overcast weather and more temperate climates. Lastly, collector domes can feature a polymer sheet containing aligned prismatic cavities, created by laser etching or joining two PMMA sheets incorporating microreplicated prisms to create nearly flat microscopic air pockets able to split and direct incoming light. Active systems forgo fixed microprism light-redirection systems in favor of a mobile PMMA parabolic mirror. A GPS control system constantly orientates the mirror to ensure maximum light harvesting, and requires no wiring as it is powered by an integrated PV cell. More collar collectors can coordinate to a master device to give synchronized operation (Power Lux).

The light-transporting device consists of an aluminum rigid tubular duct of small diameter (150–650 mm) internally coated with a material with specular reflectivity greater than 92%; the best available technologies use silver deposits or even nanothin polymer layers of photonic crystal mirrors, achieving reflectivity values as high as 99% (superreflective). The system includes straight tubes and adaptable fittings (continuous curves), which allow the conduit to follow any trajectory to connect the daylight collection point to the room to be illuminated (Fig. 8.12). Flexible reflective ducts are also available, although their ease of installation has the drawback of a noticeable decrease in performance.

Lastly, the diffuser at the end of the reflective duct spreads the conveyed light in the area to be illuminated. The diffuser's purpose is to eliminate the various bright patterns that form in the reflective pipe and affect outlet light. These patterns are called "caustics," as they can be harsh to view. Material and geometry of the diffuser must be selected carefully, since too much diffusion at the inlet may badly reduce light throughput by backscattering.[1] The diameter of the duct must be chosen in view of the geometry and the path length, as well as the extension of the surface to be illuminated. Higher internal reflectivity values enable use of smaller pipe-cross sections for the same lumen output.

Light and solar control glazing and systems 289

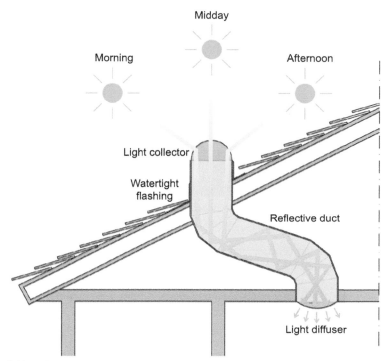

Figure 8.12 Tubular solar conveyor.

The system can be integrated with systems for forced aspiration of interior air or night-time artificial lighting.

The negligible environmental and visible impact of the small collectors, perfectly integrated in the roof system, make them particularly suitable for use in attics in urban centers in cases of both environmental protection and preservation prescriptions on the building. Tubular conveyor systems are a viable alternative when the opening of skylights and dormers on the roof is not allowed, as they are measures deemed of considerable architectural impact; in such cases, the competent authorities are issuing favorable opinions on these systems more and more frequently.

Highly significant examples of use are also noted in underground archaeological sites and fine art galleries or depositories, where exposure to directed or scattered natural light without the harmful effects of UV radiation is particularly desirable.

8.3.3 Natural mimicking artificial light

In the most extreme cases where daylight cannot enter from outside, either because of excessive distance or for hygienic reasons, a possible solution is to employ artificial light engineered to mimic natural light. One example is the CoeLux device, a smart lighting appliance capable of perfectly reproducing daylight by using nanotechnology, developed in Italy by a spin-off of Insubria University with the European Union funding.

Figure 8.13 CoeLux lighting system (switched off).
Courtesy of CoeLux s.r.l.

The system is made of three elements: a high-brightness light-emitting diode projector reproducing the spectral characteristics of sunlight; sophisticated optical systems to direct the sunrays, which allow the observer to perceive the sky and the sun at infinite distance; and finally the nanostructured material able to create the Rayleigh scattering diffusion process that happens naturally within our Earth's atmosphere.

The result is light realistic enough to trick the human brain into effectively perceiving it as natural. The CoeLux system (Figs. 8.13 and 8.14) can ideally reproduce different daylight scenarios, mimicking the light produced by the sun at 60, 30, and 45 degrees. Currently, the standard solution demanded by the market is one that reproduces the sun at an angle of 45 degrees, as this system offers greater balance between light and shade. It has been developed in two different versions: the CoeLux 45 HC has a skylight window dimension of 1700 × 850 mm, while a second, smaller version has a skylight window dimension of 1000 × 500 mm. Future enhancements may implement daylight sequencing—from dawn to dusk—and different weather

Figure 8.14 CoeLux lighting system (switched on).
Courtesy of CoeLux s.r.l.

conditions within the same skylight. Suitable applications include train and subway stations, indoor pools, fitness centers, parking garages, retail, and high-end residential. Some interesting applications are in the healthcare sector, especially in areas inherently devoid of natural light for safety reasons, such as operating theaters and radiotherapy bunkers. In such situations, the benefits of daylight are most appreciated in terms of reducing patients' stress: a CoeLux installation in Humanitas Hospital in Milan is already getting positive feedback from both patients and doctors.

8.3.4 Transparent organic light-emitting diode windows

Another promising lighting development is organic light-emitting diode (OLED)-integrated windows. OLED technology is finding its way into more and more electronic devices and appliances for displaying information and images/videos, and its characteristics make it suitable for integration in windows to give them extra functionality. OLED lamps utilize an emissive electroluminescent layer composed of an organic compound film sandwiched between two electrodes that supply it with power. Current technology employs fluorescent organic molecules, but phosphorescent OLEDs are in development to reach even higher efficiency. By integrating this assembly with a transparent indium tin oxide thin-film transistor controller, it is possible to achieve a transparent display medium. Currently transparent OLED (TOLED) prototypes have transparencies of over 85% but are almost opaque to UV wavelengths, thus acting as a UV filter as well. Window applications are manifold: window panes can let natural light in during the day and then self-illuminate at night, replacing traditional lighting fixtures and providing large, diffusing glowing surfaces for maximum visual comfort, excellent color fidelity (90% of sunlight), little temperature drawback (OLED working temperature is around 30°C), and high efficiency (up to 150 lm/W).

Further enhancements will provide high-resolution pixel matrices, turning windows into effective displays able to show videos and images or, paired with a touch-sensitive layer, act as a control input for building management systems.

Finally, researchers at Universal Display Corporation have devised a smart window that combines a TOLED layer with a reversible electrochromic mirror, achieving a multi-operational device with four operating states: with both TOLED lamp and mirror off, the windows is 70% transparent; switching TOLED on makes it glow white with a diffusing, milky light that still allows partial see-through; in the reflective state with the lamp off the window acts as a mirror; and finally, with both mirror and lamp activated, the OLED light is bounced back from the mirror into the interior, achieving much brighter, uniform light emission and avoiding light pollution to the outside (Fig. 8.15).[14]

8.4 Static solar protection glazing

Unlike simple high-thermal insulation double glazing for heat loss reduction, solar protection glazing's task is to prevent the majority of the incident radiant flow on the window from penetrating inside.

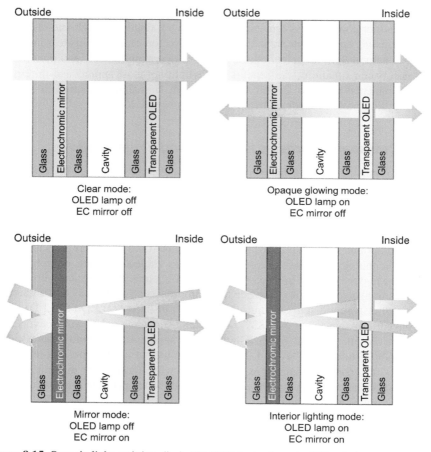

Figure 8.15 Organic light-emitting diode (OLED)/electrochromic (EC) window operation.

Solar radiation that hits a transparent material such as glass is partly reflected, partly transmitted, and partly absorbed. The spectrophotometric behavior of the glass in response to sunrays is quantified by the reflection (ρ), transmission (τ), and absorption (α) energy coefficients, which respectively indicate the ratio between the reflected, transmitted, and absorbed flow and the incident flux. The relationship between these three characteristics is:

$$\rho + \tau + \alpha = 1 \qquad [8.1]$$

The courses of these three ratios along the different wavelengths form the spectral curves of the glass. In correspondence with a particular incidence of solar radiation, these ratios depend on the color and thickness of the glass and the nature of any coating applied on the surface.

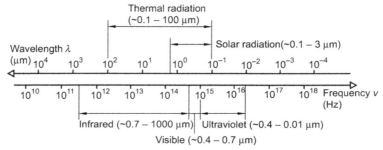

Figure 8.16 Electromagnetic wave spectrum.[15]

In particular, float glass has a high transmission in the visible range (380–780 nm), a reduced transmission in the field of near IR (780–2500 nm), and is no longer transparent to lengths below 300 nm and above 2500 nm; for these wavelengths the radiation is completely absorbed. By varying the thickness and color of the glass and applying suitable materials on the pane surface it is possible to adjust the spectral curve of the glass by reducing, for example, the transmission of IR radiation while maintaining unchanged the transparency to visible light radiation.

With reference to the different wavelengths that form incident solar radiation (Fig. 8.16), the following dimensionless variables are used to characterize the energy and luminous behavior of glazing.[16]

- Ultraviolet transmission factor τ_{uv} to indicate the proportion of radiation passing through the glass directly related to the solar spectrum with wavelengths between 280 and 380 nm.
- Visible Light transmission factor τ_v or VLT to indicate the proportion of radiation passing through the glass directly related to the solar spectrum with a wavelength range between 380 and 780 nm and visible to the human eye.
- Direct solar energy transmission factor τ_e to indicate the proportion of radiation passing through the glass directly related to the solar spectrum with wavelengths between 300 and 2500 nm.
- Transmission factor of the total solar energy, solar factor g or solar heat-gain coefficient (SHGC), given by the sum of the direct solar energy transmission factor τ_e and the factor of secondary heat exchange of glass toward the interiors q_i; the latter resulting from the transmission of heat by convection and radiation in the far IR wavelengths of the fraction of incident solar radiation that has been absorbed from the window:

$$g = \tau_e + q_i = \tau_e + \frac{a_s}{h_e} \cdot U \qquad [8.2]$$

where a_s is the coefficient of absorption of solar radiation (0.06 τ_e), h_e is the coefficient of exterior adduction (23 W/m² K), and U is the transmittance of the glass (W/m² K).

- Shading coefficient (SC), given by the ratio between the solar factor g of a transparent component and the solar factor g_s of a reference glazing constituted by a single clear glass pane 3 mm thick ($g_s = 0.87$ for normal incidence between 0 and 60 degrees):

$$SC = \frac{g}{g_s} \qquad [8.3]$$

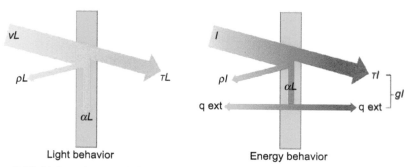

Figure 8.17 Light and energy behavior of a single glass pane.

Knowing the SC value of the transparent surface under consideration, if (I) is the solar radiation that strikes it, the solar gain through the surface is obtained by:

$$G = \text{SC} \cdot g_s \cdot I \qquad [8.4]$$

SHGC has a significant influence on heat gains in summer, when heat is transferred from the outside to the interiors of a building, and is particularly important for large windows, in which overheating in summer plays an important role.

When a glass surface is hit by solar radiation, the radiant energy (and particularly that normally hitting the glass) is only minimally absorbed or returned to the outside, due to the spectral composition of the radiation itself (Fig. 8.17). The predominant part (about 80%, depending on the glass thickness) penetrates to the interiors delimited by the window and hits the surfaces of walls and furnishings, where it is absorbed and transformed into heat. These surfaces in turn emit radiant energy. However, due to the low temperature of the emitting body, this energy has a spectral composition in which IR radiation prevails in wavelengths to which the glass is practically opaque (Fig. 8.18). Thus most solar radiant energy that has passed through the window and entered the indoor environment cannot exit by radiation through the glass, but only by conduction and adduction, with the consequence of considerably increasing the interior temperature (greenhouse effect).

Available on the market are "body-tinted glazing," "pyrolytic" coated glazing, and "selective" high-performance glazing with "magnetron"-type coatings which, compared to body tinted glass, allow better control of energy transmission without overly penalizing VLT. This property is measured using the light to solar gain (LSG) ratio index, also called the spectral selectivity index, which defines the ratio between the VLT and the SHGC: glazing with a high LSG value (selective glass) transmits a high percentage of incident visible light radiation and a small fraction of the total radiation. Selective glass on the market today has a VLT of 34–69%, an SHGC of 24–56%, and a selectivity index LSG of 1.28–2.29 (Table 8.1).

Solar protective glasses include thin-film semitransparent PV glazing, organic or inorganic, available in several transparency ratios (10–30%) and colors, and able to reduce incoming heat (SHGC of 0.29–0.41) and produce electricity at the same time (see Section 10.1 of Chapter 10).

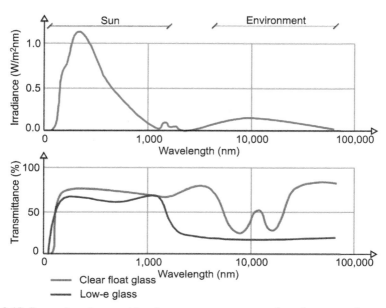

Figure 8.18 Sun and environment irradiance spectrum compared to glass transmittance.

Table 8.1 **Main selective glazing on the market with higher LSG values**

Parameters	Guardian sunguard SNX	SGG cool-lite extreme	AGC stopray ultra	Pilkington suncool
VLT (%)	62	60	60	60
SHGC	0.27	0.28	0.28	0.32
LSG	2.29	2.14	2.14	1.87
U_g (W/m² K)	1.36 (argon fill)	1.00 (unspecified fill)	1.00 (unspecified fill)	1.00 (argon fill)

The use of static solar protection glazing reduces heat load in summer while maintaining the vision through and limiting glare phenomena. However, such systems do not follow the daily solar path and weather conditions or season alternation, with the result of reduced energy harvesting during winter (especially in south-facing facades) and reduced natural light levels in the absence of direct solar radiation (in particular for east- or west-facing facades, irradiated for only half the day).

Solar protection films for retrofitting existing windows and glazed facades are also widely available. These are usually polymer or metal based, and do not reach the same performance as factory-engineered glazing, with LSG ratios up to 1.4 (70% light transmission and 0.50 SHGC). A recent development is nanotechnology-enhanced solar protection films such as 3M Ceramic Series or Hüper Optik, which employ a nanoceramic coating to ensure high selectivity and color fidelity performance, as well as better durability toward UV bleaching and corrosion in salty environments compared to polymer and metallic films.

Solar protection films are particularly convenient since they increase solar protection of interiors without having to replace the whole window or disassemble it to substitute the glass pane. Most can even be installed directly by the user after being cut to size on site.

8.5 Advanced shading systems

Among the various shading systems, external shading elements such as horizontal or vertical projecting louvers (*brise soleil*), shutters, blinds, or awnings are preferable to internal ones because they block solar radiation before it penetrates the building. *Brise soleil*, in particular, also have architectural value and can strongly characterize the external image of a building. It is important, however, to evaluate the behavior of such devices, as their presence can significantly change the performance of the openings and therefore the ventilation coefficient of rooms.

Internal elements such as curtains, blinds, or shutters work primarily by reflecting and diffusing the radiation outward, often absorbing a significant share, resulting in the release of heat in the indoor environment. These elements have good efficacy in preventing glare.

If carefully designed, outer shading systems can provide differentiated behavior over the year, excluding solar radiation in the summer while allowing access during winter. The effectiveness of such systems must be verified through the use of shading masks: the ideal screen blocks the sun path in the hours in which solar radiation is not desired. The presence of the screen, however, interferes with visual enjoyment, natural lighting, natural ventilation, and the possibility of leaning out or passage through the window. To choose the most suitable type of screen is also necessary to know its behavior in relation to noise and vibration generated by wind, facade encumbrance, ease of operation, and compatibility with the different types of opening of doors and windows.

With the development of building automation systems, ever more widespread especially in new constructions, dynamic screening systems are today able to change their geometric shape and optimize the amount of incoming solar radiation according to the climatic conditions (adaptive facades or kinetic facades).[17-21] Especially in Europe these systems are normally made using adjustable louvers or blinds integrated into a double-skin curtain wall.

Following the dynamic shading systems of the kinetic facade of the Institut du Monde Arabe in Paris by French architect Jean Nouvel (1987), there are now many examples worldwide of even more innovative solutions, such as Al Bahar Towers in Abu Dhabi (2013), designed by the engineering firms Aedas and Arup (Figs. 8.19−8.21), where special software allows the screening elements to open and close depending on the angle and power of solar rays, shaping a "thinking facade" completely self-sufficient thanks to PV panels placed on the roof and along the south facade, or the new Museu do Amanhã by Santiago Calatrava (who may be considered ventilation, privacy, and exterior view. By overlaying the individual screens, made of metal or plastic materials, the different alignment of the holes creates a kaleidoscopic effect, from defined geometric shapes to uniform and diffusing filigrees.

Adaptive Fritting consists of a double-glazing unit containing mobile plates printed with fully customizable graphic motifs. Whereas traditional printed windows have static motifs, Adaptive Fritting can vary its configuration from opaque to transparent state by aligning interior printed panels made of glass or plastic materials. The system can adapt to different aesthetic and functional requirements with geometric patterns, images, or shades, ranging in size from 0.5 to 3 m.

The Strata system consists of modular slats capable of extending and constituting a continuous surface with the surrounding elements, or contracting until they disappear inside the thin support elements. Nonrectangular screens can also be created, adaptable to irregular three-dimensional profiles. Screens can be combined horizontally or vertically, or left isolated. Slats can be completely opaque or more or less transparent, and are made of metal, plastic, or wood.

The Permea system is made from integrated panels which, by sliding on each other parallel to the surface of the building envelope, can adjust its permeability, passing from complete closure to the maximum opening; in the latter configuration all elements fall within the support structure. This way it is possible to adjust solar gains, visibility, privacy, and ventilation while protecting against dust or debris. The Permea system can also be used in sensitive targets for protection against explosions or fires.

Other innovative solutions employ water to cool the building envelope by evaporation, exploiting water vaporization heat of about 0.6 kcal/mL to reduce ambient temperature.

Air cooling by evapotranspiration is the key of the innovative Bioskin facade (Fig. 8.27), inspired by traditional Japanese cooling techniques such as *sudare* bamboo blinds and *uchimizu* water spraying, employed in the NBF Osaki in Tokyo by Japanese firm Nikken Sekkei. The eastern elevation of this 25-story highrise features filigree sunscreens made of extruded aluminum cores with a highly water-retentive terracotta shell attached via an elastic adhesive. Rainwater is collected on the roof, filtered, sterilized in storage tanks, then pumped to the porous tubes on the facade, from which it evaporates, subtracting heat from the building surface. Excess water is fed to the surrounding soil below, normalizing the water cycle and reducing the load on sewerage infrastructure. This system claims to reduce surface temperature of the building envelope by up to 12°C and its surrounding microclimate by 2°C. The southern elevation of the building also hosts solar panels to shade from solar radiation and produce electricity.

In addition, Bioskin has good potential for reducing urban heat island effect, with substantial benefits achievable if more buildings in a city use it.

Although the use of screening systems, including dynamic ones, is effective in the maintenance of solar gains in winter, reduction of heat load in summer, and elimination of glare in relation to external environmental conditions and users' needs, it may cause an excessive reduction of inside natural lighting while not allowing external vision and not being suitable for energy retrofits of existing buildings. Such systems can also have higher costs of installation, management, and maintenance.

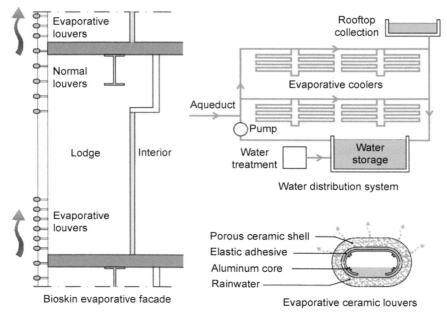

Figure 8.27 Bioskin facade system.

8.6 Conclusions and future development

Although the market now offers high-performance glazing systems for thermal insulation (triple-glazed, dual low-emissivity coating, advanced frames), the control of incident solar radiation to optimize incoming thermal and lighting flows is still particularly delicate.

Traditional static solar radiation control systems are proving unsuitable for achieving the objectives of energy efficiency and environmental well-being, and also restrain freedom of expression in the use of transparent components.

Static solar protection glazing gives effective reduction of heat loads in summer, while maintaining vision through and limiting glare phenomena. However, such systems cannot adapt to the daily solar path, changing weather conditions, or season alternation, impeding solar gain during winter and reducing daylighting in the absence of direct solar radiation. The presence of fixed or manually operated external screens interferes with visual enjoyment, natural lighting, natural ventilation, and the possibility of leaning out or passage through the window.

A more sophisticated approach using dynamic solar control systems can be effectively employed, with greater benefits from both the energetic and the visual comfort points of view. There is now a mature market for exterior dynamic shade control systems that can offer significant benefits for all buildings, regardless of their design, and that are suitable for installation in the large stock of existing buildings.

However, from an energy-efficiency perspective, dynamic shading systems can be cost-effective only in regions with higher energy cost and when a systems approach is considered.[26,27] For instance, several designers have chosen to forgo air-conditioning systems totally in moderate climates in favor of installing exterior automated shading. Buildings in hot climates with lower energy prices that still require air conditioning make installation based solely on energy efficiency harder to justify. The main reasons for installation are more often comfort, aesthetics, and other preferences. Additional research and development is needed on the one hand to reduce total system costs by enabling lower-cost motors, sensors, and controls and increasing ease of installation, and on the other hand to quantify benefits obtained better so that installation can be increasingly driven by energy efficiency in all global markets.

Lastly, in addition to the use of ultra-clear and self-cleaning glazing, natural lighting can be maximized by employing integrated light-redirection systems to bring daylight inside (anidolic mirrors, light-redirecting glazing), and artificial lighting systems that can provide efficient (TOLED) or natural-looking lighting where needed.

References

1. Smith GB, Granqvist CG. *Green nanotechnology: solutions for sustainability and energy in the built environment.* Boca Raton, FL: CRC Press; 2011.
2. Stamate M, Lazar G. Application of titanium dioxide photocatalysis to create self-cleaning materials. *Rom Tech Sci Acad Mocm* 2007;**13**(3):280−5.
3. Atwa MH, Al-Kattan A, Elwan A. Towards nano architecture: nanomaterial in architecture − a review of functions and applications. *Int J Recent Sci Res* 2015;**6**(4):3551−64.
4. Green Technology Forum. *Nanotechnology for green building.* Indianapolis, IN: Green Technology Forum; 2007.
5. Chen J, Poon C. Photocatalytic construction and building materials: from fundamentals to applications. *Build Environ* 2009;**44**:1899−906.
6. Herrmann JM. Heterogeneous photocatalysis: fundamentals and applications to the removal of various types of aqueous pollutants. *Catal Today* 1999;**53**:115−29.
7. Goffredo GB, Quagliarini E, Bondioli F, Munafò P. TiO_2 nanocoatings for architectural heritage: self-cleaning treatments on historical stone surfaces. *J Nanoeng Nanosyst* 2013;**228**(1):2−10.
8. Jelle BP, Hynd A, Gustavsen A, Arasteh D, Goudey H, Har R. Fenestration of today and tomorrow: a state-of-the-art review and future research opportunities. *Sol Energy Mater Sol Cells* 2012;**96**:1−28.
9. Ohama Y, Van Gemert D. *Application of titanium dioxide photocatalysis to construction materials.* New York: Springer; 2011.
10. Kischkoweit-Lopin M. An overview of daylighting systems. *Sol Energy* 2002;**32**(2):77−82.
11. IEA. *Daylight in buildings. A source book on daylighting systems and components. A report of IEA SHC Task 21/ECBCS annex 29.* Washington: IEA-SHC; 2000.
12. DOE (U.S. Department of Energy). *Windows and building envelope research and development: roadmap for emerging technologies.* Washington, DC: DOE Energy Efficiency & Renewable Energy; 2014.
13. Casini M. *Costruire l'ambiente. Gli Strumenti e i metodi della progettazione ambientale.* Milano: Edizioni Ambiente; 2009.

14. Universal Display Corporation. *Novel smart windows based on transparent phosphorescent OLEDs Final Technical Report*. Ewing, NJ: UDC; 2005.
15. ISO 9288. *Thermal insulation — heat transfer by radiation — physical quantities and definitions*. 1989.
16. EN 410. *Glass in building. Determination of luminous and solar characteristics of glazing.* 2011.
17. Loonen RCGM, Trčka M, Cóstola D, Hensen JLM. Climate adaptive building shells: state-of-the-art and future challenges. *Renew Sustain Energy Rev* 2013;**25**:483—93.
18. Pesenti M, Masera G, Fiorito F, Sauchelli M. Kinetic solar skin: a responsive folding technique. *Energy Procedia* 2015;**70**:661—72.
19. Miao Z, Li J, Wang J. Kinetic building envelopes for energy efficiency: modeling and products. *Appl Mech Mater* 2011;**71—78**:621—5.
20. Razaz ZE. Sustainable vision of kinetic architecture. *J Build Appraisal* 2010;**5**(4):341—56.
21. Heiselberg P, editor. *IEA ECBCS Annex 44 integrating environmentally responsive elements in buildings — expert guide part 1: responsive building concepts*. IEA; 2009.
22. Tzonis A, editor. *Santiago Calatrava's creative process. Fundamentals*, vol. 1. Basel: Birkhauser; 2001.
23. Tzonis A, Lefaivre L. In: *Santiago Calatrava's creative process. Sketchbooks*, vol. 2. Basel: Birkhauser; 2001.
24. Jodidio P. *Santiago Calatrava complete works 1979—2009*. Cologne: Taschen; 2009.
25. Drozdowski Z. The adaptive building initiative: the functional aesthetic of adaptivity. *Archit Des* 2011;**81**(6):118—23.
26. Adriaenssens S, Rhode-Barbarigos L, Kilian A, Baverel O, Charpentier V, Matthew Horner M, et al. Dialectic form finding of passive and adaptive shading enclosures. *Energies* 2014;**7**:5201—20.
27. Xu X, van Dessel S. Evaluation of an active building envelope window system. *Build Environ* 2008;**43**(11):1785—91.

Dynamic glazing

9.1 Passive dynamic glazing

The need to balance diverse needs from the energy and lighting points of view is leading to the use of next-generation products, such as chromogenic transparent materials that allow selective and dynamic control of the thermal energy and incident light and have the ability to change their optical properties in response to a light, electrical, thermal, or chemical stimulus.

Chromogenic materials belong to the category of smart materials, a new class of highly innovative materials able to perceive stimuli from the external environment (such as mechanical stress, temperature variations, humidity, pH, electromagnetic fields, and solar radiation) and react immediately by modifying independently and reversibly their mechanical, physical—chemical, or electrical properties or geometrical characteristics, adapting to changing environmental conditions. They include self-cleaning, shape-memory, and phase-change materials, piezoelectric, photovoltaic (PV), electrochromic (EC), photochromic, thermochromic materials, etc.

The use of transparent chromogenic materials in architecture allows transparent envelopes with variable performance, called smart windows, dynamic glazing, or switchable glazing, able to optimize the energy behavior of buildings and at the same time meet the comfort needs of users. Intelligent glazing can be used in a wide range of building products, such as windows, doors, skylights, and partitions, and is easily integrated inside high-performance insulated glass units. Expectations for demand growth for dynamic glass are very high. In 2013 the market for smart windows was worth over $1.5 billion, and it is expected to reach more than $5.8 billion by 2020 with an estimated compound annual growth rate of 20% (2014—2020).[1]

Based on the mode of operation, intelligent glass is distinguished in two main categories: passive control, or self-regulating, and active control, adjustable to users' needs.

Passive dynamic systems do not require an electrical stimulus for their operation, but respond independently to the presence of natural stimuli such as light (photochromic glass) or heat (thermochromic and thermotropic glazing). Compared with active systems they are easier to install and more reliable, with the drawback that they cannot be controlled by the user on request.[2,3]

9.1.1 Photochromic glazing

Photochromic glass is able to modify its transparency properties autonomously in relation to incident light intensity. This ability is due to the presence in the glass paste of organic or inorganic compounds which act as "optical sensitizers," such as metal halides (chloride and silver bromide) reactive to ultraviolet (UV) light, or plastics which absorb the sun's energy according to the output color spectrum variation. When photochromic glass is directly exposed to solar radiation, the difference in spectral absorption between the energy layers of glass and additional substances leads to the formation of a reversible process of intense coloring. The response rate to environmental changes is in the order of a few minutes, and generally the passage from the tinted state to the clear one takes twice as long. These differences in response time can lead to problems in case of sudden and frequent changes in external brightness or cast shadows on the building that can cause uneven and unsightly areas of light and shadow. Furthermore, following the chromatic transition photochromic glazing becomes absorbent rather than reflective, with possible glass pane overheating phenomena, which may lead to rupture by thermal shock in the event of intense solar radiation.

Currently, the main use of these products is as glass for the optical and car industries. The diffusion of photochromic glasses in architecture is hampered by the still high cost, the complexity of the technological system, the inability for the user to control performance directly, the difficulty in obtaining a uniform distribution of photochromic substances inside the glass pane, and the gradual loss of the reversibility of the process over time. Still, the technological problems have significantly reduced in recent years, allowing extended slab sizes and improved stability over time.

9.1.2 Thermochromic glazing

Thermochromic (TC) glazing (Pleotint, Ravenbrick, Solarsmart, etc.) can autonomously modify its optical properties according to the external surface temperature, which determines a chemical reaction or a phase transition between two different states. The material remains transparent when there are temperatures lower than the transition value, and becomes opaque at higher temperatures (Fig. 9.1). The transition temperature interval is generally between 10°C (maximum transparency) and 65°C (minimum transparency). The properties of thermochromism can be observed in a wide range of organic and inorganic compounds and in films of metal oxides, such

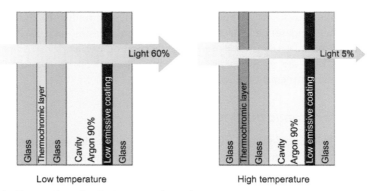

Figure 9.1 Thermochromic glazing operating scheme.

as vanadium oxide: by switching from semiconductor to metallic state, this acquires a reflective behavior highly sensitive in the infrared (IR) zone.

At present the most promising technological solution for the deployment of this glazing is the use of thermochromic materials directly in plastic film of polyvinyl butyral (PVB) with a thickness of 1.2 mm, introduced to the market in late 2010. Since PVB is one of the products mostly used for the manufacture of laminated safety or acoustic glass, this solution allows the best integration in manufacturing processes and the possibility of providing a higher-quality product at a reduced cost. The typical ranges of light transmission and solar heat gain in correspondence with transparent and opaque states when coupled with a clear glass with switching times in the order of a few minutes are reported in Table 9.1. Beyond the lack of user control, one disadvantage is that TC glazing may not reach the temperature required for switching to a dark state even in the presence of solar radiation, with the drawback of not eliminating glare for users. Duration is guaranteed for an operating time of at least 20 years, and it is cheaper than active control dynamical systems (return on investment <4 years).

9.2 Active dynamic glazing

Active dynamic systems can be controlled directly or connected to a computerized building management system to respond to changes in external (temperature, solar radiation) or internal (temperature, artificial and natural lighting levels, heat gains, presence of people) climatic conditions or the needs of users. This allows adjustment of the intensity of penetrating visible and IR radiation without the use of screening systems, and significantly reducing energy consumption for air conditioning and lighting (savings are estimated at more than 20%). The most advanced systems on the market integrate with PV systems for total electrical self-sufficiency, in addition to possibilities of remote control via smartphones and independent adjustment of different panels of the same window (light zoning) or even different zones of the same panel, up to becoming real imaging displays with touchscreen technology (advanced technology windows).

Electrically controllable active systems include electrochromic (EC) glass, suspended particle devices (SPDs), liquid crystal devices (polymer-dispersed liquid crystals—PDLCs), and more recently, still experimental glazing devices based on microblinds (micro electrical mechanical systems—MEMS) or with special nanotechnological coatings. Each technology has different characteristics, performance, and costs, making it more suitable for determined applications or requirements (privacy, switching speed, solar-gain reduction, etc.).[4–6]

9.2.1 Suspended particle devices

SPDs (Isoclima, Vision Systems, Innovative Glass, Hitachi Chemicals, etc.) consist of a double sheet of glass within which is located a layer of thin laminate of suspended particles, similar to rods, immersed in a fluid, and placed between two electrical conductors of transparent thin plastic film. When power is turned on, the suspended rod particles align, light passes through, and the SPD smart-glass panel clears. When

Table 9.1 Main thermochromic glazing on the market

Temperature	Visible light transmission (VLT) (%)	Solar heat-gain coefficient (SHGC)	Ug	VLT (%)	SHGC	Ug	VLT (%)	SHGC	Ug
	Pleotint Suntuitive clear			Innovative Glass Solarsmart			Ravenbrick Ravenwindow		
Low	60	0.37	1.36	55	0.36	1.36	33	0.28	1.36
High	13	0.17	Argon	5	0.12	Argon	5	0.18	Argon

Dynamic glazing

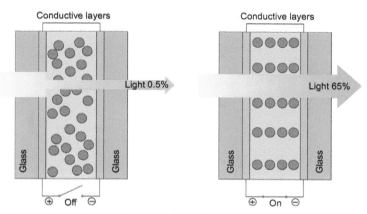

Figure 9.2 Suspended particle devices operating scheme.

the power is switched off, suspended particles are instead randomly oriented, blocking the light, and the glass appears dark (or opaque) blue or, in more recent developments, gray or black. SPD glass can lighten or darken, allowing instantaneous control of the amount of light and heat passing through (Fig. 9.2). SPD smart glass, when dark, can block up to 99.4% of visible radiation. Finally, SPD glass protects from harmful UV rays when switched on or off.

The typical ranges of light transmission and solar heat gain in correspondence of transparent and opaque states are respectively VLT 65–0.5% and SHGC 0.57–0.06, with switching times of some seconds. The very high switching speed, total controllability by the user, and the fact that the obscured state coincides with the device powered off make SPD glass particularly suitable for the automotive (side and rear windows, transparent sunroofs), marine (windows, skylights, portholes, partitions, and doors), and aviation sectors (more than 30 different models of aircraft have SPD windows installed).

The device requires about 100 V AC to operate from the off (tinted) state to the transparent one, and can be modulated to any intermediate state. Power requirements are 5 W/m^2 for switching and 0.55 W/m^2 to maintain a state of constant transmission. With further research, operating voltages may drop to about 35 V AC. New suspensions are also being developed to obtain colors other than blue (green, red, and purple) and a greater variation in the solar factor. At the moment glass panes can be up to 1524 × 3048 mm (length any size) and are available in several shapes, both planar and curved. Durability and optical solar properties have not been verified in the long term as the products are new to the market, but the high cost remains a problem (to date it is the most expensive dynamic glass on the market).

9.2.2 PDLC devices

Liquid crystal devices (Scienstry, Polytronix, Essex Safety Glass, Switchglass, Smartglass International, Magic Glass, Dream Glass, Sonte, etc.) consist of a double sheet of glass containing a PDLC package composed of a polymer matrix film sandwiched

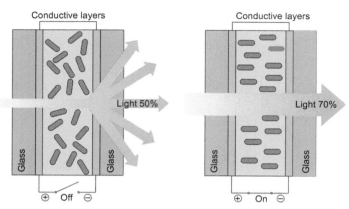

Figure 9.3 Polymer-dispersed liquid crystals device operating scheme.

between two electrical conductors on transparent thin plastic films. Within this interlayer, tiny liquid crystal spheres with a diameter of the same order of magnitude as the wavelength of visible radiation are dispersed.

In the absence of electrical stimulus, the liquid crystals have a disordered arrangement and crossing light rays undergo random diffractions, making glazed elements appear white and translucent; when an electric field is applied, liquid crystals align in the same direction, ensuring the transparency of the panels (Fig. 9.3). The degree of transparency can be controlled by the voltage applied. The light transmittance of liquid crystal glazing in the active state does not normally exceed 70% and in the off state it is about 50%, although appropriate dyes may be added to darken the device in the off state. Liquid crystal systems, while able to spread direct incident solar radiation optimally, do not block it enough to obtain a significant reduction of the solar factor, usually between 0.69 and 0.55. Furthermore, compared to EC systems, liquid crystals systems are not bistable and require a constantly applied electric field for correct operation, resulting in continuous consumption of electrical energy (about 5–10 W/m^2 of surface operating between 65 and 110 V AC). PDLC systems are mainly used for the construction of interior or exterior partitions in applications that usually require privacy, such as shop windows, meeting rooms, intensive care areas, bathroom and shower doors, or transparent walls to use as temporary projection screens. PDLC devices are also available in rolls as an adhesive, bespoke intelligent film to apply to existing glazing, even with integrated WiFi control.

9.2.3 Electrochromic devices

EC glazing (Sage, View, EControl-Glas etc.) exploits the properties of some materials to change the parameters of transmission, reflection, and absorption of solar radiation according to an electrical stimulus adjustable by an external user. Variation of the properties of these elements is attributable to the addition or extraction of mobile ions from the EC layer: when the electric field is activated, the introduced ions react, generating compounds which alter the coloring of the material (Figs. 9.4–9.6).

Figure 9.4 Electrochromic façade (clear).
Courtesy of EControl-Glas GmbH.

Figure 9.5 Electrochromic façade (half-tinted).
Courtesy of EControl-Glas GmbH.

The central part of an EC device is an ion conductor (or electrolyte, usually $LiAlF_4$) sandwiched between two layers, respectively constituted by an electrochromic nanofilm (also called electrode, commonly tungsten trioxide [WO_3] or niobium pentoxide [Nb_2O_5]) and an accumulation layer (counter-electrode, $Li_xV_2O_5$). The two outer layers are made of transparent conductive oxides such as In_2O_3 or SnO_2, and the electron accumulation layer and transparent conductor may be incorporated in a single layer. When electric potential difference is applied between the two transparent conductors, Li^+ ions extracted from the accumulation layer pass through the conductor layer and are led into the EC layer, thereby changing its optical properties—in the case of WO_3, Li + ions induce light absorption (cathodic coloration).[7] The insertion

Figure 9.6 Electrochromic façade (fully tinted).
Courtesy of EControl-Glas GmbH.

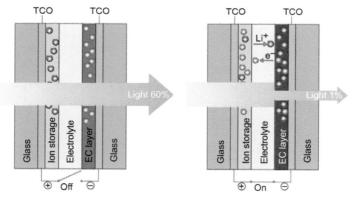

Figure 9.7 Electrochromic (EC) glazing operating scheme.

and extraction of lithium ions is balanced by an equal and opposite migration of electrons. Conversely, when the electrical stimulus is turned off, the ions are extracted from the EC layer, bleaching it, and, through the conductive layer, are deposited into the storage layer, making the device transparent again; at the same time the electrons flow back to the EC layer via an external circuit (Fig. 9.7).

Other EC materials, such as nickel oxide hydroxide (NiOOH), iridium oxide (IrO$_2$), or vanadium pentoxide (V$_2$O$_5$), instead become tinted in their oxidized state—ie, when they lose ions—and exploit anodic coloration.

An anodic EC layer can in fact replace the ion accumulation layer to enhance coloring intensity and modulation: in the clear state, ions stay in the anodic layer and electrons stay in the cathodic one, allowing both to be bleached; upon voltage application, ions and electrons trade places, reducing the cathode and oxidizing the anode, and both layers darken.[7]

Table 9.2 Main electrochromic products on the market

Tinted state	Guardian Sunguard EC (low emissive)			View Dynamic Glass		
	VLT (%)	SHGC	Ug (W/m²K)	VLT (%)	SHGC	Ug (W/m²K)
0	50	0.34	1.10 (argon)	58	0.46	1.64 (argon)
1/3	35	0.24		40	0.29	
2/3	18	0.13		20	0.16	
3/3	3	0.06		3	0.09	
Tinted state	EControl-Glas (low emissive)			SGG Sageglass		
	VLT (%)	SHGC	Ug (W/m²K)	VLT (%)	SHGC	Ug (W/m²K)
0	55	0.40	1.10 (argon)	60	0.41	1.64 (argon)
1/3	–	–		18	0.15	
2/3	–	–		6	0.10	
3/3	15	0.12		1	0.09	

Available glazing (Table 9.2) typically has green or blue colors in relation to the EC materials most widely used (eg, tungsten oxide varies its color from transparent to blue) and the degree of transparency can be modulated in intermediate states from clear (off device) to completely tinted (Figs. 9.8 and 9.9). Light transmission varies from 60% in the transparent state to 1% when opaque. SHGC is between 0.46 and 0.06 (Fig. 9.10).

The amount of energy required by the system to switch between the different color states is minimal (2.5 Wp/m²) and, thanks to the property of EC materials to possess a bistable configuration, even less energy is required to maintain the desired tinted state (less than 0.4 W/m²).

Figure 9.8 Electrochromic dynamic glazing.
Courtesy of EControl-Glas GmbH.

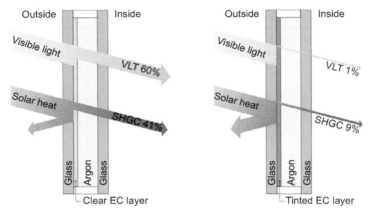

Figure 9.9 Electrochromic glazing control states.

If the device is working properly, the change of properties of the glass is almost perfectly uniform across the entire surface. Darkening starts from the edges, moving inward, and is a slow process, ranging from several seconds to some minutes depending on panel size. Switching speed is also linked to the glass temperature. The coloring process typically takes a little longer than the clearing process. In moderate or warm climates, a 90 × 150 cm window usually takes 5—10 min to accomplish at least 90% of its coloring cycle. The time increases in conditions of low temperature, but the need to control the coloring of the glass is less likely. The gradual change of light transmission is advantageous, however, because it allows the occupants to adapt naturally to changes in light levels without discomfort or distraction. EC glass provides visibility even in the darkened state, and thus preserves visible contact with the outside environment.

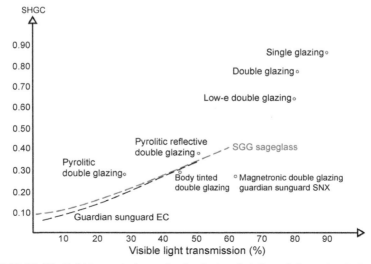

Figure 9.10 Visible light transmission and solar factor of static and dynamic glazing.

Figure 9.11 Electrochromic dynamic glass control states.
Courtesy of Vetrotech Saint-Gobain International AG.

Technological room for improvement in EC glass concerns increasing the number of control states (currently four; Fig. 9.11) and the switching speed, increasing opacity in the tinted state to improve privacy, and further reducing the already limited energy consumption.

Akin to EC glazing are gasochromic windows, in which tinting is achieved by exposing the EC layer to oxygen or hydrogen introduced in the window cavity.[8] The EC layer is made of a thin film (less than 1 μm) of WO_3 covered with platinum: exposed to diluted hydrogen it undergoes chemical reduction and darkens, reverting back to a clear state if diluted oxygen is introduced. Coloring gases are produced by an electrolyzer, which may serve multiple windows through a gas distribution system. Visible transmittance can vary between 0.10 and 0.59, with an SHGC range of 0.12−0.46. Transmittance levels of less than 0.01 for privacy or glare control are possible. Compared to traditional EC glazing, gasochromic glass does not need transparent electrodes or an ion-conducting layer, resulting in simpler construction. However, water is needed in the WO_3 films to allow rapid transport of the hydrogen, and may escape in higher temperatures, so care must be taken in the tungsten sputtering process to incorporate the right amount of water and make it stable up to temperatures above 100°C. Gasochromic technology is best suited for triple-glazed windows, since one cavity is required for the gasochromic operation, and is not suitable for low-emissivity coating or argon/krypton filling.

Among the different dynamical active control systems, EC devices are particularly interesting and are currently considered the most suitable and promising chromogenic technology for the control of radiant energy through transparent components of the

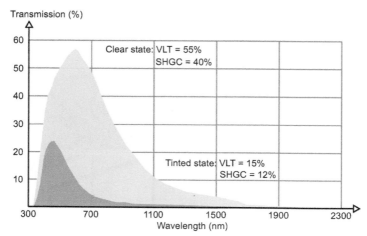

Figure 9.12 Spectral transmission of Electrochromic glass in different tint states (EControlglas).

building envelope (to switch from clear to completely tinted, a 200 m² EC glass facade needs about the same energy used to power a single 60 W lightbulb). Compared to SPD and PDLC glass, EC glazing has lower power consumption for both switching and keeping the desired tint state. It also gives excellent protection from solar (SGHC values varying from 0.46 to 0.06) and UV radiation (Fig. 9.12) while always allowing vision through (unlike PDLCs), and, to date, has better and proven durability, guaranteed up to over 30 years (one of the first installations of EC glass in the Desert Regional Medical Center in Palm Springs, California, dates from 2003 and the glass is still operative today), along with lower costs compared to SPD and PDLC devices.

Compared to the first introductions, registered in the early 2000s, new EC products allow larger glass pane sizes (up to 1524 × 3048 mm); are available in different shapes (circular, triangular, trapezoidal, etc.) and colors; offer the possibility of adjustment between four different tinted stages; allow independent modulation of individual panels (light zoning) and implementation of up to three control zones, even with different shapes, on the same pane (Sageglass Lightzone) to maximize daylight and optimize visual comfort even in case of partial shading of the facades; can be self-powered by window-integrated PV systems (Sageglass Unplugged) without the need for supply and control wiring, with easier installation in particular for replacing existing windows; are controllable via WiFi by smartphones; and have considerably higher durability, guaranteed for more than 30 years (Fig. 9.13).

There are now several hundred examples of buildings that have EC glazing installed around the world and in different climatic zones, from educational, commercial, and public buildings to hotels, hospitals, and worship facilities. According to the US Department of Energy, the goal of achieving zero-energy or positive-energy buildings is not feasible without the use of dynamic glazing.[9]

Figure 9.13 Chabot College electrochromic glazing.
Courtesy of Vetrotech Saint-Gobain International AG.

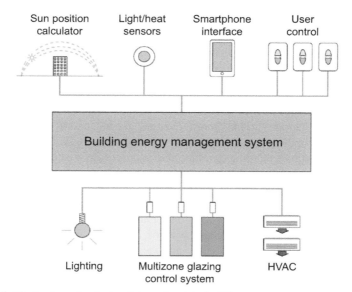

Figure 9.14 Electrochromic glazing integration into building management system.

The main advantages of using EC glass are, in particular:

- reduction of up to 60% of artificial lighting needs by increasing light transmission through the windows, resulting in increased visual comfort for occupants and reduced energy costs;
- the ability to adjust the lighting levels in indoor environments while maintaining transparency and exterior vision, thus occupants can enjoy outdoor views both during the day and at night (even in its darkest state, with less than 2% of VLT, the glazing is still transparent);
- the reduction of summer and winter heating, ventilation, and air-conditioning requirements thanks to the ability to control heat gains from solar radiation, with consequent reduction of energy costs for management and installation due to the use of plants of smaller size (up to 25% less power needed) (Fig. 9.14);

- the elimination of both internal and external solar control screens, with consequent reduction of installation, maintenance, and, in the case of powered mobile systems, management costs;
- the protection of materials and furnishings from direct solar radiation, reducing discoloration and degradation due in particular to UV solar radiation, which is entirely blocked by the EC glazing;
- greater freedom in architectural design, allowing an increase in the glazed surface to opaque surface ratio of the envelope (window-to-wall ratio) without affecting the building's energy performance, and the use of glass components in situations where this would normally compromise environmental comfort;[10]
- potential application in all cases of upgrading energy efficiency of existing buildings;
- the ability to achieve high scores in building environmental certification systems such as US LEED (Leadership in Energy and Environmental Design) and UK BREAAM (Building Research Establishment Environmental Assessment Methodology) (at the end of 2015 the project at the 3.0 University Place building in Philadelphia, featuring EC facades by Sageglass on its entire building shell, was the world's first commercial office building to obtain LEED V4 Platinum precertification).

In this picture it is apparent that the use of dynamic windows can bring numerous benefits in terms of energy efficiency, environmental comfort, and architectural quality of buildings. Static solutions with selective glass and fixed or mobile screens do not allow optimization of solar gains and light conditions during the year, thus limiting the size of glazed components in the design. Solutions with automated dynamic sunscreens coupled with building automation systems offer excellent energy performance, but have high installation, maintenance, and management costs, and hinder the view from the inside to the outside. In contrast, dynamic glazing, and in particular EC glass, allows adjustment of the amount of incoming light and heat (Fig. 9.15), according to the effective need, creating a building envelope able to adapt fully to the weather conditions (climate-adaptive building shells) and improving the building's overall performance in every kind of climate, especially in hot and Mediterranean countries (Table 9.3).

EC systems also prove more convenient compared to static systems, thanks to the significant reduction in energy consumption for artificial lighting and air conditioning all year round, and to automated screens systems, due to lower installation (absence of internal and external screening systems and related motion devices, ability to use lower-power air-conditioning systems) and management costs.

Despite these indisputable advantages, the costs for installing EC glass, and dynamic glazing in general, still remain too high for more widespread application in residential and commercial sectors. The extra cost of dynamic windows compared to traditional insulating glass units is around €215/m^2, with a payback period of 26—33 years for residential buildings and 57—61 years for commercial buildings.[9] Beyond the high cost, which should drop with increasing market penetration and improvement of production processes, another factor hampering widespread use is the still limited information between professionals and consumers, due to a lack of standardization in the technology.

9.2.4 Emerging technologies

Among emerging dynamic technologies, intelligent glass with integrated microblinds and nanocrystal-based EC materials, as well as hue control and energy harvesting

Figure 9.15 Electrochromic dynamic glazing control states. Courtesy of View Inc.

switchable glazing, are of particular interest for possible future applications in architecture.

Developed by the National Research Council, Canada, Micro-Blinds sun-shielding glass is composed of inorganic prestressed curling electrodes, invisible to the naked eye (size 100 μm) and able to unwind following a weak electrostatic stimulus.[11] These MEMS are manufactured by depositing on the glass slab a magnetron layer not dissimilar to a low-emissivity coating, and subsequently patterning it by laser. At the current experimental stage their performance is comparable to conventional dynamic electric control glass; however, they do not require expensive indium-tin oxide conductive layers, and have activation and deactivation times in the order of milliseconds. MEMS can be made with materials of different properties, allowing, for example, favoring UV-ray resistance or, in contrast, permeability for healthier indoor environments, or using highly reflective materials to improve performance further in the shielding (unwound) configuration. Development is now focusing on laser etching modes and creating market-sized products.

As regards nanotechnological EC materials, researchers at the University of Berkeley have developed a novel material consisting of indium-tin oxide nanocrystals embedded in a glassy matrix of niobium oxide (Fig. 9.16).[12] With an electrical pulse, the resulting compound allows independent control of visible light and near-IR radiation, blocking unwanted energy intakes while maintaining the possibility of best exploiting natural light. Further enhancement of this technology at Cockrell School of Technology at the University of Texas has structured the two components into a single porous film with optimized pathways for electronic and ionic charges, resulting in the possibility of control of up to 90% of near-IR radiation and 80% of visible light, with switching times in the order of few minutes.[13]

Research is also focusing on self-powered shading technologies, like that developed by researchers at Nanyang Technical University.[14] The prototype is able to change its color from transparent to dark blue just through exposure to oxygen present in the air, by using a layer of Prussian blue pigment with EC characteristics, while at the same time accumulating electric energy as a genuine transparent battery. Prussian blue can become transparent if it receives electrons from a donor, lowering its iron

Table 9.3 Main active and passive chromogenic technologies on the market[4]

Properties	Passive systems	Dynamic glazing Active systems		
	TC	EC	SPD	PDLC
Optical and thermal performance				
Clear state	Low temperature	Off	On	On
Dark state	High temperature	On	Off	Off
VLT (clear)	60%	60%	65%	Up to 75%
VLT (dark)	5%	1%	0.5%	50%
SHGC (clear)	0.37	0.46	0.57	0.69
SHGC (dark)	0.12	0.06	0.06	0.55
UV transmission (clear)	0%	0.4%	0.1%	0.5%
UV transmission (dark)	0%	0%	0.1%	0.5%
Privacy in dark state	No	No	Limited	Yes
Number of light control levels from clear to dark	No	Typically 4 states	Unlimited	2 (transparent and frosted)
Continuous states between dark and clear	Yes	Yes	Yes	No
Light zoning	n/a	Yes	Yes	Yes
Operating temperature	From −20 to 160°C	From −20 to 70°C	From −40 to 120°C	From −20 to 70°C
Configuration options				
Maximum size	1651 mm × any length	1524 × 3048 mm	1524 mm × any length	1828 × 3567 mm

Shapes	Any shape, including curved	Rectangle, square, trapezoid, triangle	Any shape, including holes anywhere and curved	Any shape, including holes anywhere and curved
Colors	Blue, green, bronze, gray	Blue, green	Typically blue	Clear, bronze, gray, green tint
Electrical properties				
Operating voltage	n/a	12 V DC	65–110 V AC	65–110 V AC
Power requirement for state transition	n/a	2.5 W/m^2	5 W/m^2	5–10 W/m^2
Power requirement for state maintenance	n/a	0.4 W/m^2	0.55 W/m^2	5–10 W/m^2
Switching speed	Several minutes	Typically 3–5 min to reach 90% of range	Typically 1–3 s	Instantaneous (40 ms)
Control	No	Wall switch, remote control, movement sensor, light and temperature sensor, timer	Wall switch, remote control, movement sensor, light and temperature sensor, timer	Wall switch, remote control, movement sensor, light and temperature sensor, timer
Integration with building management system	n/a	Yes	Yes	Yes
Costs and durability				
Cost	Lowest	Medium	Highest	High
Durability	>20 years	>30 years	>20 years	>10 years

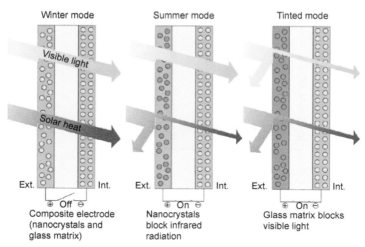

Figure 9.16 Nanotechnological electrochromic glazing, Berkeley University.

content. By connecting a cathode of indium-tin oxide coated with Prussian blue to an aluminum anode, characterized by high electropositivity, through a liquid electrolyte, the glazing clears in just 4 s, while the electrode discharges. Decoupling the connection and exposing the electrodes to oxygen, the battery is recharged by oxidizing the iron present in the cathode, reconstituting the dark blue tone. Once recharged, the smart window can become clear again to let light in, and simultaneously supply low-power electrical devices such as light-emitting diodes.

Finally, an extremely promising technology devised by researchers at the University of Cincinnati in collaboration with Hewlett Packard and EMD/Merck combines control of both transmission and hue of entering light (Fig. 9.17).[15] Unlike the traditional scheme with two electrodes separated by a conductive fluid, found in SPD, PDLC, and EC technologies, this introduces a third electrode disposed on a honeycomb mesh, creating a grid of pixels where a colloidal solution of two complementary colors (red/cyan, green/magenta, blue/yellow) is dispersed. By applying small voltages to each of the three electrodes, the colored particles are drawn respectively toward the upper electrode, the inferior one (characterized by replicated polymer micropits capable of compacting pigments, inhibiting chromatic spread), or the perimeter one, allowing passage from a neutral dark state (both colors in uniform dispersion) to a neutral clear state (colors compacted on the perimeter electrode and in the lower electrode micropits), and from a cold hue (blue particles dispersed, yellow particles compacted around the perimeter electrode) to a warm one (yellow particles dispersed, blue particles accumulated around the perimeter electrode), or any combination these four states, with unprecedented versatility and switching times in the order of 10 s.

Depending on the bicolored dispersions employed, it is also possible to get windows able to adjust light transmittance and opacity independently (combining black particles with light diffusing white particles) or light and thermal transmittance (dispersion of black particles and light particles absorbing IR radiation). Finally, the

Dynamic glazing

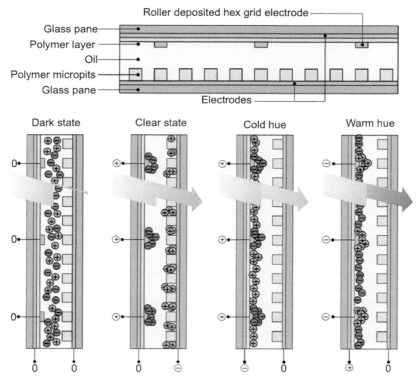

Figure 9.17 Light and hue control dynamic glass, Cincinnati University.

production process is relatively simple, allowing large-size roll-to-roll manufacturing at a competitive cost, even in films to apply on existing windows.

9.3 Conclusions and future trends

The use of dynamic windows can bring numerous benefits in terms of energy efficiency, environmental comfort, and architectural quality of buildings, and can play a pivotal role in robust market development for zero-energy buildings.

Conventional static solar control solutions such as selective glass and fixed or mobile shading devices do not permit the optimization of solar gains and daylighting across the entire year, effectively limiting the size of glazed components in the design phase. Dynamic shading solutions integrated with building automation systems are very effective, but are expensive to install, operate, and maintain, and impede outside view.

In contrast, dynamic glazing such as ECs allow free control of incoming light and heat, giving a climate-adaptive building shell fully adjustable to any weather condition, with improved overall performance for the building in all climates, and even more pronounced in warmer countries. Dynamic glazing systems are also more convenient than automated screen systems due to lower installation and management costs, and static systems, as they reduce artificial lighting and air conditioning consumption across the entire year.

The critical aspects are still the high cost of the products, although this is destined to drop with increasing market penetration and the improvement of production processes, and the limited information between professionals and consumers, due to a lack of standardization in the technology.

Rather than improving the already good performance, it therefore appears more important to reduce the cost of dynamic glass drastically, focusing on materials, production process improvement, and easier installation.

Nanotechnology will play a large role in widening the whole dynamic windows market, as nanoparticle-conductive films are expected gradually to substitute indium-tin oxide coatings, allowing the abandonment of expensive and hard-to-supply raw materials and implementation of cheaper production processes such as roll-to-roll manufacturing.

To ensure wide market penetration of dynamic glass, the 2025 target should be to achieve less than €65/m^2 extra cost, allowing a payback time of 10−12 years for residential buildings and 21−22 years for commercial ones.[9]

In existing buildings, in addition to high costs, spread is still hindered by the need to replace the entire window and install extensive wiring. It is therefore essential to promote research in the field of dynamic high-performance films to be used to retrofit existing windows, and in the development of internal self-powering systems and built-in remote control via WiFi for windows (internet of things).

References

1. Markets and Markets. *Smart glass market by technology (electrochromics, SPD, liquid crystal, microblinds, thermographic and photochromics), by application (architectural, transportation, solar power generation, electronics & others), by geography − global forecast to 2014−2020*. Dallas, TX: Markets and Markets; 2014.
2. Granqvist CG, Green S, Niklasson GA, Mlyuka NR, von Kræmer S, Georén P. Advances in chromogenic materials and devices. *Thin Solid Films* 2010;**518**:3046−53.
3. Saeli M, Piccirillo C, Parkin IP, Binion R, Ridley I. Energy modeling studies of thermochromic glazing. *Energy Build* 2010;**42**:1666−73.
4. Casini M. Smart windows for energy efficiency of buildings. *Int J Civ Struct Eng* 2015;**2**(1):230−8.
5. Beevor M. *Smart building envelopes*. Department of Engineering: University of Cambridge; 2010.
6. Baetens R, Jelle BP, Gustavsen A. Properties, requirements and possibilities of smart windows for dynamic daylight and solar energy control in buildings: a state-of-the-art review. *Sol Energy Mater Sol Cells* 2010;**94**:87−105.
7. Makhlouf ASH, Tiginyanu I. *Nanocoatings and ultra-thin films*. Cambridge: Woodhead Publishing Limited; 2011.
8. Smith GB, Granqvist CG. *Green nanotechnology: solutions for sustainability and energy in the built environment*. Boca Raton, FL: CRC Press; 2011.

9. DOE (U.S. DEPARTMENT OF ENERGY). *Windows and building envelope research and development: roadmap for emerging technologies*. Washington, DC: DOE Energy Efficiency & Renewable Energy; 2014.
10. Sbar NL, Podbelski L, Yang HM, Pease B. Electrochromic dynamic windows for office buildings. *Int J Sustain Built Environ* 2012;**1**:125−39.
11. Lamontagne B, Barrios P, Py C, Nikumb S. The next generation of switchable glass: the micro-blinds. *Proc Glass Perform Days* 2009:637−9.
12. Llordés A, Garcia G, Gazquez J, Milliron DJ. Tunable near infrared and visible-light transmittance in nanocrystal-in-glass composites. *Nature* 2013;**500**:323−6.
13. Kim J, Ong GK, Wang Y, Leblanc G, Williams TE, Mattox TM, et al. Nanocomposite architecture for rapid, spectrally-selective electrochromic modulation of solar transmittance. *Nano Lett* 2015;**15**(8):5574−9.
14. Wang J, Zhang L, Yu L, Jiao Z, Xie H, Lou XW, et al. A bi-functional device for self-powered electrochromic window and self-rechargeable transparent battery applications. *Nat Commun* 2014;**5**(4921):1−7.
15. Mukherjee S, Hsieh WL, Smith N, Goulding M, Heikenfeld J. Electrokinetic pixels with biprimary inks for color displays and color-temperature-tunable smart windows. *Appl Opt* 2015;**54**(17):5603−9.

Energy-generating glazing

10.1 Advanced photovoltaic glazing

Among the different ways of integrating photovoltaic (PV) technologies into the building envelope, those related to glazed components are particularly interesting for the possibility of combining energy production, protection from solar radiation, daylight control, and color quality.[1]

In addition to the more traditional glass—glass PV systems, manufactured by inserting crystalline cells inside two glazed panes, it is now possible to create transparent closures using PV technologies based on semitransparent thin film made from inorganic (a-Si, CdTe) or organic (organic photovoltaic—OPV, dye-sensitized solar cells—DSSCs) material, or innovative systems based on microspheres (solar sphelar), glass prisms (photovoltaic glass units—PVGUs), and transparent luminous solar collectors (TLSCs).

To ensure maximum yield of window-integrated PVs, external glass layers that cover solar cells must be as transparent as possible to let all the light radiation in. Highly transparent low-iron glass panes, however, still reflect a part of the incoming radiation, in particular solar rays coming at an angle or diffused light, preventing it from reaching the photosensitive surface beneath. For this reason it is best to resort to glass treated with antireflective coatings: best available technologies employ nanocoatings able to achieve 97% energy transmittance levels, boosting electric energy yield by up to 6% compared to traditional ultra-clear glass, and some have hydrophilic self-cleaning properties too.

Light-gathering enhancement solutions also include coatings able to capture light coming at an angle and efficiently diffuse it toward the photoactive surface, increasing installation flexibility in relation to suboptimal orientation or inclination of the solar modules. Among these, particularly interesting is one devised by researchers from Arabian and Taiwanese universities which employs a hierarchical structure of fused silica nanorods arranged in a honeycomb pattern of nanowalls to increase conversion efficiency greatly. In particular, the subwavelength feature of the nanorods and the efficient scattering ability of the nanowalls enhances efficiency of a silicon cell by 5.2% at normal incidence, and up to 46% at an incident angle of 60 degree. The nanostructured surface also exhibits excellent self-cleaning characteristics: the water drop contact angle is reduced to 124 degree, making drops roll away with dirt and enabling 98.8% of efficiency to be maintained after six weeks of outdoor exposure.[2]

10.1.1 Crystalline silicon photovoltaic glazing

Crystalline silicon (c-Si) cells are obtained from thin slices of silicon (wafers) 160–240 μm thick, cut from a single crystal or a block. The type of crystalline cell produced depends on the silicon wafer manufacturing process. The main types of crystalline cells are:

- monocrystalline
- polycrystalline or multicrystalline
- ribbon and sheet-defined film growth (ribbon/sheet c-Si).[3]

Single-crystal silicon PV cells are formed with wafers manufactured using expensive single-crystal growth methods such as the Czochralski technique, and have commercial efficiencies of 17–20% with a record of 25.6% in laboratory conditions obtained by Panasonic (2014) by laminating amorphous silicon (a-Si) layers on top of the monocrystalline silicon wafer (patented as HIT—heterojunction with intrinsic thin-layer) and bringing electrode contacts to the back of the cell (Table 10.1). Their standard color ranges from dark blue to nearly black, and they usually have a "pseudosquare" shape due to the wafer being cut from a cylindrical ingot.

Multicrystalline silicon cells, usually formed with multicrystalline wafers manufactured from a cheaper cast solidification process, are more popular as they are less expensive to produce but are marginally less efficient, with average conversion efficiency around 14–18%. Record efficiency of 21.25% in laboratory conditions was achieved by Trina Solar (2015) by exploiting surface passivation of the cell, to extend the charge carriers' average lifetime. Polycrystalline cells' standard color ranges from blue to dark blue, showing their typical crystalline pattern.

Ribbon and sheet-grown silicon cells forgo wafers sawn from ingots for respectively drawing molten silicon or epitaxially depositing it around a monocrystalline seed. These processes can be considerably cheaper, as there is no material waste from wafer cutting. The record efficiency for epitaxial crystal cells was set by Solexel at 21.2% in 2014.

The color appearance of c-Si solar cells does not depend strictly on the material, but is instead determined by the thickness of the antireflective coating applied. Solar cells are coated with a layer of silicon nitride with thickness optimized to achieve highest

Table 10.1 Crystalline silicon solar cells' maximum efficiency

Crystalline silicon solar cells technology	Maximum efficiency (2015)	
	Company	Value (%)
Monocrystalline heterostructure (HIT) cells	Panasonic	25.60
Monocrystalline cells	Sunpower	25.00
Epitaxial crystal cells	Solexel	21.20
Polycrystalline cells	Trina Solar	21.25

efficiency. By varying the coating thickness is possible to achieve solar cells with the desired color appearance, albeit with the inherent trade-off of some loss of efficiency.[3]

Glass—glass PV modules (Fig. 10.1) are realized by encapsulating c-Si solar cells inside two glass panes with transparent resin (polyvinyl butyral). The modules can be used both for single-glazed windows and inside double or triple glazing. The distance between the solar cells determines the power of the module and its transparency degree (Figs. 10.2—10.4); the latter, in addition to spacing the cells, can be achieved by microperforated PV cells.

In particular, the main parameters which characterize glass—glass PV panels are as follows.

- Size and shape of the glazing element: glass—glass PV modules can be rectangular, trapezoidal, and triangular, both flat and curved.
- Distribution pattern (texture): the cells can be arranged in different patterns, provided that they allow preparation of the electrical connections.
- Distance between the cells (degree of transparency): by varying the distance between the cells it is possible to control interior daylight, and therefore the characteristics of visual permeability between the inside and the outside. This system also ensures multifunctionality of the PV component, which can be employed as a sunscreen element.
- Type of cell: in manufacturing glass—glass modules it is possible to use cells characterized by different technologies (monocrystalline and polycrystalline silicon), sizes, shapes, colors, and number of active surfaces (single or double-sided cells).

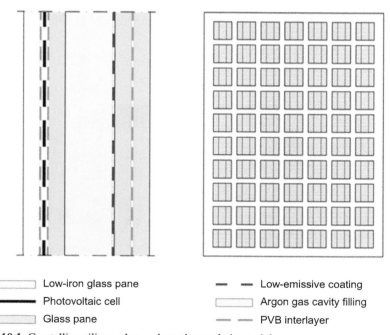

Figure 10.1 Crystalline silicon glass—glass photovoltaic modules.

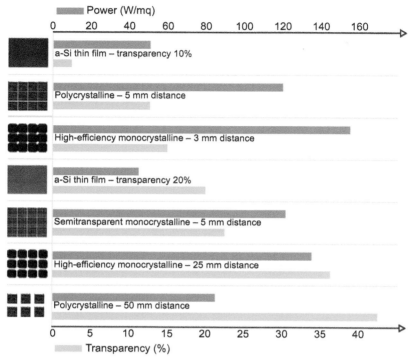

Figure 10.2 Yield and transparency comparison between thin-film and crystalline silicon PV modules.

Figure 10.3 Crystalline silicon glass—glass photovoltaic glazing.
Photo by the author.

Figure 10.4 Berlin Hauptbahnhof crystalline silicon PV glazing.
Photo by the author.

Table 10.2 Polycrystalline silicon glass—glass modules

Cells/m²	Transparency (%)	Power (Wp)
12	70	48
18	55	72
24	41	96
30	27	120
36	12	144

Table 10.2 shows some reference power values of polycrystalline silicon glass—glass modules according to the different degrees of transparency. Assuming a transparency of 41% (distance between the cells of 32.5 mm), the module has a peak power of 96 Wp/m².

Concerning the profile of the glazed elements, crystalline PV modules are usually only available as flat panels because of the difficulties of bending PV cells, which are extremely fragile and can break easily because of their reduced thickness (about 0.2 mm).

For this reason, curved or flexible PV products were until recently made only with thin-film technology. Today some companies, such as Spain-based Vidursolar, have overcome these difficulties with a new production process that allows manufacture of curved crystalline solar cells and glass—glass PV modules with a

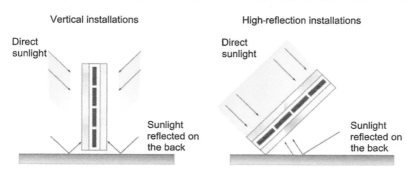

Figure 10.5 Bifacial photovoltaic cells.

Table 10.3 **Bifacial photovoltaic cell yield**

Contribution of solar radiation on back of panel	0%	5%	10%	15%	20%	25%	30%
Power (Wp/m^2)	160	168	183	190	198	213	220
Module efficiency (%)	16.1	16.8	17.6	18.3	19.0	19.8	20.5

curvature radius up to 5,000 mm. This new process gives even more possibilities for the integration of PV technology in buildings, and greatly expands the repertoire of solutions available to designers in urban areas.

PV cells may also have one photoactive side or two. Bifacial cells (Fig. 10.5) exploit both the direct solar radiation that over the course of the day affects the front and back faces and the indirect radiation reflected from the surfaces underlying or adjacent to the modules, allowing a greater production of electricity, up to over 200 Wp/m^2 (Table 10.3).

Back-face cells contribute to produced energy depending on the surrounding surfaces: low-reflectance surfaces (eg, grass) raise production by up to +18%, whereas high-reflectance surfaces (eg, sand) contribute up to +35%. Snow can increase energy production up to +80%.

Bifacial cells' cost premium is reasonable compared to traditional products, since they share most of the manufacturing process; this, together with their greater efficiency, is favorable to their market penetration. The world's largest installation of bifacial cells is under way in Chile at La Hormiga solar plant, whose 9,180 bifacial modules total 2.5 MWp and have expected yearly production of 5.78 GWh.

10.1.2 Semitransparent thin-film PV glazing

Thin-film cells employ materials with high light absorption values, able to capture solar energy effectively even at extremely reduced thickness. They are usually constituted of 5—10 different layers to form the p-n junction, integrate electrodes, and

Figure 10.6 Semitransparent amorphous thin-film photovoltaic glazing.
Courtesy of Onyx Solar Group llc.

provide resistance reduction, deposited on a low-cost substrate such as glass, stainless steel, or polymer. Thin-film manufacturing techniques allow photoactive semiconductors to be deposited over larger areas, using layers of microscale thickness to obtain monolithically interconnected modules which can then be laser cut into multiple thin cells.[1] In particular, a-Si solar cells are manufactured by depositing a 1 μm thick silicon layer on glass substrates (Fig. 10.6).[2]

Additionally, flexible cells can be obtained by depositing the photosensitive material on a thin polymer film; they can be integrated into textile building components or consumer end-products.

Available second-generation thin-film technologies include:

- amorphous silicon (a-Si)
- multijunction thin-silicon film (a-Si/μc-Si)
- cadmium telluride (CdTe)
- copper, indium, gallium, (di)selenide/(di)sulfide (CIGS)
- copper, zinc, tin, sulfide (CZTS)
- gallium arsenide (GaAs).

The efficiency of commercially available thin-film solar cells is between 10% and 16% on average, with best laboratory values over 20% (Table 10.4).

Compared to crystalline technology, thin-film solar cells have a higher energy yield with equal installed power. This is due to their lower dependence on operating temperature, which can reach 70°C in summer months, along with a better response when the diffused-light component is higher and irradiance levels are lower, such as in overcast weather, resulting in a yearly energy yield up to 15% higher than c-Si solar cells under actual operating conditions. This is even more pronounced in installations in shaded, highly polluted or heavily irradiated areas. For these reasons thin-film solar cells are also less affected than crystalline solar cells by suboptimal direct sun radiation exposure, allowing greater flexibility in architectural integration. However, thin-film solar

Table 10.4 Thin-film solar cells' maximum efficiency

PV inorganic thin-film technology	Maximum efficiency (2015)	
	Company	Value (%)
GaAs cells	Alta devices	28.8
Copper indium gallium selenide (CIGS) cells	ZSW	21.7
CdTe cells	First Solar	21.5
Amorphous Si:H (stabilized)	AIST	13.6
CZTS/Se cells	IBM/Solar frontier	12.6

cells' lower conversion efficiency means that a larger installed surface is required to achieve the same peak power.

Thin-film technology today allows semitransparent PV (STPV) films that can be applied inside two glass panes or on the inner side of external glass of double or triple glazing (position 3) (Fig. 10.7).

In addition to producing electricity, STPV thin film allows sunlight in, avoiding ultraviolet (UV) and infrared (IR) radiation, and permits vision through the glass. Different degrees of transparency (normally 10%, 20%, or 30% up to fully opaque) can be chosen depending on the degree of solar protection or light transmission required. Glass can have any shape (circular, trapezoidal, triangular) and can also have a curved profile. The chromatic effect may be neutral or colored through the application of interlayers of the desired color (Figs. 10.8 and 10.9).

Inorganic thin-film technology is usually based on a-Si or CdTe. CIGS technology holds the efficiency record for thin film (23.3% for National Renewable Energy Laboratory (NREL) concentrator cells), but is unsuitable for semitransparent application. The power of the module depends on the technology used, which determines the efficiency of conversion of solar radiation, and on the degree of transparency of the film (Fig. 10.10). Increasing transparency decreases electric power (Table 10.5).

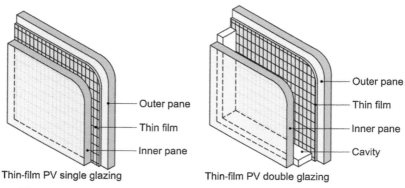

Figure 10.7 Semitransparent amorphous silicon thin-film photovoltaic glazing.

Energy-generating glazing

Figure 10.8 Semitransparent colored amorphous silicon thin-film photovoltaic glazing. Courtesy of Onyx Solar Group llc.

Figure 10.9 Semitransparent amorphous silicon thin-film colored photovoltaic pavement. Courtesy of Onyx Solar Group llc.

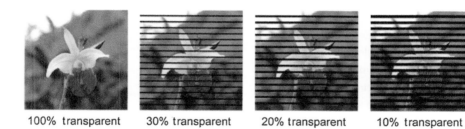

100% transparent   30% transparent   20% transparent   10% transparent

Figure 10.10 Amorphous silicon thin-film photovoltaic glazing transparency.

Table 10.5 **Semitransparent amorphous silicon thin-film PV glazing (Onyx solar)**

Transparency	0%	10%	20%	30%
Peak power (Wp/m^2)	62.5	44.4	38.9	33.3
Visible light transmission (VLT) (%)	0.2	10.8	17.3	28.4
Solar heat-gain coefficient (SHGC)	22	29	34	41
Light to solar gain (LSG) ratio	0.00	0.37	0.50	0.70
UV transmission (%)	0.0	1.5	1.5	4.7
Ug (W/m^{2}K)	1.1 (argon filling and low-emissivity coating)			

10.1.3 Organic photovoltaic glazing

Developments of PV technology are represented by organic photovoltaics, also known as third generation photovoltaics, consisting of organic solar cells (OPV) and of dye photoelectrochemical cells (DSSCs) which, thanks to the ability to exploit diffused light, can be applied on any part of the building envelope with any exposure, and even in interiors.

10.1.3.1 Organic solar cells

Organic solar cells are formed by one or more layers of photoactive polymers placed between two metallic electrodes; the solar light absorbed by the polymers triggers a transition of electrons and electronic gaps pairs (excitons) from the donor layer to the receptor, generating a difference in electrical potential between the two electrodes. The excitons are split in separated charges by a two-material heterojunction.

In organic—organic composites the heterojunction consists of a polymer acting as the gaps transporter and a fullerene derivative to transport electrons (Fig. 10.11). In organic—inorganic hybrids it consists of a light-absorbing polymer to transport gaps, while the electrons are transported by nanostructured inorganic materials such as titanium dioxide (TiO$_2$).

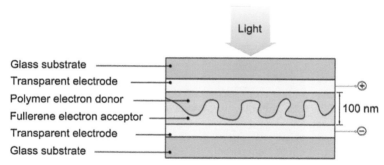

Figure 10.11 Hybrid organic solar cells' structure.

These materials are characterized by high optical absorption, allowing the use of very thin flexible films suitable for plastic substrates. Economical high-output processes such as roll-to-roll manufacturing are viable, as is spray application: fully sprayed polymer solar cell modules conform to virtually any kind of substrate and automated spray processes such as the one devised by CHOSE Laboratory, completely performed in air, allow fabrication of fully spray-coated modules on glass substrates. Conversion efficiency is still limited to 1%, though, and transfer on plastic substrates is still in progress.[4]

Conversion efficiency reached 11.5% in 2015, compared to initial values of around 2—3% in the early 2000s.

Compared to traditional inorganic semiconductors such as silicon, in polymer-based cells much more photon energy is lost during conversion, due to the inherently lower short-circuit current density and open circuit voltage. Research on molecular engineering is thus focusing on increasing conversion performance by improving the transport of electrons between the several layers, and increasing the exchange surface of charges between the donor and acceptor with "bulk"-type heterojunctions where two or more polymers are mixed together while maintaining separate phases. New polymers with improved conversion efficiency are also being researched, such as the naphthobisoxadiazole-based polymer devised by researchers at Kyoto University: this innovative material, using oxygen atoms to replace sulfur in key molecular positions, yields higher open circuit voltages and lower photon energy loss (0.5 eV compared to 0.7—1.0 eV in conventional polymer systems), reaching a conversion efficiency of 9%.

Another important field is increasing the quantum efficiency of OPV materials, which today absorb only blue photons and do not exploit the energy contained in red ones. Furthermore, OPV cells can take any color required, allowing wide possibilities of building integration.

Lastly, a more disruptive theory states that by switching the metal contacts on the top and bottom of the cell to generate an electric field on top of the cell instead of the bottom, the electron/holes pairs (excitons) could be separated without the need for complex bulk heterojunctions, enabling simpler and more effective single organic semiconductor PV cells.[5]

Researchers working at Hong Kong Polytechnic University have created semitransparent perovskite solar cells (PSCs) with graphene top electrodes.[6] These cells have a claimed power conversion efficiency of around 12% when illuminated from fluorine-doped tin oxide (FTO) bottom electrodes or the graphene top electrodes, and developers foresee the potential to produce them at below $0.06/W, more than 50% less than conventional silicon solar cells. The already excellent conductivity of graphene was further enhanced by coating the electrode with a patina of poly(3,4-ethylenedioxythiophene) (PEDOT):PSS conductive polymer that also acts as an adhesion layer to the perovskite during the lamination process. Furthermore, multilayering graphene through chemical vapor deposition to create transparent electrodes reduces their surface electric resistance while retaining transparency. Lastly, performance is further optimized by improving the contact between the top graphene electrodes and the hole transport layer (spiro-OMeTAD) on the perovskite films. The mechanical flexibility of graphene and the simplified preparation of the cells make this device suitable for mass production via direct printing or using a roll-to-roll process.

10.1.3.2 Dye-sensitized solar cells

DSSC PV cells, also known as Grätzel cells after their inventor, belong to the category of hybrid solar cells, since they use components of both organic and inorganic types. Generally these cells employ a layer of microporous TiO_2 as the anode, impregnated with both inorganic and organic photosensitive pigments (dye sensitized), and platinum or tin oxide enriched with fluorspar as the cathode (Figs. 10.12–10.14). These

Figure 10.12 DSSC prototype.
Courtesy of Solaronix SA.

Figure 10.13 DSSC multicell module.
Photo by the author.

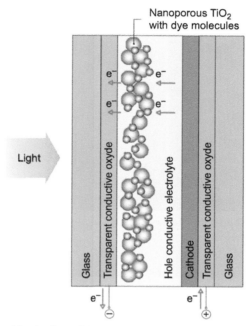

Figure 10.14 Dye-sensitized solar cells' structure.

are enclosed by two transparent conductive oxide layers such as FTO, which can be directly integrated on supporting glass as coatings.

Among inorganic dyes, most used is ruthenium, but researchers from Lund University have successfully replaced it with cheaper and more available iron sensitizers with a quantum yield of 92%[7]; organic dyes are usually plant-derived anthocyanins or animal-derived porphyrins.

The TiO_2 anode is a key player in determining the cell efficiency, and is thus the focus of extensive research: employing titanium aerogel as a photoanode support has shown excellent performance, claiming an efficiency improvement of 16%.[8] Another potential development aims at increasing the light-harvesting efficiency by adding sensitized molecular "antennas" to the TiO_2 anode, constituting an integrated trinuclear complex. The antennas are able to absorb different wavelengths efficiently, then funnel them to the central semiconductor fragment, effectively broadening the useful solar energy spectrum in a way akin to multijunction solar cells.[9]

Similarly, an organic gel developed by the University of Connecticut acts as a wavelength-expanding antenna, capturing blue photons, conventionally unused by solar cells, and lowering their energy levels via an artificial photosynthesis process. The device is made of biodegradable materials such as organic dyes, coconut fat acids, and cow blood proteins; it is inexpensive, but still not stable enough to be incorporated in manufacturing processes. In addition to DSSCs, this gel can be integrated in consumer silicon solar cells, noticeably enhancing their efficiency.

Anode and cathode are separated by a liquid electrolyte based on organic compounds. The photons pass through the electrode and hit the photosensitive pigments,

Figure 10.15 SwissTech Congress Center DSSC facade.
Courtesy of Solaronix SA.

exciting the electrons until they transfer to the titanium anode and are extracted, generating electrical current. The electrolyte, after extracting return electrons from the cathode, refurnishes the dye with the electron it has just lost to the TiO_2, allowing the dye solar cell to transform sunlight continually into electric current.

Conversion efficiency is 11.9% (2014), compared to around 8–9% in the early 2000s. A first important example of architectural integration of DSSCs is the new SwissTech Congress Center of Ecole Polytechnique Fédérale in Lausanne (Figs. 10.15–10.17), the institute where Professor Michael Grätzel developed the first prototypes of these particular cells in 1991. The total area of the intervention, implemented in collaboration with Romande Energie, is 300 m^2, covered by 1,400 modules,

Figure 10.16 SwissTech Congress Center DSSC facade.
Courtesy of Solaronix SA.

Energy-generating glazing 341

Figure 10.17 SwissTech Congress Center DSSC facade.
Courtesy of Solaronix SA.

50 × 35 cm in size, divided into five different shades of red, green, and orange according to the design by artists Daniel Schlaepfer and Catherine Bolle.

Due to the use of liquid electrolytes, cells with DSSC technology have thermal stability problems. At low temperatures the electrolytes may freeze, blocking the production of power and potentially causing physical damage. In contrast, high temperatures cause the expansion of the liquid, making the sealing of modules difficult. In addition, the electrolyte solutions contain volatile organic compounds, harmful to humans, requiring special care in sealing to prevent leakage due to galvanic corrosion. Replacing the liquid electrolyte with a solid one (solid-state DSSCs—ss-DSSCs) has been a major focus of research so far, since the use of a solid electrolyte eliminates these problems but decreases the efficiency of total conversion.

Whereas in traditional DSSC PV cells the layer of TiO_2 is immersed in a liquid electrolyte, the solid-state prototypes employ a layer of microporous TiO_2 that incorporates within its pores the solid material that transports the electronic gaps; the latter can be constituted by organic materials or p-type semiconductors that promise conversion efficiency up to over 10%. Michael Grätzel himself has announced the development of ss-DSSCs with an efficiency of 15%, achieved using hybrid perovskite $CH_3NH_3PbI_3$ as pigment, obtained by the successive deposition of PbI_2 and CH_3NH_3I solutions directly on the layer of TiO_2.[10]

Another method to obtain ss-DSSCs is evaporation of the liquid electrolyte, leaving behind a solid, hole conducting structure able to transport positive charges. This effect was discovered by researchers at Uppsala University, who noted how exhausted DSSCs continued to produce electricity even after the copper-based electrolyte depleted. Furthermore, the efficiency of these "zombie solar cells" in some cases increased up to 8%. Laboratory testing is under way to reproduce this aging process and measure obtained ss-DSSC resistance.[11]

Finally, a further development of DSSCs is constituted by quantum dot solar cells (QDSCs), which are again based on the Grätzel architecture but forgo light-sensitizing dyes in favor of low-band-gap semiconductor quantum dots such as CdS, CdSe, PbS, or Sb_2S_3. These can be carefully tuned in their band gap and light response by simply changing their nanocrystal size. A mesoporous TiO_2 layer and liquid/solid electrolytes act similarly as in conventional DSSCs: the TiO_2 layer is made photoactive by coating with quantum dots by chemical bath or electrophoretic deposition or successive ionic layer adsorption and reaction.

10.1.3.3 Manufacturing and development scope

Conventional semiconductor industry methods for silicon cell manufacture require high temperatures, doping, vacuum deposition, and other expensive processes, and are thus restricted to large industrial compounds, while OPV film panels and DSSCs, can be manufactured by fast and low-cost roll-to-roll deposition (Fig. 10.18) and printing techniques, with consequent reduction of production time and costs, opening the market to small and medium-sized enterprises.[3]

In particular, the organic solar cell printing technique has seen continuous research over recent years, with notable results such as the Melbourne-based CSIRO Laboratory method that is able to print liquid crystals cheaply on a polymer substrate to give flexible, moldable, semitransparent A3-sized solar cells that can be used pretty much anywhere, with energy conversion efficiency reaching 9.3% in 2015 and an expected cost of less than $10/m.

Support layers can be thin and flexible and can be integrated with OLED (organic light-emitting diode) and photovoltachromic technology, creating "smart" glass surfaces capable of interacting with the environment and working, depending on the need, as a PV panel, an absorbing screen, or a light. During the day they can act as

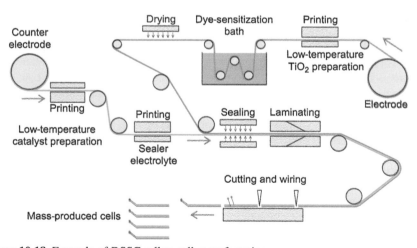

Figure 10.18 Example of DSSC roll-to-roll manufacturing.
Drawn from Luque A, Hegedus S. *Handbook of photovoltaic Science and engineering*. West Sussex: Wiley; 2011.

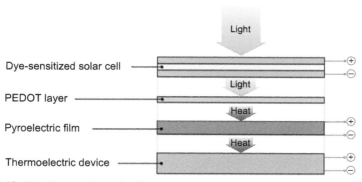

Figure 10.19 Hybrid solar/thermal cell.

"tents" that produce energy, while at night they turn into light sources to substitute for lightbulbs. At Expo 2015 in Milan German Belectric demonstrated its OPV cells, manufactured by a potentially inexpensive, continuous multistep printing process employing machines akin to those used in currency printing to achieve the bespoke, visually customizable cells that characterized the German pavilion canopies. During the day the cells produce and store electricity (5% efficiency even with diffused light), which is then used at night to provide illumination.

Furthermore, the simple and compact assembly of organic solar cells makes them suitable for integration in hybrid energy cells to combine electricity production from the conversion of different energy forms. For instance, researchers have recently created a hybrid cell capable of converting both light and thermal energy by combining a DSSC, a PEDOT conductive polymer and pyroelectric film, and finally a thermoelectric device (Fig. 10.19).[12] Light that is not absorbed by the solar cell passes through and warms the PEDOT layer below; this heat is then converted into additional electricity by the pyroelectric film and the thermoelectric device, boosting photoconversion efficiency by 20%.

The main objective of current research is improving the efficiency of solar energy/electricity conversion, and the development of a technology that allows further reduction in production costs while improving cell stability over time—today the technology is still hampered by degradation phenomena due to a gradual decrease of conversion efficiency over the years.

10.1.4 Spherical cell photovoltaic glazing

Another example of integration of PVs in transparent glazing is provided by the spherical solar cells produced by Sphelar-Power (Figs. 10.20 and 10.21).[13] Unlike traditional silicon cells, the modules are composed of small spheres, 1—2 mm diameter, capable of absorbing solar rays from every direction and thus being effective over the course of the whole day and able to exploit light diffused by the sky or reflected by the surroundings (Fig. 10.22).

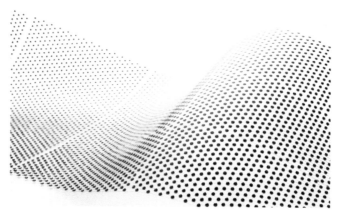

Figure 10.20 Sphelar-Power PV glazing.
Courtesy of Sphelar-Power Corporation.

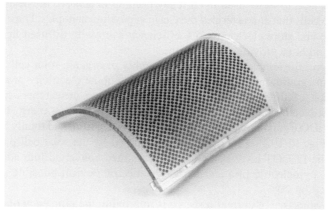

Figure 10.21 Sphelar-Power PV glazing.
Courtesy of Sphelar-Power Corporation.

The spherical cells are obtained from fragments of silicon and do not require cutting of expensive wafers with consequent waste of material. The silicon is melted and separated into droplets of uniform size that, once solidified, form the P-type core of the PV cell. On their surface, a thin layer is then grown into which phosphorus is diffused at high temperature, thus constituting a PN junction with the electrodes on the two opposite poles. Finally, the cells are wired to each other by a flexible mesh (Fig. 10.23).

The small size of the spheres allows integration with glass of any shape and size, even with a curved profile. Increasing or decreasing the concentration of the cells can vary the degree of transparency of the glass at will, or create special graphical effects (Table 10.6).

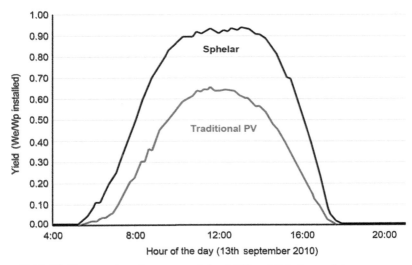

Figure 10.22 Yield comparison between traditional PV and Sphelar module. Courtesy of Sphelar-Power Corporation.

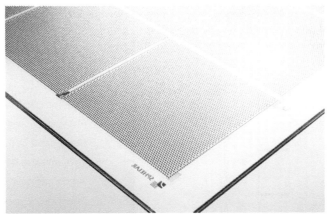

Figure 10.23 Sphelar-Power semitransparent PV module. Courtesy of Sphelar-Power Corporation.

Table 10.6 Yield of spherical cell PV glazing[14]

Cell density (%)	11.1	13.1	16.5	19.0	27.5
Power (W/m^2)	13.1	15.0	18.8	21.5	31.0

10.1.5 Prismatic optical cell photovoltaic glazing

Another interesting building-integrated PV application is the Photovoltaic Glass Unit (PVGU) (Fig. 10.24), developed by Guardian Glass and Pythagoras Solar, which combines the production of electrical energy with the optimization of daylight and solar gain thanks to a system of optical prismatic cells placed on the second position of the double-glazing unit, behind the low-iron external pane, and able to divert the direct solar radiation toward PV cells arranged perpendicularly to the glass surface. This solution allows diffused natural light inside, preventing glare while at the same time allowing vision outside. This glazing is characterized by excellent selectivity (VLT = 49%, SHGC = 14%, LSG index = 3.50) and performance (maximum power 120 Wp/m^2, module efficiency 12%).

10.1.6 Transparent luminous solar collectors

Developed by researchers at Michigan State University, TLSCs (Fig. 10.25) are composed of organic salts that absorb specific UV and IR wavelengths in the nonvisible light spectrum, and then reemit them by luminescent effect as invisible IR radiations.[15] These radiations are then guided to the window frame, where thin strips of conventional PVs convert them into electrical energy (efficiency up to 10%). Solar radiation falling in the visible range remains almost unchanged, giving the TSLC a transparency up to 90%. Nontoxic and easy-to-find raw materials, and a production process based on vacuum deposition, analogous to other high-performance films, make this technology attractive for large-scale markets.[15] In Holland a consortium led by the University of Eindhoven has tested highway noise barriers that integrate

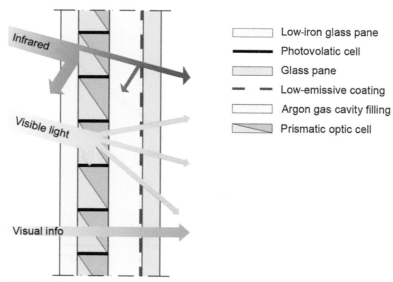

Figure 10.24 PVGU operating scheme.

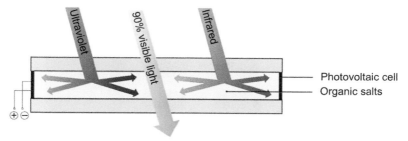

Figure 10.25 Transparent luminous solar collector.

colored TLSCs, validating that 1 km of energy-generating barriers can supply up to 50 households with electric power. Currently research is addressing a downside of one of the most used reemitting materials, inorganic salts, namely their tendency to readsorb some of the radiation emitted: researchers from Los Alamos National Laboratory and the University of Milan Bicocca are trying to substitute them with quantum dots custom engineered for a larger separation between adsorption and emission bands (Stokes shift).[16] Research focuses on eliminating toxic heavy metals such as cadmium from quantum dot composition, and optimizing their absorption and emission to avoid any color casting of transmitted light.

TSLCs are gaining interest among a variety of operators, in both building and energy sectors, for their promising properties of good efficiency even with diffused light, low investment costs, and increased durability due to reduced cell overheating as a result of selective concentration. Italian energy company Eni has focused in its research center on TLSC development, with 16 patents of suitable fluorescent colorants deposited and a prototype application in Rome: an electric bike recharging station featuring 192 photoactive yellow glass panes totaling 60 m^2 and 500 Wp of installed power.

10.2 Bioadaptive glazing

An innovative form of integration between the transparent envelope and renewable energy systems is the so-called "bioadaptive facade," made of glazed photobioreactors (PBRs) containing microalgae, which can serve as a sunscreen and at the same time continuously generate the biomass needed for the production of biogas to power air conditioning and sanitary hot water in the building. An example is BIQ House in Hamburg by Arup (2013) (Figs. 10.26 and 10.27).[17]

The system can find application as movable *brises soleil*, as cladding of opaque closures, or directly as a glass roof functioning as a dynamic screen against solar radiation: the increase of the incident solar radiation in fact raises algae concentration and consequently their shading capacity.

These bioreactors normally have a size of 2.5 × 0.7 m and a thickness of about 8 cm, and contain, in addition to algae, water and nutrients (nitrogen, phosphorus, and sulfur)

Figure 10.26 BIQ House, Arup, Hamburg.
Courtesy of Arup.

Figure 10.27 BIQ House bioadaptive facade.
Courtesy of Arup.

(Fig. 10.28). Together with nutrients, the culture is continuously added through the piping system with both carbon dioxide (CO_2), necessary for the photosynthesis process, and compressed air to ensure homogeneous distribution of the algae within the reactor and their regular growth. The CO_2 can be obtained from sodium bicarbonate ($NaHCO_3$) or sodium carbonate (Na_2CO_3), or directly from the atmosphere through

Figure 10.28 BIQ House algae reactors.
Courtesy of Arup.

the use of exchanging resins. The production of 1.0 ton of algae biomass requires 1.8 tonnes of CO_2. PBRs are mounted on a steel structure that also houses piping and wires.

When the reactor is exposed to the sun, the algae, mostly unicellular small organisms about 5 μm in size, increase their mass and absorb the nutrients and CO_2, producing H_2 gas. The growth speed of the algae depends on the amount of solar radiation received during the day, and thus on the latitude, exposure, and inclination of the PBR.

Photosynthesis requires about 0.75 L of water for every kg of algae biomass. If the lipid content of microalgae is around 50%, then 1 L of biogas requires 1.5 L of water. Under stress, microalgae tend to clump together, and therefore frequent cleaning of the reactors is needed by the addition of special products ($KOH + K_2CO_3$).

The optimal growth temperature for most algae is usually around 20–30°C with a pH of 7–9; growth stops below 5°C and above 35°C. To keep the water at a constant temperature, excess heat is collected by exchangers and used for the production of sanitary hot water, with a conversion efficiency of the radiant energy of approximately 38% (compared to 60–65% for common solar panels).

Every 7–10 days the excess biomass is extracted from the reactor and transported to facilities for the production of methane biogas, which is in turn used to power the boiler of the building for winter heating (Fig. 10.29). The conversion efficiency of solar radiation into biomass is about 10%, thus allowing a production of thermal energy from the biogas of around 30 kWh per square meter of bioadaptive facade per year.

Annual energy yield of the system in terms of hot water and thermal energy from the biogas combustion obviously depends on the size of the reactors and the amount of solar radiation received, in turn a function of the latitude, orientation, and tilt of the reactors. Table 10.7 shows the production data of BIQ House in Hamburg (latitude 53°24′2″ north, longitude 10°1′10″ east, solar radiation 940 kWh/m^{2}yr).

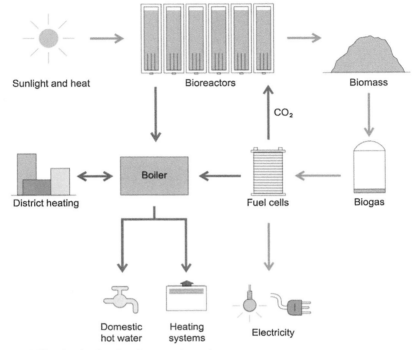

Figure 10.29 Bioadaptive facade operating scheme.

Table 10.7 **BIQ House: energy yield of a 200 m² bioadaptive façade operating 300 days/year**[17]

Biomass production	4.5 kg/m²yr
Methane gas production from biomass (biogas)	3 m³/m²yr
Total energy from biogas	32.4 kWh/m²yr
Net energy from biogas	22.7 kWh/m²yr
Thermal solar energy	150 kWh/m²yr

Since some microalgae have the ability to emit light in certain circumstances, biologists at the University of San Diego were able to create colored algae by adding phosphorescent proteins to their cells.[18] Colors are blue, cyan, green, yellow, and orange. Artificial LED (light-emitting diode) lights can be used as a light source for growing microalgae, especially in street furniture and night-lighting structures. Biochemist Pierre Calleja is working on algae-powered streetlights. The lamps absorb CO_2 from the atmosphere to feed the algae, and through the process of photosynthesis the batteries are recharged during the day and at night use stored energy to light up the city.

10.3 Conclusions and future trends

The role of glass surfaces as energy-generating components is important for building integration of renewable energy sources and achieving the objective of zero-energy buildings.

STPVs employ different PV cell technologies, from arrays of isolated opaque c-Si cells to transparent thin-film PV modules, and are suitable for windows and skylight integration in both commercial and residential buildings.

Thanks to both inorganic (second generation) and organic (third generation) thin-film technology, PV systems can now be fully integrated in flat or curved windows and even in ethylene tetrafluoroethylene (ETFE) cushions. In addition to electricity generation, STPV glazing may be used for solar heat control while providing adequate daylight and vision through.

OPVs are a future technology with great potential, and will facilitate the ever-wider use of solar technology in coming years. In the mid to long term, both OPVs and DSSCs are projected to achieve higher profitability at a lower levelized cost of electricity, including a shorter energy return time and a better ecological balance. This is due to their mainly inexpensive raw materials and manufacturing processes, which include spraying, standard printing, and roll-to-roll techniques allowing cheap large-scale production, with manufacturing costs expected to reach €0.50/W by 2020. The major challenges for this sector are the low device efficiency—the current record efficiency is 11.9% for DSSCs, achieved by Sharp in 2013, and 11.5% for organic cells, set by Hong Kong University of Science and Technology (HKUST) in 2015—and their instability in the long term, as organic materials have issues with high temperatures and UV light degradation. In particular, DSSCs employ a liquid electrolyte and are thus prone to leakage or rupturing.

Further developments of PVs concern the application of quantum dots to solar cells technologies: these are nanometer scale crystals of conventional semiconductors, which band gap can be carefully tuned by changing their particle size to closely match the desired wavelength without the need of efficiency-degrading coatings or dopants.

Quantum Dots Solar Cells' (QDSCs) structure and operation is similar to that of DSSCs, with conversion efficiencies up to 9.9% (obtained in 2015 by the University of Toronto), and their unique characteristics make them suitable for extremely varied applications. For instance, quantum dot colloidal solutions can be spray applied to curved or flexible substrates, creating a "PV paint," although with lower efficiency values—the University of Toronto recorded 7.2% efficiency in 2014.[19] Theoretical potential efficiency of a single-size QDSC is as high as 63.2%, while multiple particle sizes would act as a multijunction solar cell with several band gaps, achieving a theoretical limit efficiency of 86.5%.

Lastly, Perovskite Solar Cells (PSCs) are the latest research trend in the field of cost-efficient PVs. They are based on semiconductor compounds structured as the actual perovskite mineral, or calcium titanium oxide, and usually consist of hybrid tin or more commonly lead halide materials, such as methylammonium lead halide. PSCs can operate as a conventional thin-film cell or as a DSSC, and can even be combined with conventional c-Si cells to obtain a hybrid tandem device. Their promising efficiency due to a convenient band-gap range and cheap manufacturing costs means they have rapidly advanced in recent years, with laboratory efficiency increasing from 3.8% in 2009 to a record value of 20.1% obtained in 2014 by Korea Research Institute of Chemical Technology (KRICT) researchers by combining common methylammonium lead halide perovskite with formamidinium lead iodide. Furthermore, perovskite's electronic structure is well suited to light absorption, the PSCs conduct the subsequently mobilized electrical charges well, and perovskite crystals are very easy to process from base solutions. A viable manufacturing process is for instance to coat a substrate with a solution containing the electrode materials, effortlessly obtaining a dense layer of perovskite crystals after evaporation. Perovskite also lends itself well to spray application, with the potential to make a number of substrates energy active at a reduced cost, and print processing.

Economic studies state that PSCs may have an energy payback time as short as two to three months, compared to more than two years required by c-Si and six months for CdTe solar cells.[20]

At present, however, perovskite-based cells still suffer from high instability in ambient conditions, as the organic molecule part of their hybrid structure quickly degrades when exposed to elements. Concerns about the use of toxic lead are an additional issue: using tin as a stabilizer is a less efficient alternative, though several studies are researching this field. Research samples are also still in the order of 0.1 cm in size, although possible solutions to improve scalability are under way. By inserting the PSC between two heavily doped extraction layers, a NiMgLiO semiconductor and titanium oxide, very rapid carrier extraction was achieved from photoactive perovskite even with 10−20 nm thick layers, avoiding pinholes and eliminating local structural defects over large areas. This way it was possible to obtain a 1 cm^2 cell able to withstand 1,000 h of light soaking thanks to the robust inorganic nature of the external layers.[21]

References

1. Mcevoy A, Markvart T, Castaner L, editors. *Practical handbook of photovoltaics, fundamentals and applications*. Waltham (MA): Architectural Press; 2012.
2. Lin C, Tsai M, Wei W, Lai K, He H. Packaging glass with a hierarchically nanostructured surface: a universal method to achieve self-cleaning omnidirectional solar cells. *ACS Nano* 2016;**10**(1):549−55.
3. Luque A, Hegedus S. *Handbook of photovoltaic science and engineering*. West Sussex: Wiley; 2011.
4. La Notte L, Mineo D, Polino G, Susanna G, Brunetti F, Brown TM, et al. Fabrication of fully-Spray-processed organic photovoltaic modules by using an automated process in air. *Energy Technol* 2013;**1**(12):757−62.
5. Raya B, Baradwaj AG, Khana MR, Boudourisb BW, Alama MA. Collection-limited theory interprets the extraordinary response of single semiconductor organic solar cells. *Proc Natl Acad Sci* 2015;**112**(36):11193−8.
6. You P, Liu Z, Tai Q, Liu S, Yan F. Efficient semitransparent perovskite solar cells with graphene electrodes. *Adv Mater* 2015;**27**(24):3632−8.
7. Harlang TCB, Liu Y, Gordivska O, Fredin LA, Ponseca Jr CS, Huang P, et al. Iron sensitizer converts light to electrons with 92% yield. *Nat Chem* 2015;**7**:883−9.

8. Lin C, Wei T, Lee K, Lu S. Titania and Pt/titania aerogels as superior mesoporous structures for photocatalytic water splitting. *J Mater Chem* 2011;**21**:12668−74.
9. Evans RC, Douglas P, et al. *Applied photochemistry*. Dordrecht (NL): Springer; 2013.
10. Burschka J, Pellet N, Moon S, Humphry-Baker R, Gao P, Nazeeruddin MK, et al. Sequential deposition as a route to high-performance perovskite-sensitized solar cells. *Nature* 2013;**499**:316−9.
11. Freitag M, Daniel Q, Pazoki M, Sveinbjörnsson K, Zhang J, Sun L, et al. High-efficiency dye-sensitized solar cells with molecular copper phenanthroline as solid hole conductor. *Energy & Environ Sci* 2015;**8**:2634−7.
12. Park T, Na J, Kim B, Kim Y, Shin H, Kim E. Photothermally activated pyroelectric polymer films for harvesting of solar heat with a hybrid energy cell structure. *ACS Nano* 2015;**9**(12):1183039.
13. Taira K, Nakata J. Catching rays. *Nat Photonics Technol Focus* 2010;**4**(9):602−3.
14. Taira K. Sphelar cells save on silicon, capture light from all paths. *JEI Dempa* February 2007:26−7.
15. Zhao Y, Meek GA, Levine BG, Lunt R. Near-infrared harvesting transparent luminescent solar concentrators. *Adv Opt Mater* 2014;**2**(7):606−11.
16. Meinardi F, Mcdaniel H, Carulli F, Colombo A, Velizhanin AK, Makarov NS, et al. Highly efficient large-area colourless luminescent solar concentrators using heavy-metal-free colloidal quantum dots. *Nat Nanotechnol* 2015;**10**:878−85.
17. International Building Exhibition Hamburg. *Smart material house BIQ Hamburg*. IBA Hamburg GmbH; 2013.
18. Rasala BA, Barrera DJ, Ng J, Plucinak TM, Rosenberg JN, Weeks DP, et al. Expanding the spectral palette of fluorescent proteins for the green microalga Chlamydomonas reinhardtii. *Plant J* 2013;**74**(4):545−56.
19. Kramer IJ, Moreno-Bautista G, Minor JC, Kopilovic D, Sargent EH. Colloidal quantum dot solar cells on curved and flexible substrates. *Appl Phys Lett* 2014;**105**.
20. Gong J, Darling SB, You F. Perovskite photovoltaics: life-cycle assessment of energy and environmental impacts. *Energy & Environ Sci* 2015;**8**:1953−68.
21. Chen W, Wu Y, Yue Y, Liu J, Zhang W, Yang X, et al. Efficient and stable large-area perovskite solar cells with inorganic charge extraction layers. *Science* 2015;**350**(6263):944−8.

Index

'*Note:* Page numbers followed by "f" indicate figures and "t" indicate tables.'

A

Active dynamic glazing, 307–323
  electrochromic devices, 310–318, 311f–317f, 319f, 320t–321t
  emerging technologies, 318–323, 322f–323f
  PDLC devices, 309–310, 310f
  suspended particle devices, 307–309, 308t, 309f
Adaptive building systems, 300, 300f
Advanced insulating materials
  aerogels
    appliances, 148
    building applications, 140–148, 141f
    floors and roofs, 142
    insulating products, 132–140, 133f–134f, 134t, 135f, 135t–136t, 137f–138f, 138t, 139f
    low-temperature heating, 147, 148f
    origin and properties, 127–130, 128f–129f, 130t
    perimeter walls, 142–146, 142f–144f, 144t, 145f–146f
    preparation method, 130–132, 131f
    tensile membranes, 147–148
    thermal bridge correction, 146–147, 146f–147f
  biobased insulating materials, 160–167, 161t–162t, 163f, 163t, 164f–166f
  future trends, 174–175
  transparent insulating materials, 167–174, 167f–170f, 170t, 171f, 172t, 173f–174f
  vacuum insulating panels, 149–160, 149f–150f, 152f–153f
    building applications, 159–160
    floors and roofs, 159
    future developments, 158–159
    modified atmosphere insulation panels, 158
    other uses, 160
    specifications and performance, 154–159, 154f, 155t, 157t
    walls, 159–160, 160f
Advanced low-emission glazing, 249–255, 250f, 251t, 252f, 253t, 254f–255f
Advanced photovoltaic glazing, 327–347
  crystalline silicon photovoltaic glazing, 328–332, 328t, 329f–331f, 331t, 332f
  dye-sensitized solar cells (DSSC), 338–342, 338f–341f
  manufacturing and development scope, 342–343, 342f–343f
  organic photovoltaic glazing, 336–343
  organic solar cells, 336–338, 336f
  prismatic optical cell photovoltaic glazing, 346, 346f
  semitransparent thin-film PV glazing, 332–334, 333f–334f, 334t, 335f, 336t
  spherical cell photovoltaic glazing, 343–344, 344f–345f, 345t
  transparent luminous solar collectors, 346–347, 347f
Advanced pitched roofs, 121, 122f
Advanced shading systems, 296–301, 297f–300f
  adaptive building systems, 300, 300f
  Bioskin facade system, 301, 302f
  dynamic shading systems, 296–297, 297f, 303
Advanced/traditional pitched roofs., 122f
Advanced window frames, 258–260, 259f, 259t
Aerogel blanket inclusion, 148

Aerogels
    appliances, 148
    building applications, 140–148, 141f
    floors and roofs, 142
    insulating products, 132–140, 133f–134f, 134t, 135f, 135t–136t, 137f–138f, 138t, 139f
    low-temperature heating, 147, 148f
    origin and properties, 127–130, 128f–129f, 130t
    perimeter walls, 142–146, 142f–144f, 144t, 145f–146f
    preparation method, 130–132, 131f
    tensile membranes, 147–148
    thermal bridge correction, 146–147, 146f–147f
Aerogips panels, 145f–147f
Aeropan Basic/Spaceloft fiber-reinforced aerogel blankets, 132–133, 134f
Aeropan insulation system, 142–143, 142f–144f
Aerowall, 144–146
Airloy, 139–140, 140t
Algae-based insulation, 164–165, 165f
Algae-photobioreactors facade systems, 347–350, 348f–350f, 350t
Antireflective glazing, 279–280, 280f
Applications/products, PCM, 197–215
    ceilings, 204–209, 205f–206f, 206t, 207f–208f
    exterior walls/roof linings, 211–212
    floors, 209–211, 210f–211f
    glazing, 212–214, 212f, 213t
    interior insulation, heavyweight construction buildings with, 201f, 202–204, 203f
    mechanical systems, 215
    structural concrete, 213f–214f, 214–215
    walls, inner lining of, 198–202, 198f, 200t

B
BASF Micronal, 192–193, 192f–193f
Biobased insulating materials, 160–167, 161t–162t, 163f, 163t, 164f–166f
Bioskin facade system, 301, 302f
Blue smoke, 127
Building automation system (BAS), 47

Building automation technologies and building automation control systems (BAT/BACS), 19
Building energy consumption, 3–6
Building information modeling (BIM) systems, 8–9
Building-integrated photovoltaics (BIPV), 21–27, 22f, 24t, 25f–26f, 77–78
Building management system, 46, 46t
Building materials
    classification, 55–59, 55t, 56f
        architecture properties, 57t
        ceramic materials, 56
        composite materials, 56
        metallic materials, 55
        polymeric materials, 56
        zero-energy buildings, 58
    3D printing, 94–100
        applications, 96
        Canal House printed sections, 99f
        Cool Bricks, 97–98, 98f
        Metal Objects, 95
        QuakeColumn, 97–98, 97f
        WASProject, 100, 100f
    future trends, 101–102
    nanotechnology, 59–84
        adhesives, 78–79
        ceramics, 77
        chemical vapor deposition process, 67, 67f
        classification, 63–65, 63f, 64t
        concrete, 75–76
        definition and concepts, 60–63, 61t
        energy/environmental and construction sectors, applications in, 68–71, 69t
        environmental monitoring and control systems, 83–84
        glazing, 77–78
        graphene, 73–75
        insulating materials, 78
        lighting, 80–81, 80f
        manufacturing processes, 65–68, 66f–67f
        metals, 76
        nanoplates, 64, 65f
        nanoproducts for architecture, 71–84, 72t–74t
        nanoscience, 68–69

nanostructured materials, 65, 66f
oil-absorbent nanoporous polymers and aerogels, 70
paints, 79—80
patents classification, 69t
photovoltaics, 81—83, 82f
physical bottom-up methods, 68
plastics, 76—77
quantum size effects, 62
soil depuration, 70
smart materials
  applications, 86t—87t
  classification, 85t
  energy-exchanging materials, 90—94, 91f—94f
  property-changing materials, 84—90, 89f

C

Carbon-neutral buildings, 42
Cavity insulation, 117
Cavity walls, 117
Ceilings, 204—209, 205f—206f, 206t, 207f—208f
Chemical sensors, 70
Climate-adaptive building shell, 43
Closed-cavity facade (CCF), 263
CoeLux lighting system, 290f
Composite aerogels, 130
Cool Bricks, 242, 243f
Cool roofs, 219—229, 221f
  emerging technologies, 225—229
  energy-saving benefits, 220
  fictitious temperature concept, 219—220
  products and specifications, 222—225, 223f—225f, 226t
  standards and regulations, 221—222
  urban heat island (UHI) effect, 219
Crystalline silicon (c-Si), 23
  photovoltaic glazing, 328—332, 328t, 329f—331f, 331t, 332f

D

Design
  building information modeling (BIM) systems, 8—9
  building-integrated photovoltaics, 21—27, 22f, 24t, 25f

Energy Performance of Buildings (2010/31/EU Directive), 5—6
future trends, 49—50
green buildings, 27—38
  products, 28—35, 29f—30f, 31t, 33f—35f
  rating systems, 35—38, 36f—37f
smart buildings, 38—49, 39t—41t
  building management system, 46, 46t
  internet of things (IoT), 48—49
  smart envelope, 42—44, 45t
  smart systems, 44—49, 46t
smart/sustainable and inclusive buildings, 6—9
zero-energy buildings (ZEBs), 10—27
  building energy equipment system scheme, 19, 19f
  definition and concepts, 10—12, 12t
  sociocultural-level measures, 20—21
  strategies, 12—21
  technical and construction-level measures, 16—17
  technological-level measures, 17—20
  typological-level measures, 14—16, 15t
Digital Building Operating Solution (Di-BOSS), 47
Double-skin glazing system, 261
3D printing, 94—100
  applications, 96
  Canal House printed sections, 99f
  Cool Bricks, 97—98, 98f
  Metal Objects, 95
  QuakeColumn, 97—98, 97f
  WASProject, 100, 100f
Dye-sensitized solar cells (DSSC), 338—342, 338f—341f
Dynamic glazing
  active dynamic glazing, 307—323
    electrochromic devices, 310—318, 311f—317f, 319f, 320t—321t
    emerging technologies, 318—323, 322f—323f
    PDLC devices, 309—310, 310f
    suspended particle devices, 307—309, 308t, 309f
  future trends, 323—324
  passive dynamic glazing
    photochromic glazing, 305—306
    thermochromic glazing, 306—307, 306f

E

Electrochromic dynamic glazing, 313f
Electrochromic facade, 311f–312f
Electrochromic (EC) glazing operating scheme, 312f
Electromagnetic wave spectrum, 293, 293f
Emerging technologies, building skin, 225–229
   lenticular cool roof, 229
   PCM color coatings, 229
   radiative sky cooling, 225–228, 227f
   thermochromic cool roofs, 228–229
Energy-generating glazing
   advanced photovoltaic glazing, 327–347
      crystalline silicon photovoltaic glazing, 328–332, 328t, 329f–331f, 331t, 332f
      dye-sensitized solar cells (DSSC), 338–342, 338f–341f
      manufacturing and development scope, 342–343, 342f–343f
      organic photovoltaic glazing, 336–343
      organic solar cells, 336–338, 336f
      prismatic optical cell photovoltaic glazing, 346, 346f
      semitransparent thin-film PV glazing, 332–334, 333f–334f, 334t, 335f, 336t
      spherical cell photovoltaic glazing, 343–344, 344f–345f, 345t
      transparent luminous solar collectors, 346–347, 347f
   bioadaptive glazing, 347–350, 348f–350f, 350t
   future trends, 351–352
EnergyPlus, 196
Environment-adaptive skin facades, 239–243
   Cool Bricks, 242, 243f
   Hydroceramic, 242, 242f
   Hydromembrane, 241, 241f
   passive deployable insulation, 240
   Self-activated building envelope regulation (SABER), 239–240
   Thermobimetal, 240
   TiO_2 photocatalyst evaporative shell, 243
   water-reacting facade, 240
ESP-r, 197
ETFE. *See* Ethylene tetrafluoroethylene (ETFE)
Ethylene tetrafluoroethylene (ETFE), 266–274, 267f–274f
External cladding insulation, 142–143
External thermal insulation composite system (ETICS), 115–117, 159–160

F

Fiber-reinforced aerogel blanket (FRAB), 136–140
   aerogel batting, 135f
   hydrophobic behavior, 138f
Fictitious temperature concept, 219–220
Fire-resistant glazing, 265–266, 266f
Floors, 209–211, 210f–211f
Forest Stewardship Council (FSC), 31
Frozen smoke, 127
Functional model/building design, PCM, 193–197
   adequate solar protection and nighttime ventilation, 195
   behavior, 194–195
   EnergyPlus, 196
   ESP-r, 197
   heating periods, 195
   LHTES, 193
   passive systems, 194
   solar direct gain, 195
   TRNSYS software, 197
   winter heating, solar thermal energy for, 196

G

Gel dehydration, 131–132
Glazed double-skin facades, 260–264, 260f–262f, 264f
Glazing, 212–214, 212f, 213t
   advanced low-emission glazing, 249–255, 250f, 251t, 252f, 253t, 254f–255f
   advanced shading systems, 296–301, 297f–300f
      adaptive building systems, 300, 300f
      Bioskin facade system, 301, 302f
   advanced window frames, 258–260, 259f, 259t
   antireflective glazing, 279–280, 280f
   dynamic glazing. *See* Dynamic glazing

Index 359

ethylene tetrafluoroethylene transparent closures, 266−274, 267f−274f
fire-resistant glazing, 265−266, 266f
future development, 274−275, 302−303
glazed double-skin facades, 260−264, 260f−262f, 264f
heating glazing, 264−265, 265f
light-redirection and optical systems, 284−291, 284f
  light-redirection louver systems, 285−287, 285f−286f
  natural mimicking artificial light, 289−291, 290f
  transparent organic light-emitting diode windows, 291, 292f
  tubular solar conveyors, 287−289, 287f−289f
monolithic aerogel insulating glazing, 257−258, 258f
self-cleaning glazing, 280−283, 281f
  superhydrophilic photocatalytic glazing, 282−283, 283f
  superhydrophobic nanotechnological glazing, 282, 282f
static solar protection glazing, 291−296, 292f−295f, 295t
suspended film glazing, 255−256, 256f
vacuum insulating glass (VIG), 256−257, 257f
Green architecture, 160
Green buildings, 27−38
  products, 28−35, 29f−30f, 31t, 33f−35f
  rating systems, 35−38, 36f−37f
Green walls, 232f−234f
  air purification, biofilters/active living walls for, 234
  air quality, 230
  hydroponic system, 234
  leaf area index (LAI), 230
  module-based green wall system, 231
  sphagnum, 231
  street canyons, 230
  Treebox Rain Garden's system, 232
  vertically integrated greenhouse, 236−238, 238f

H
HDPE. See High-density polyethylene (HDPE)

Heating glazing, 264−265, 265f
Heating/ventilation and air-conditioning (HVAC), 115−116
Heat-reflective paints, 123
Heat removal fluid (HRF), 27
Hemp fibers, 165−166
High-density polyethylene (HDPE), 193
High luminous transmission, 130
High-temperature solar applications, 148
Hydroceramic, 242, 242f
Hydromembrane, 241, 241f

I
Insulated glass unit (IGU), 250
Insulating materials
  classification and thermal properties, 110−114, 111t−112t
  functional model and building facade applications, 114−123, 115f
    advanced pitched roofs, 121, 122f
    cavity walls, 117
    inner face, insulation placed on, 116−117
    outer face, insulation placed on, 115−116
    thermal reflective surfaces, 121−123, 123f
    ventilated walls, 117−121, 118f−120f
  future trends, 124−125
  heat transfer physics, 107−110, 107f, 109t
Integration, PV, 22
Interior insulation, heavyweight construction buildings with, 201f, 202−204, 203f
Internal glazing, 260
Iron oxide (Fe_3O_4), 151

K
Kinetic Facade, 296−297
Knudsen effect, 127−128, 129f

L
Light-emitting diode (LED), 273
Light-redirection louver systems, 285−287, 285f−286f
Light-redirection/optical systems, 284−291, 284f
  light-redirection louver systems, 285−287, 285f−286f
  natural mimicking artificial light, 289−291, 290f

Light-redirection/optical systems (*Continued*)
  transparent organic light-emitting diode windows, 291, 292f
  tubular solar conveyors, 287–289, 287f–289f
Light-transporting device, 288
Low-e coating, 252–253
  high-performance, 254
  spectral transmittance, 254f
Low-tech biotech innovative building approach, 32–33

M

Macroencapsulation, 188–189, 189f
Massive timber, 30
Mechanical systems, 215
Metal frames, 258–259
Microencapsulation, 189–193, 190f–191f, 191t, 192f–193f
Monolithic aerogel insulating glazing, 257–258, 258f
Monolithic silica aerogel, 128f–129f
Mushrooms root (mycelium) insulating materials, 161–163

N

Nanocoatings, 71–73
Nanocomposite concrete, 75–76
Nano-enhanced concrete products, 75
Nanofilms, 71–73
Nanotechnology, 59–84
  adhesives, 78–79
  ceramics, 77
  chemical vapor deposition process, 67, 67f
  classification, 63–65, 63f, 64t
  concrete, 75–76
  definition and concepts, 60–63, 61t
  energy/environmental and construction sectors, applications in, 68–71, 69t
  environmental monitoring and control systems, 83–84
  glazing, 77–78
  graphene, 73–75
  insulating materials, 78
  lighting, 80–81, 80f
  manufacturing processes, 65–68, 66f–67f
  metals, 76
  nanoplates, 64, 65f
  nanoproducts for architecture, 71–84, 72t–74t
  nanoscience, 68–69
  nanostructured materials, 65, 66f
  oil-absorbent nanoporous polymers and aerogels, 70
  paints, 79–80
  patents classification, 69t
  photovoltaics, 81–83, 82f
  physical bottom-up methods, 68
  plastics, 76–77
  quantum size effects, 62
  soil depuration, 70
Natural mimicking artificial light, 289–291, 290f

O

Oil-absorbent nanoporous polymers, 70
OLED. *See* Organic light-emitting diode (OLED)
Organic light-emitting diode (OLED), 291, 292f
Organic photovoltaic glazing, 336–343
Organic solar cells, 336–338, 336f
Organisation for Economic Co-operation and Development (OECD), 4–5

P

Passive deployable insulation, 240
Passive dynamic glazing
  photochromic glazing, 305–306
  thermochromic glazing, 306–307, 306f
Perimeter walls, 142–146
Permea system, 301
Perovskite solar cell, 337, 341, 351–352
Phase-change materials
  building applications and products, 197–215
  ceilings, 204–209, 205f–206f, 206t, 207f–208f
  exterior walls/roof linings, 211–212
  floors, 209–211, 210f–211f
  glazing, 212–214, 212f, 213t
  interior insulation, heavyweight construction buildings with, 201f, 202–204, 203f
  mechanical systems, 215
  structural concrete, 213f–214f, 214–215
  walls, inner lining of, 198–202, 198f, 200t

charge/discharge cycle, 183f
classification and technical specifications, 184—188, 184f
  inorganic phase-change materials, 185—186
  organic phase-change materials, 186—188
cold accumulators, 183
eutectics, 187
fatty acids, 187
functional model and building design, 193—197
  adequate solar protection and nighttime ventilation, 195
  behavior, 194—195
  EnergyPlus, 196
  ESP-r, 197
  heating periods, 195
  LHTES, 193
  passive systems, 194
  solar direct gain, 195
  TRNSYS software, 197
  winter heating, solar thermal energy for, 196
future trends, 215—216
incongruent melting, 186
latent heat and fusion points, 184—185, 185t
nanomaterial application, 188
packaging and encapsulation methods, 188—193
  macroencapsulation, 188—189, 189f
  microencapsulation, 189—193, 190f—191f, 191t, 192f—193f
salt hydrates, 185—186
segregation, 186
sodium sulfate decahydrate, 186
subcooling, 186
thermal conductivity, 187
thermal energy storage systems (TES), 183—184
thermal mass and latent heat storage, 179—184, 179f—180f, 180t—181t
transitions diagram, 182f
Photocatalytic materials, 88
Photoluminescent materials, 91
Photovoltaic glazing, 327—347
  crystalline silicon photovoltaic glazing, 328—332, 328t, 329f—331f, 331t, 332f
  dye-sensitized solar cells (DSSC), 338—342, 338f—341f
  manufacturing and development scope, 342—343, 342f—343f
  organic photovoltaic glazing, 336—343
  organic solar cells, 336—338, 336f
  prismatic optical cell photovoltaic glazing, 346, 346f
  semitransparent thin-film PV glazing, 332—334, 333f—334f, 334t, 335f, 336t
  spherical cell photovoltaic glazing, 343—344, 344f—345f, 345t
  transparent luminous solar collectors, 346—347, 347f
Piezoelectric pavement, 93—94, 93f—94f
Plug-in hybrid/electric vehicles, 19
Polymer-dispersed liquid crystals devices, 310, 310f
Polymethylmethacrylate (PMMA), 168, 171
Polyvinyl butyral (PVB), 307
Polyvinyl chloride (PVC), 258
Prismatic optical cell photovoltaic glazing, 346, 346f
Prism solar protection panels, 284

Q

QLED. *See* Quantum light-emitting diode (QLED)
Quantum light-emitting diode (QLED), 80—81, 80f
Quantum dot solar cells (QDSCs), 341—342, 351

R

Recycled cellulose, 166
Recycled jeans insulation, 166, 166f
Recycled textile insulation materials, 166
Reflective microspheres, 123f
Renewable energy, 18
Resilient design, 40

S

SABER. *See* Self-activated building envelope regulation (SABER)
SCENIHR. *See* Scientific Committee on Emerging and Newly Identified Health Risks (SCENIHR)

Scientific Committee on Emerging and Newly Identified Health Risks (SCENIHR), 60
Self-activated building envelope regulation (SABER), 239–240
Self-cleaning glazing, 280–283, 281f
  superhydrophilic photocatalytic glazing, 282–283, 283f
  superhydrophobic nanotechnological glazing, 282, 282f
Self-powered shading technologies, 319–322
Semitransparent PV (STPV) products, 25
Semitransparent thin-film PV glazing, 332–334, 333f–334f, 334t, 335f, 336t
Shape-memory alloy, 88–90, 89f
Silica aerogel light transmission, 132
Silicon carbide (SiC), 151
Silicon nitride (SiNx), 152
Silicon oxides (SiOx), 152
Skin
  cool roofs, 219–229, 221f
    emerging technologies, 225–229
    energy-saving benefits, 220
    fictitious temperature concept, 219–220
    products and specifications, 222–225, 223f–225f, 226t
    standards and regulations, 221–222
    urban heat island (UHI) effect, 219
  environment-adaptive skin facades, 239–243
    Cool Bricks, 242, 243f
    Hydroceramic, 242, 242f
    Hydromembrane, 241, 241f
    passive deployable insulation, 240
    Self-activated building envelope regulation (SABER), 239–240
    thermobimetal, 240
    TiO_2 photocatalyst evaporative shell, 243
    water-reacting facade, 240
  future trends, 243–244
  green walls, 232f–234f
    air purification, biofilters/active living walls for, 234
    air quality, 230
    hydroponic system, 234
    leaf area index (LAI), 230
    module-based green wall system, 231

sphagnum, 231
street canyons, 230
Treebox Rain Garden's system, 232
vertically integrated greenhouse, 236–238, 238f
Smart buildings, 38–49, 39t–41t
  building management system, 46, 46t
  internet of things (IoT), 48–49
  smart envelope, 42–44, 45t
  smart systems, 44–49, 46t
Smart envelope, 42–44, 45t
Smartlight device, 286
Smart materials
  applications, 86t–87t
  classification, 85t
  energy-exchanging materials, 90–94, 91f–94f
  property-changing materials, 84–90, 89f
Smart systems, 44–49, 46t
Soil depuration, 70
Solar heat-gain coefficient (SHGC), 293–294
Sol–gel preparation, 130
Solid smoke, 127
Solid-state lighting technologies, 80
Spaceloft fiber-reinforced aerogel blankets, 133, 134f, 134t, 136t, 138t, 141, 141f
Spherical cell photovoltaic glazing, 343–344, 344f–345f, 345t
Static solar protection glazing, 291–296, 292f–295f, 295t
Strata system, 301
Structural concrete, 213f–214f, 214–215
Supercritical drying process, 136, 139
Superhydrophilic photocatalytic glazing, 282–283, 283f
Superhydrophobic coatings, 282
Superhydrophobic nanotechnological glazing, 282, 282f
Suspended film glazing, 255–256, 256f
Suspended particle devices (SPD), 307–309, 308t, 309f

T
Tenuous solid skeleton, 131
Tessellate system, 300–301
Tetraethyl orthosilicate (TEOS), 130
Tetramethyl orthosilicate (TMOS), 130

Thermal mass/latent heat storage, 179–184, 179f–180f, 180t–181t
Thermal reflective coatings, 123f
Thermal reflective surfaces, 121–123, 123f
Thermal transmittance, 108
Thin dielectric films, 253
TIM. *See* Transparent insulating materials (TIM)
TiO_2 photocatalyst evaporative shell, 243
Titanium dioxide (TiO2), 32, 282–283
Transparent insulating materials (TIM), 167–174, 167f–170f, 170t, 171f, 172t, 173f–174f
Transparent luminous solar collectors, 346–347, 347f
Transparent organic light-emitting diode windows, 291, 292f
TRNSYS software, 197
Tubular optical systems, 287
Tubular solar conveyors, 287–289, 287f–289f

V

Vacuum insulating glass (VIG), 256–257, 257f
Vacuum insulating panels (VIP), 149–160, 149f–150f, 152f–153f
  building applications, 159–160
  floors and roofs, 159
  future developments, 158–159
  modified atmosphere insulation panels, 158
  other uses, 160
  specifications and performance, 154–159, 154f, 155t, 157t
  walls, 159–160, 160f
Vapor-open application, 152–153
Vehicle-to-building/vehicle-to-grid technology, 47
Ventilated walls, 117–121, 118f–120f
VIG. *See* Vacuum insulating glass (VIG)
VIP. *See* Vacuum insulating panel (VIP)
Volatile organic compounds (VOC), 136–138

W

Water-reacting facade, 240
White photovoltaics, 81–82, 82f
Wood foam, 163–164

X

X-aerogels, 139

Z

Zero-energy buildings (ZEBs), 10–27
  building energy equipment system scheme, 19, 19f
  definition and concepts, 10–12, 12t
  sociocultural-level measures, 20–21
  strategies, 12–21
  technical/construction-level measures, 16–17
  technological-level measures, 17–20
  typological-level measures, 14–16, 15t